GLENN B. WIGGINS is curator in the Department of Entomology, Royal Ontario Museum, and professor in the Department of Zoology, University of Toronto.

The Trichoptera or caddisflies, one of the orders of aquatic insects, are important because their larvae are involved in the food chains of most types of freshwater habitats. Caddisfly larvae are also remarkable among insects for the variety of retreats, nets, and portable cases they construct.

This is the first comprehensive reference work on the larval stages of the genera of Nearctic Trichoptera, and contains much new information not previously available. Keys for identification of the 18 families and all but six of the 142 genera now recognized are supplemented by Anker Odum's drawings of larval morphology and case structure for each genus; outlines of present knowledge of distribution, taxonomy, and biology are also given. An introductory section treats the biology, morphology, classification, and methods of field study for larval Trichoptera. The work is a fundamental reference for freshwater biologists and ecologists, systematic entomologists, and students in courses dealing with aquatic invertebrates throughout North America.

GLENN B. WIGGINS

Larvae of the North American Caddisfly Genera (Trichoptera)

Illustrated by Anker Odum

UNIVERSITY OF TORONTO PRESS
TORONTO BUFFALO LONDON

Toronto Buffalo London
Reprinted 1978
Printed in Canada

Canadian Cataloguing in Publication Data

Wiggins, Glenn B., 1927–
Larvae of the North American caddisfly genera
(Trichoptera)

Bibliography: p.
Includes index.
ISBN 0-8020-5344-0

1. Caddis-flies – North America. 2. Larvae – Insects. 3. Caddis-flies – North America – Identification. I. Title.

QL517.1.A1W53 595.7′45′097 C77-001277-9

This book is dedicated to Professor Herbert H. Ross

I consider it as of the utmost importance fully to recognise that the amount of life in any country, & still more that the number of modified descendants from a common parent, will in chief part depend on the amount of diversification which they have undergone, so as best to fill as many & as widely different places as possible in the great scheme of nature.

Charles Darwin

Charles Darwin's Natural Selection (p. 234) ed. R.C. Stauffer 1975

Contents

Preface

This is a reference work to the identity, structure, and biology of larvae of the North American caddisfly genera. More precisely perhaps, it is a stage in the evolution of such a reference, for a definitive work even at the generic level is still well beyond the information now available. The book is the result of a project I began some years ago to increase knowledge about larval Trichoptera in North America. Systematic collections on which it is based were brought together in the course of more than 150,000 miles of travel in field expeditions through many parts of Canada and the United States, especially in the western mountains where the fauna is highly diverse yet little explored. Associations between larval and adult stages were established for some 350 species, approximately 30% of the 1,200 or so species of Trichoptera now known in the two countries; over 200 of these were established for the first time.

Although the information made available is still far short of enabling one to produce keys for the identification of North American caddisfly larvae to the species level, it represents a substantial advance at the generic level. Since the genus in the Trichoptera, as in most groups, represents an ecological as well as a morphological type, it provides a useful and an incisive base for generalization. To date, 142 Nearctic genera are recognized and larvae have been identified for all but 6 (4%); diagnostic characters for larvae of 24 genera are given here for the first time, and from this general project characters for an additional 16 genera were originally published elsewhere. Having gained a better understanding of the range of larval characters covered by most genera, I have been able to establish more precise diagnoses for them and consequently have introduced many new characters into the generic keys. Ultimately, of course, keys for identification of larval Trichoptera to the species level will provide the most effective aid to workers in freshwater biology, but, in general, this level of precision is not possible in North America because sufficient basic data have not yet been assembled.

I am conscious of my good fortune in having the support of several institutions and agencies in this project. The Royal Ontario Museum has provided the working facilities for my studies, and the documented collections on which they are based are part of the research materials of the Department of Entomology. In the early years of my field work,

1962-68, when intensive surveys were made in the western United States, financial support was received from the National Science Foundation. Operating grants from the National Research Council of Canada have supported the project in recent years. Funds from the Fisheries Research Board of Canada and from the Canadian National Sportsmen's Show have also been received. Substantial grants toward the cost of publishing this book were received from the National Research Council of Canada and the Publications Fund of the University of Toronto Press. For all of this support I am profoundly grateful.

Figures for the book were prepared by Mr Anker Odum, until recently the scientific illustrator in our department. They are sufficient evidence of Mr Odum's considerable ability in this field and of his own fascination with insects; I shall add only that I have worked closely with him in an attempt to ensure that every drawing conveyed as much accurate morphological information as possible.

Individual acknowledgment to all of the persons who have contributed in some way to this project is not possible here, but I am most grateful for the co-operation, generosity, and encouragement that I have everywhere received. To Mr Toshio Yamamoto of our department I owe special gratitude for his knowledgeable and enthusiastic assistance in field work and innumerable other aspects of the project. Several others who were members of our field expeditions have contributed substantially to the growth of the collections on which the work is based: D. Barr, H.E. Frania, L.H. Kohalmi, B.P. Smith, I.M. Smith, and the late R.S. Scott. Professor Rosemary Mackay of the University of Toronto has been helpful in resolving ecological matters and problems of many kinds. It is this sense of community in acquisition and analysis of the collections which underlies use of the pronoun *we* throughout.

My professional colleagues have responded generously to requests for specimens, information, and comment. It is a pleasure to acknowledge this co-operation from N.H. Anderson, D.G. Denning, O.S. Flint, J.C. Morse, V.H. Resh, H.H. Ross, F. Schmid, S.D. Smith, J.D. Unzicker, and J.B. Wallace.

Students in Aquatic Entomology at the University of Toronto and at the Lake Itasca Biology Sessions of the University of Minnesota (1970, 1972, and 1974) helped me to appreciate that keys for identification of insects are too often inconclusive and inadequately illustrated. In affording me opportunities to test earlier versions of the present keys, they also helped to confirm my suspicion that when effort is invested in producing extensively illustrated keys, learning can be at least easier and is more often exciting.

Services made available through the Royal Ontario Museum are invaluable in a work of this kind, and this book represents supporting contributions of many persons: those involved in the entire museum process leading to deposition of adequately documented specimens in the research collection, secretaries and photographers in hours of careful work during preparation of the manuscript, and librarians for their efforts to maintain the reference base essential for research in systematics. I wish also to acknowledge the assistance of J.C.E. Riotte for translation of important reference works, Sharon Hick for compiling the literature citations, Zile Zichmanis for art work, and Felix Bärlocher for analysis of larval food. In acknowledging these contributions, I extend my appreciation to those directors and trustees of the Royal Ontario Museum whose support my work has had over the years of my curatorial appointment. I hope that they will see in this book another example of the ROM's 'record of nature through countless ages.'

In matters relating to publication, Lorraine Ourom and Ian Montagnes of the University of Toronto Press have given me advice and encouragement.

Facilities for some of the field work were made available by several institutions: American Museum of Natural History, Southwestern Research Station, Portal, Arizona; University of Minnesota, Lake Itasca Forestry and Biological Station; Oregon State University, Corvallis; Queen's University Biological Station, Chaffeys Locks, Ontario; University of California Sagehen Creek Research Project, Truckee; University of Montana Biological Station, Yellow Bay, Flathead Lake; Harkness Research Laboratory, Ontario Ministry of Natural Resources, Algonquin Provincial Park.

My wife and children shared many of the longer field expeditions, and more recently contended with my long preoccupation during preparation of the manuscript. I have appreciated their companionship and assistance as well as their forbearance.

Finally, I acknowledge a debt to those who have gone before me. Identification of the North American caddisfly larvae was at best uncertain before appearance of H.H. Ross' *Caddis Flies or Trichoptera of Illinois* in 1944, and was again improved at the generic level by his contribution to the revised edition of Ward and Whipple (ed. Edmundson 1959). Taxonomic refinement by Ross, D.G. Denning, F. Schmid, O.S. Flint, and others has continued to develop the classification of the Nearctic Trichoptera into a most useful scientific asset. In offering the present work as a further step in the evolution of a fundamental faunal reference for larval Trichoptera on this continent, I am mindful that it builds upon the work of others.

It is my hope that this book will have some part in the development of freshwater biology, bringing nearer the time when the diversity of freshwater communities can be adequately understood, and ultimately preserved. It has been my privilege to explore freshwater habitats over much of North America; in studying creatures seldom seen by man, I have delighted in sensing the timeless grandeur of natural processes that shaped them as they are. Perhaps this book will also bring to others more of those uniquely human insights that come from seeing and comprehending the life with which we share this planet.

G.B.W.
March 1976

GENERAL SECTION

Introduction

The caddisflies, or Trichoptera,* are a relatively small order of insects widely distributed on almost every habitable land mass. Estimates have placed the group in the range of 10,000 species, but the rate at which species new to science continue to be discovered suggests that this figure may be conservative (Malicky 1973). On the whole, and especially in North America, these insects have made little impact on popular attention. That is because adult caddisflies are mostly rather small brown or grey moth-like insects and are active mainly at night; and also because larval caddisflies live only in water, where they are well enough concealed to escape notice by casual observers. Caddisflies are, however, common; in fact, their larvae are abundant in freshwater habitats, which in itself means these insects are of some importance as components of freshwater systems. But because caddisflies have evolved the means to exploit resources of the full range of freshwater habitats from cold springs, through streams, rivers, and marshes, to the shorelines and depths of lakes, and to temporary pools, they are significant in transfer of energy through the trophic levels of most aquatic systems. This broad diversification is the reason both for their abundance as freshwater animals and for their importance in freshwater biology.

Diversification in their larval stages has also brought about a close dependence of genera and species of Trichoptera on particular features of the aquatic environment. This relationship has led to an important application of caddisflies and of benthic organisms in general in freshwater biology, for the presence of certain species in a body of water can be taken as evidence that conditions within it are natural and largely undisturbed. Benthic

*The name *Trichoptera* is derived from the Greek *trichos* ('hair') and *pteron* ('wing'), in reference to the dense hair covering on the wings of the adults. The origin of *caddisfly* is obscure, but according to Hickin (1967) it dates in reference to these insects at least to Izaak Walton's *Compleat Angler* (1653; 'cod-worm or caddis'); in reference to cotton or silk materials, various versions of the word date to 1400 and *cadysses* appeared in Shakespeare's *The Winter's Tale* (1611) in this sense. Although connection between the two usages has not yet been established, Hickin states that itinerant vendors of bits of material fastened pieces to their clothing as an advertisement and were called *cadice men*, suggesting a parallel with the case-making larvae.

communities of insects contain, then, important indicators of the level of pollution and perturbation within a body of water, and here the Trichoptera are especially significant because there are relatively large numbers of species in most freshwater habitats.

Since fresh waters are one of the most basic natural resources of all, the communities of organisms that live within them and maintain their ecological harmony are subjects of high priority for scientific study. Even so, freshwater biologists have always had problems in elucidating the precise ecological role of insect species in a community largely because of difficulty in identifying their larval stages. Insufficient taxonomic work on the larvae of caddisflies, and indeed of all aquatic insects, has always been a severe barrier to the progress of freshwater biology (Wiggins 1966; Hynes 1970).

Apart from their role in aquatic communities, caddisfly larvae are remarkable creatures in their own right. Most biologists are aware that the larvae of these insects build shelters, but few have ever seen the exquisite seine-like caddis nets fastened in thousands to the rocks of rivers and streams everywhere; not many appreciate that the bizarre cases carried about by caddis larvae represent a way of using silk production to increase respirational efficiency, thereby enabling caddisflies to exploit the resources of lentic habitats. The variety of structures built by these larvae ought to have established them among the most fascinating insects to be found, but there is little evidence for that happening. To all but a few specialists, the diversity in caddisfly larval behaviour is largely unappreciated.

OBJECTIVES

There is, then, a useful place for a reference work on North American caddisfly larvae. This book could have taken several forms, but I have chosen to organize the information with the needs of three categories of users in mind.

For freshwater biologists and ecologists, I have attempted to provide the means for precise identification of families and genera of caddisfly larvae in aquatic communities in North America, along with some ecological information.

For university courses concerned with various aspects of aquatic biology, I have tried to make the book suitable for student use. Because freshwater communities are enormously diverse and consistently available, their value in teaching is coming to be recognized more and more. Taxonomic diversity in these communities may be itself an object in teaching, or a means to demonstrate ecological principles. In either case, discovering how many different types of organisms there are in a freshwater community can be a new and exciting experience for students, but one that easily becomes frustrating because of identification keys that are too often vague and inadequately illustrated. It is also my hope that the book will draw more attention to use of caddisfly larvae in student experiments concerned with case-making behaviour or with feeding; caddis larvae are well suited to these studies, but little used. Thus, for students I have attempted to make generic types more than disconnected sets of morphological characters, and to make them come alive as whole organisms adapted in elegant ways to particular niches of the freshwater community.

Finally, for systematists, I have tried to produce a reference that would serve as an atlas of gross morphology of the trichopteran larval types in North America. It is well established that data from larval morphology are essential in assessing systematic relationships of Trichoptera (and other insects) and for advancing hypotheses concerning their phylogeny, but much of this morphological information is widely dispersed or is not available in the literature. Within the constraints of design and space established by the first two objectives, compromise is often made in this third one, with the result that illustrations of smaller structures in many groups have not been included.

GEOGRAPHIC LIMITS

The genera covered are all those currently recognized within the Nearctic region, excluding the islands of the Caribbean. The species totals for each genus are, however, those known to occur north of the border between Mexico and the United States approximately up to the end of 1975.

ORGANIZATION AND METHODS

This book is primarily a reference work. To make the information more easily accessible to those not familiar with the classification of Trichoptera, families are arranged in alphabetical sequence and numbered consecutively from 1 to 18. Within each family genera for which larvae are known are arranged alphabetically and numbered consecutively such that the reference number 10.36 specifies Limnephilidae (family no. 10), *Neophylax* (limnephilid genus no. 36); individual illustrations on each plate are lettered consecutively with A assigned to the habit drawing of the larva, B to the illustration of head and thorax, and C to the principal figure of the case, where feasible.

A rather large part of the book is devoted to illustrations because this is the most effective way of communicating structural information. Fine distinctions in form and proportions of body parts are useful in identification, and can be effectively conveyed only by illustration. The stylized arrangement of the legs in the habit figures is made to permit a clear view of the relative lengths of segments and of setal arrangement. Abdominal gills are drawn to reveal their segmental position. In the figure of head and thorax, all parts are shown in full dorsal view, although they would not be seen simultaneously in that relationship in life. Illustrations of cases and retreats are provided in some number because these structures provide excellent characters for generic recognition in the field, and also because, being among the extraordinary things built by any animal, they deserve to be better known.

The General Section is concerned with various aspects of existing knowledge of the Trichoptera. Systematic relationships and a classification of the Nearctic families, subfamilies, tribes, and genera are outlined under Classification and Phylogeny; Biological Considerations represents an attempt to place aspects of habitat, case-building, respiration, feeding, and life cycles in an evolutionary context. Both of these chapters offer useful ways of thinking about caddisflies, and provide an evolutionary basis for the work that would otherwise be obscured by the alphabetical arrangement. Structural parts of caddisfly larvae are considered in the chapter on Morphology. Under Techniques are discussed methods of collecting, rearing, and preserving larval specimens.

Introduction

In the Systematic Section, the key to families includes those recognized in the Nearctic region. General features of each family are outlined under the family heading; an indication of the broad relationships among genera can be obtained from the groupings in subfamilies or tribes, although all of these taxa are arranged alphabetically. For each genus for which larvae are known, a summary of essential features is provided. Under Distribution and Species is given the general world distribution of the genus; distribution in North America is based largely on records in the literature, although some from the ROM collection were not published previously. The number of species known north of Mexico up to the end of 1975 is provided, but in many genera these totals will, of course, be increased as more faunal work is done. Literature references are given for larval diagnoses and descriptions at the species level. By indicating for each genus the number of species known as larvae, both in the literature and in our collections, I have shown the base on which the present generic diagnosis rests. In many cases this base represents only a small part of the total number of species known, and users of this book should be aware that larvae yet to be discovered will probably render some of the diagnoses incomplete for particular genera.

Information summarized for each genus under Morphology pertains largely to features diagnostic for the genus, or to general features that would help to confirm an identification made by using the key. The length given for the larva is usually that of the largest specimen examined, or occasionally of one recorded in the literature; the measurement is a straight line on the long axis of the body from the anterior-most margin of the head, usually as positioned in the habit figures of larvae, to the posterior edge of the anal proleg and does not include protruding setae. The larva illustrated is not necessarily the largest one examined, but wherever possible, and in fact in almost all genera, the larva illustrated is a final instar. Magnifications given in the figure captions for habit drawings of larvae in most families are based on a straight-line measurement of the entire larva, as indicated above; but in the families Hydropsychidae, Philopotamidae, Polycentropodidae, Psychomyiidae, and Rhyacophilidae, where preserved larvae are usually strongly curved and difficult to measure, the magnification is based on length of the head capsule from the posterior margin to the anterior border of the frontoclypeal apotome, excluding labrum and mandibles.

Under Case, or Retreat for those groups in which the larval structure is fixed in one place, are outlined essential features of these larval constructions. Illustrations of portable cases include an end-on view of the posterior end of the case, an important functional and diagnostic feature of case-making behaviour. The case length given is the maximum for material I have studied or for those recorded in the literature, and is not necessarily the length of the case illustrated or the case of the largest larva examined.

Generic data summarized under Biology are drawn from our collections and field observations, supplemented wherever possible by pertinent literature references; these references should always be consulted for additional information because I have made only the most summary kind of statement. Food studies by others have been cited whenever available, but for many genera there is no information on food in the literature. For some of these, temporary slide mounts of the entire gut content were prepared, and examined under magnifications up to 400X. Visual estimates were made of proportions of components on the slide; usually the guts of three larvae were examined, the number of specimens expressed in parentheses in the text. It is my intention only to provide the basis for some statement of items ingested where none has been made previously; I have no illusion about the quantitative and seasonal basis of the samples.

Miscellaneous comments under Remarks often include references to important taxonomic reviews of adult stages of North American species in the genus under consideration. I have felt that these sources would not be known to many users of this book, but could be useful if adults were associated with larvae. Where no reference is given to works dealing with taxonomy of the adults, either under this heading or elsewhere, it can be assumed that the most useful general review for that genus is still to be found in the classic *Caddisflies, or Trichoptera, of Illinois* by H.H. Ross (1944).

Taxonomic and nomenclatorial designations are not included in this work, but most will be found in the *Trichopterorum Catalogus*, of which 15 volumes have been published to date (Fischer 1960-73). The taxonomic index to the present work lists all taxa mentioned, and includes synonyms established since 1944.

USE OF KEYS

In making generic identifications with this book, users who do not recognize the families of Trichoptera at sight should first identify specimens with the key to families; the generic illustrations and general outline for a family should be used to corroborate placement made through the key, and the appropriate key to genera can then be entered with assurance. At the generic level, reliance on the illustrations alone can lead to misidentification because the plates do not necessarily include every condition of a character represented within a genus. Users should also note that the diagnostic characters are based on the final instar; some diagnostic characters are probably effective for at least the later of the subterminal instars, but others, especially those relating to setae and gills, can change at each instar. Diagnostic characterization of subterminal larval instars even at the generic level is not yet possible for Trichoptera in any precise way. It should also be noted that larval stages for six genera have still to be discovered: four in the Limnephilidae and one in each of the Polycentropodidae and Psychomyiidae. These deficiencies are specified in the general sections treating the families concerned. Larval associations for some genera are proposed tentatively on the basis of circumstantial evidence, but the qualifications are stated in the generic section.

Classification and Phylogeny

Comparative data now available on families of Trichoptera, and drawn from considerations of morphology, behaviour, and ecology of all stages support their assembly into five natural groups (Fig. 1): free-living forms, saddle-case makers, purse-case makers, net-spinners or retreat-makers, and tube-case makers. Morphological similarity among members of the first three of these groups is so strong that usually they are grouped together; the superfamily Rhyacophiloidea (Ross 1967) appropriately denotes this relationship and is used in the present work. Assignment of the net-spinners or retreat-makers to the superfamily Hydropsychoidea, and the tube-case makers to the Limnephiloidea, also proposed by Ross (1967), is useful and is adopted here.

In an earlier system of classification Martynov (1924, 1930) recognized two suborders: the Annulipalpia comprising those families listed in the classification which follows under Hydropsychoidea and Rhyacophiloidea; and the Integripalpia, equivalent to the Limnephiloidea here. These categories are no longer as useful as perhaps they once were, partly because of the differing view that the Rhyacophilidae, Glossosomatidae, and Hydroptilidae (i.e. the Rhyacophiloidea) should be placed in the Integripalpia rather than the Annulipalpia (Ross 1956, 1967); most workers, however, are largely agreed that the three superfamily groups are natural units. But the major disadvantage of the Integripalpia-Annulipalpia classification system in whatever way the two groups are defined is that because these two categories approximate only grossly the five basic biological groups of the Trichoptera, they are not useful for generalization – an important function of higher classification. The system of three superfamilies overcomes most of these shortcomings, and seems largely adequate for theoretical considerations as well. A recent comprehensive review of the Trichoptera by Malicky (1973) also adopts the higher classification of three superfamilies. On this basis, then, the classification of the Nearctic families and genera follows; bibliographic data for most of the genera are available in the *Trichopterorum Catalogus* (Fischer 1960–73), and for more recent names in the generic summaries of the present work. Families, subfamilies, and tribes are listed in a general trend from primitive to derived, in accordance with current concepts where available; genera are listed alphabetically.

Superfamily **HYDROPSYCHOIDEA**

Family	Subfamily	Genus
Philopotamidae	Philopotaminae	Dolophilodes *Ulmer* Wormaldia *McLachlan*
	Chimarrinae	Chimarra *Stephens*
Psychomyiidae	Psychomyiinae	Lype *McLachlan* Psychomyia *Latreille* Tinodes *Curtis*
	Paduniellinae	Paduniella *Ulmer*
	Xiphocentroninae	Xiphocentron *Brauer*
Polycentropodidae	Polycentropodinae	Cernotina *Ross* Cyrnellus *Banks* Neureclipsis *McLachlan* Nyctiophylax *Brauer* Polycentropus *Curtis* Polyplectropus *Ulmer*
	Dipseudopsinae	Phylocentropus *Banks*
Hydropsychidae	Arctopsychinae	Arctopsyche *McLachlan* Parapsyche *Betten*
	Diplectroninae	Aphropsyche *Ross* Diplectrona *Westwood* Homoplectra *Ross* Oropsyche *Ross*
	Hydropsychinae	Cheumatopsyche *Wallengren* Hydropsyche *Pictet* Potamyia *Banks* Smicridea *McLachlan*
	Macronematinae	Leptonema *Guérin-Méneville* Macronema *Pictet*

Superfamily **RHYACOPHILOIDEA**

Family	Subfamily	Genus
Rhyacophilidae	Rhyacophilinae	Himalopsyche *Banks* Rhyacophila *Pictet*
	Hydrobiosinae	Atopsyche *Banks*
Glossosomatidae	Glossosomatinae	Anagapetus *Ross* Glossosoma *Curtis*
	Agapetinae	Agapetus *Curtis*
	Protoptilinae	Culoptila *Mosely* Matrioptila *Ross* Protoptila *Banks*
Hydroptilidae	Ptilocolepinae	Palaeagapetus *Ulmer*
	Hydroptilinae	Agraylea *Curtis* Dibusa *Ross* Hydroptila *Dalman* Ochrotrichia *Mosely* Oxyethira *Eaton* Stactobiella *Martynov*
	Orthotrichiinae	Ithytrichia *Eaton* Orthotrichia *Eaton*
	Leucotrichiinae	Alisotrichia *Flint* Leucotrichia *Mosely* Zumatrichia *Mosely*
	Incertae sedis	Mayatrichia *Mosely* Neotrichia *Morton*

Superfamily **LIMNEPHILOIDEA** (Limnephilid branch)

Family	Subfamily	Genus
Phryganeidae	Yphriinae	Yphria *Milne*
	Phryganeinae	Agrypnia *Curtis* Banksiola *Martynov* Fabria *Milne* Hagenella *Martynov* Oligostomis *Kolenati* Oligotricha *Rambur* Phryganea *Linnaeus* Ptilostomis *Kolenati*
Brachycentridae		Adicrophleps *Flint* Amiocentrus *Ross* Brachycentrus *Curtis* Eobrachycentrus *Wiggins* Micrasema *McLachlan* Oligoplectrum *McLachlan*
Limnephilidae	Dicosmoecinae	Allocosmoecus *Banks* Amphicosmoecus *Schmid* Cryptochia *Ross* Dicosmoecus *McLachlan* Ecclisocosmoecus *Schmid* Ecclisomyia *Banks* Ironoquia *Banks* Onocosmoecus *Banks*
	Pseudostenophylacinae	Pseudostenophylax *Martynov*
	Limnephilinae	TRIBE STENOPHYLACINI Chyranda *Ross* Clostoeca *Banks* Desmona *Denning* Hydatophylax *Wallengren* Philocasca *Ross* Pycnopsyche *Banks*

Superfamily **Limnephiloidea** continued

Family	Subfamily	Genus
Limnephilidae cont'd		TRIBE LIMNEPHILINI
		Anabolia *Stephens*
		Arctopora *Thomson*
		Asynarchus *McLachlan*
		Clistoronia *Banks*
		Grammotaulius *Kolenati*
		Halesochila *Banks*
		Hesperophylax *Banks*
		Lenarchus *Martynov*
		Leptophylax *Banks*
		Limnephilus *Leach*
		Nemotaulius *Banks*
		Philarctus *McLachlan*
		Platycentropus *Ulmer*
		Psychoronia *Banks*
		TRIBE CHILOSTIGMINI
		Chilostigma *McLachlan*
		Chilostigmodes *Martynov*
		Frenesia *Betten & Mosely*
		Glyphopsyche *Banks*
		Grensia *Ross*
		Homophylax *Banks*
		Phanocelia *Banks*
		Psychoglypha *Ross*
	Apataniinae	TRIBE APATANIINI
		Apatania *Kolenati*
	Neophylacinae	Farula *Milne*
		Neophylax *McLachlan*
		Neothremma *Dodds & Hisaw*
		Oligophlebodes *Ulmer*
	Goerinae	TRIBE LEPANIINI
		Goereilla *Denning*
		Lepania *Ross*
		TRIBE GOERINI
		Goera *Stephens*
		Goeracea *Denning*
		Goerita *Ross*

Superfamily **Limnephiloidea** continued

Family	Subfamily	Genus
Limnephilidae cont'd	Incertae sedis	Imania *Martynov* Manophylax *Wiggins* Moselyana *Denning* Pedomoecus *Ross* Rossiana *Denning*
Lepidostomatidae		Lepidostoma *Rambur* Theliopsyche *Banks*

Superfamily **LIMNEPHILOIDEA** (Leptocerid branch)

Family	Subfamily	Genus
Beraeidae		Beraea *Stephens*
Sericostomatidae	Sericostomatinae	Agarodes *Banks* Fattigia *Ross & Wallace* Gumaga *Tsuda*
Odontoceridae	Odontocerinae	Marilia *Müller* Namamyia *Banks* Nerophilus *Banks* Parthina *Denning* Psilotreta *Banks*
	Pseudogoerinae	Pseudogoera *Carpenter*
Molannidae		Molanna *Curtis* Molannodes *McLachlan*
Helicopsychidae		Helicopsyche *von Siebold*
Calamoceratidae		Anisocentropus *McLachlan* Heteroplectron *McLachlan* Phylloicus *Müller*

Superfamily **Limnephiloidea** continued

Family	Subfamily	Genus
Leptoceridae	Leptocerinae	TRIBE ATHRIPSODINI Ceraclea *Stephens* TRIBE LEPTOCERINI Leptocerus *Leach* Nectopsyche *Müller* TRIBE MYSTACIDINI Mystacides *Berthold* Setodes *Rambur* Triaenodes *McLachlan* TRIBE OECETINI Oecetis *McLachlan*

Phylogeny of the families of Trichoptera has been investigated intensively by Ross (1956, 1967), and ancestral and derivative characters outlined for larvae and adults. It is sufficient for general orientation to summarize his interpretation: the Hydropsychoidea and Rhyacophiloidea were considered to have been derived from ancestral Trichopteran stock in a basal dichotomy, with their most primitive living families the Philopotamidae and Rhyacophilidae, respectively; and the Limnephiloidea were considered to have been subsequently derived from rhyacophiloid ancestors, extant families arising from two general lineages - the limnephilid branch and the leptocerid branch. Riek (1970) proposed, however, that larvae of the Hydropsychidae are the most generalized of the Trichoptera, and that some of the case-making families, essentially those of the leptocerid branch, were probably derived independently of the others from different types of retreat-making ancestors.

Biological Considerations

ANCESTRAL HABITAT

The five basic groups of caddisfly families (Fig. 1) discussed above under classification and phylogeny are all represented in cool, running waters. Further, it has been shown from morphological evidence that in families represented in both lotic and lentic waters, the genera with more ancestral characters occur in cool, lotic habitats and those with derived character states occur in warm, lentic sites (Ross 1956). Accordingly, cool, running waters are believed to be the ancestral habitat in which progenitors of the Trichoptera first became aquatic, and the habitat in which differentiation into the five basic groups probably occurred. These groups represent essentially different methods of living and feeding by larvae, and it is likely that their differences evolved as more efficient means of exploiting available niches in lotic habitats. It is a common observation that, apart from the Diptera, the Trichoptera are usually more numerous in species and more diverse biologically in a given lotic habitat than other aquatic insect orders; this can be seen as an indication of the effectiveness of basic ecological diversification in the Trichoptera. That all caddisfly larvae produce silk was undoubtedly an asset in their ecological diversification because silk adds, in effect, a whole new dimension to behavioural evolution; the situation is analogous to the role of silk in the very marked ecological diversification of spiders.

HABITAT DIVERSITY

It will be seen from Figure 1 that families of North American Trichoptera differ markedly in the ability of their members to exploit warm, lentic habitats: thus all 18 (100%) North American caddisfly families are represented in cool, running waters; 17 (95%) in warm, lotic sites; 7 (38%) in standing waters of lakes and marshes, excluding those warm-lotic families whose larvae also live along wave-washed shores of lakes (i.e. Hydropsychidae, Sericostomatidae, and Helicopsychidae); and only 3 families (17%) in temporary pools, a habitat that for caddisfly larvae can be considered a more demanding extension of permanent, lentic waters.

In Figure I it will be recognized that rigid distinction between habitat categories is for the most part impossible; the dotted lines between them are to be interpreted as broad bands of intergrading conditions. This series of habitat types grades through decreasing water currents and increasing water temperatures; and current was shown to be critical in shortening the diffusion path by which oxygen actually becomes available for respiration by caddis larvae (Jaag and Ambühl 1964). Consequently, it would seem that the general trend of the series of habitat types in Figure I leads to steadily increasing selection pressure for greater respirational efficiency. If this premise is correct and enhanced respirational efficiency is one of the requirements for caddisfly larvae living in lentic waters, it can hardly be coincidence that in five of the seven families with representatives in lentic waters, the larvae are tube-case makers: Leptoceridae, Molannidae, Phryganeidae, Limnephilidae, and Lepidostomatidae. The two exceptions are the Polycentropodidae and the Hydroptilidae.

RESPIRATION

That the portable tube-case of Trichoptera enhances respiratory efficiency is an idea shared by students of these insects at least since Dodds and Hisaw (1924) and Milne (1938), and possibly others before them. The principle is that dorsoventral abdominal undulation or ventilation by the larva brings a steady current of water through the anterior opening and out the posterior opening, a flow that can be easily detected by an observer; thus abdomen and gills are bathed in a current of continuously renewed water (Fig. II). The tubular case serves then as a conduit for a channelled flow of water, enabling the larva to create its own current. According to this theory, the three humps of abdominal segment I serve to provide a space between the larva and the sides of the case, allowing the respiratory current to bathe all sides of the abdomen. It should be added here that at least some larvae can reverse the direction of water flow (Merrill and Wiggins 1971; Tindall 1963).

Supporting evidence for the respiratory advantage of the tube-case has come from several sources. Jaag and Ambühl (1964) found that larvae of the limnephilid genus *Anabolia*, when inside their cases, were able to survive at lower oxygen concentrations than those without cases because they were able to remove more oxygen from the water. Several workers have shown that a rise in temperature or a decrease in dissolved oxygen results in increased rate and amplitude of ventilatory movement of case-bearing caddisfly larvae (Van Dam 1938; Fox and Sidney 1953); Feldmeth (1970) found that in two species of *Pycnopsyche* the rate of ventilation increased as current speed decreased, indicating that abdominal ventilation compensates for low current velocity. Philipson (1954) demonstrated that the frequency of ventilation movements in *Hydropsyche instabilis* and *Polycentropus flavomaculatus* decreased with increasing current. Larvae of rheophilic genera such as *Rhyacophila* and the philopotamid *Wormaldia* do not normally ventilate by abdominal movements (Philipson 1954), although under duress larvae of these two families and of the Glossosomatidae are said to ventilate (Tomaszewski 1973). This is, of course, still a picture painted with broad strokes, but the evidence indicates that the most rheophilic caddis larvae are dependent upon stream current for renewal of freshly oxygenated

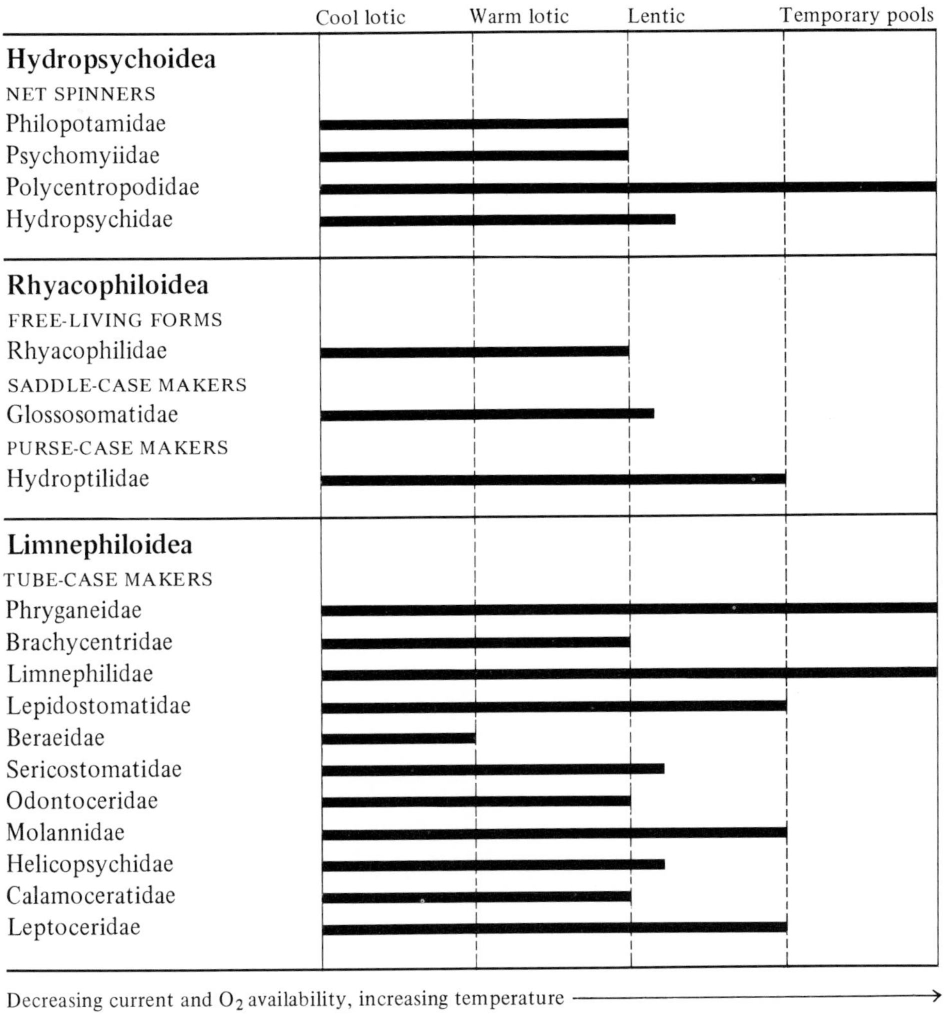

Cool lotic Warm lotic Lentic Temporary pools

Hydropsychoidea

NET SPINNERS

Philopotamidae

Psychomyiidae

Polycentropodidae

Hydropsychidae

Rhyacophiloidea

FREE-LIVING FORMS

Rhyacophilidae

SADDLE-CASE MAKERS

Glossosomatidae

PURSE-CASE MAKERS

Hydroptilidae

Limnephiloidea

TUBE-CASE MAKERS

Phryganeidae

Brachycentridae

Limnephilidae

Lepidostomatidae

Beraeidae

Sericostomatidae

Odontoceridae

Molannidae

Helicopsychidae

Calamoceratidae

Leptoceridae

Decreasing current and O_2 availability, increasing temperature →

1 Habitat diversity in families of North American Trichoptera

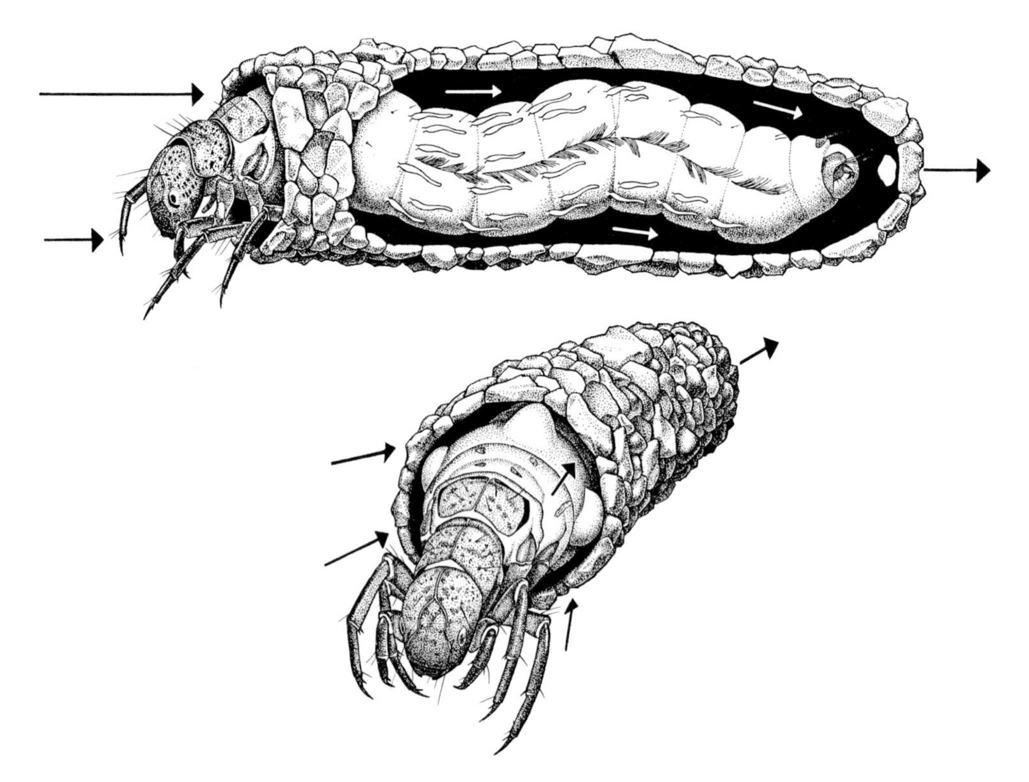

II Circulation of water generated by abdominal ventilation through case of typical case-making caddis larva

water, and less rheophilic larvae can create their own current by abdominal ventilation at rates geared to acquire the necessary oxygen with the lowest energy expenditure.

Although the Hydropsychoidea are largely rheophilic, certain members of the Polycentropodidae are tolerant of lentic waters (Fig. I). Some species of *Polycentropus*, for example, live on lake bottoms with *Chironomus* larvae that are provided with haemoglobin; and some *Polycentropus* live in temporary pools, where larvae make a fixed tubular retreat of silk (Fig. 15.5F) within which they ventilate by abdominal undulation (Wiggins 1973a).

Since larvae of a few purse-case hydroptilids occur in lentic waters and ventilate in the normal way, there is little doubt that the portable purse-case with its ventral slits can serve as an effective tubular conduit for a respiratory current. Among Nearctic hydroptilids at least, it is worth noting that species of *Oxyethira* probably occupy the most lentic habitats; their cases are not the typical purse with slits, but a flattened flask with openings confined to front and rear (Fig. 7.11B). One may wonder to what extent the ability of these larvae to exploit lentic waters is correlated with a more tube-like case architecture.

We see then (Fig. I) that representatives from all three superfamilies of Trichoptera have invaded lentic waters independently, and that certain of the Limnephiloidea and Hydropsychoidea have further succeeded in colonizing temporary pools. Even though the superfamilies differ biologically in many ways, the means by which their members have adapted to lentic waters are similar in principle; thus, the basic caddisfly larval tube, either portable or fixed, can be regarded as a device to increase respirational efficiency, thereby helping to open lentic habitats to exploitation by these insects. It is altogether remarkable that an increase in respirational efficiency has been achieved as a consequence of silk production.

FEEDING

The role of caddisfly cases in respiration should not obscure other functions because the structures built by all caddisfly larvae are, of course, critically important in feeding, too. Many of the net-spinning Hydropsychoidea build a fixed retreat or shelter in which the larva lies while food particles of appropriate size are collected from the current by its silken capture net. In a functional sense, these retreat-makers are filter feeders (Cummins 1973), and their food includes small fragments of plant and animal materials, faeces of other invertebrates, algal cells, and in some instances other invertebrates. Food of the Philopotamidae is very fine particulate materials; larvae of the Hydropsychidae ingest small invertebrates as well, and over all are considered omnivorous; within the Polycentropodidae some genera are specialized for fine-particle collecting, but others are highly predacious. In the Psychomyiidae, detrital particles, associated microflora, and algae are also collected, but probably largely from the substrate surface.

Among the Rhyacophiloidea, most larvae of the free-living Rhyacophilidae are predators, but not entirely so for some *Rhyacophila* are herbivorous (Thut 1969). It seems likely that the flat-bottomed, tortoise-like, portable case of the highly rheophilic Glossosomatidae evolved as a sheltering dome enabling these larvae to move on to the exposed upper surfaces of rocks where they could scrape growths of diatoms and the fine organic particles that collect there. All three thoracic legs in glossosomatid larvae are approximately the same size (Fig. 4.4A), as they are in the Rhyacophilidae and Hydropsychoidea, a condition that can be considered primitive for the Trichoptera. But in some of the Hydroptilidae and in all of the Limnephiloidea, the legs are progressively longer from front to rear; and except for highly specialized swimmers such as *Leptocerus*, these larvae are generally adept at walking with the last two pairs of legs (Tomaszewski 1973). Therefore, for feeding, the evolutionary significance of portable cases - of saddle, purse, or tube design - is that larvae were able to move more actively in search of energy sources. Many of the Hydroptilidae became specialized for feeding on filamentous algae, although a few groups feed on periphyton and fine organic particles. A few of the Limnephiloidea - some *Oecetis* and *Ceraclea* - became predators; some became diatom feeders as in the Apataniinae, Neophylacinae, and Goerini.

For the Limnephiloidea, development of respirational independence allowed the exploitation of food unrelated to current. Many genera, particularly in the Limnephilidae and Lepidostomatidae, became largely dependent on dead plant materials torn from leaves and wood fragments which support dense growths of fungi and bacteria; the microorganisms evidently have much to do with the palatability and nutrient value of these materials (Bärlocher and Kendrick 1973, 1975). Larvae feeding in this way have come to be known as shredders (Cummins 1973), playing one of the most generally important ecological roles in fresh waters, for in this manner leaves and other large pieces of plant materials are reduced to the fine organic particles utilized by collectors. Information available to date suggests that omnivorous feeding is the general condition for larvae in most genera of the Phryganeidae, Molannidae, Leptoceridae, and Odontoceridae.

These general aspects of feeding are summarized in Table 1, based on, but amplifying for Trichoptera, the useful classification of trophic categories for aquatic insects generally, developed by Cummins (1973). This must be seen only as a general overview; several families appear in more than one category on the basis of different genera, although the generic examples listed are not necessarily complete for a given family. A further complication arises in that the food of certain species can change seasonally or as the larvae grow larger; *Banksiola crotchi* became almost totally predacious in the final instar after feeding on detrital materials up to that point (Winterbourn 1971b). Other detailed analyses of food ingested by larval Trichoptera reveal a strong tendency for opportunistic feeding within single species, evidently limited mainly by the size of the pieces a larva is capable of ingesting from among those available; one example is the finding by Cummins (1973) that larvae of *Glossosoma nigrior* fed largely on periphyton in a Pennsylvania stream but on detritus in a Michigan stream. Trophic generalism is, in fact, the pattern of feeding by aquatic insect larvae over all (Cummins 1973). For Trichoptera, then, the central point emerges that in evolution leading to marked divergence in ways of obtaining food, there has been little restrictive specialization in the kind of food that can be utilized.

CASE-MAKING

Although individual shelters and cases are constructed by larvae in certain groups of beetles and moths, the variety of these structures is much greater in the Trichoptera than in any other single group of insects. This is a considerable asset in studying the evolution of case-making behaviour. After making comparative analyses of the steps involved in the various constructions by Trichoptera and cocoon-making by Lepidoptera, Ross (1964) and Malicky (1973) suggested a close basic similarity in this behaviour in the two orders. Analysis of case-building behaviour in the Trichoptera on a more detailed level has been undertaken by several workers, such as Hansell (1972, 1974) and Hanna (1960).

Ability of caddisfly larvae to recognize and re-enter their own cases was found by Merrill (1969) to be considerably different in several families; she also found that sensors on the anal claws are important in regulating the maximum length to which a case is built (Merrill 1965). Behaviour leading to construction of portable tubular cases appears to have arisen independently at least twice, once in the Limnephiloidea and once in the hydroptilid genera *Neotrichia* and *Mayatrichia* of the Rhyacophiloidea. In fact, Riek (1970) suggested that the limnephiloid families have apparently arisen more than once from different types of non-case-making ancestors.

Although the cases and retreats built by larval caddisflies are extremely important in respiration and feeding, consideration of their structural details in relation to microhabitat is illuminating in other respects. For the most part these structures are lined with silken strands, and plant and rock pieces are incorporated on the outside. But one can find examples of ways in which this basic behaviour has evolved to solve many engineer-

TABLE 1
Classification of trophic categories for North American Trichoptera (based on Cummins 1973)

Method of feeding		Dominant food	North American Trichoptera
Shredders		Herbivores feeding on living vascular hydrophytes and filamentous algae	Brachycentridae (*Eobrachycentrus, Micrasema*) Hydroptilidae (Hydroptilinae) Leptoceridae (*Triaenodes*) Phryganeidae
		Detritivores feeding on pieces of decomposing vascular plant tissue and associated microflora	Beraeidae Calamoceratidae Lepidostomatidae Limnephilidae (Dicosmoecinae, Lepaniini, Limnephilinae, Pseudostenophylacinae) Odontoceridae Phryganeidae Sericostomatidae
Collectors	Filter or suspension feeders	Detritivore-herbivores feeding on fine organic particles and living algal cells	Brachycentridae (*Brachycentrus, Oligoplectrum*) Hydropsychidae Philopotamidae Polycentropodidae (*Neureclipsis, Phylocentropus*)
	Substrate surface feeders	As above	Brachycentridae Leptoceridae Psychomyiidae
Scrapers		Herbivore-detritivores feeding on periphyton and fine organic particles	Glossosomatidae Helicopyschidae Hydroptilidae (*Ithytrichia, Leucotrichiinae*) Leptoceridae Limnephilidae (Apataniinae, Goerini, Neophylacinae) Molannidae Limnephilidae (Dicosmoecinae)
Predators		Carnivores feeding on whole animals or large parts	Hydropsychidae Leptoceridae (*Ceraclea, Oecetis*) Molannidae Phryganeidae Polycentropodidae (*Nyctiophylax, Polycentropus*) Rhyacophilidae

ing problems: streamlining, ballast, buoyancy, structural rigidity, camouflage, internal water circulation, external water resistance, regulation of mesh size in nets to obtain an effective compromise between current speed and fine-particle filtering, protection from predators that would swallow the case and from those that would intrude, and so on. Some of these solutions in relation to rapid currents were explored by Dodds and Hisaw (1925); comments on many of the others are offered in appropriate places in the systematic part of the present work. Again, the significance of silk production in increasing the diversity of solutions, and hence of niches exploited, seems clear. It can be added that this is a very rich area for experimental studies in behaviour and adaptive evolution, and one that is little explored.

The often-repeated idea that caddisfly species can be distinguished by the cases they build has only slight support. By and large, case architecture is characteristic at the generic level. If in a particular area a genus is represented by only one species, the case is likely to be of diagnostic value, but in a series of several congeneric species case types are generally not diagnostic for the species.

LIFE CYCLES

Most caddisflies in temperate latitudes complete one generation each year, passing through five larval instars, a pupal stage, and a winged adult stage; the time required for completion of the actual metamorphosis, i.e. from separation of the larval cuticle to eclosion or emergence of the adult from the pupal skin, is very generally of the order of 3 weeks. For purposes of discussion it is not inappropriate to regard this as the generalized condition, characterized by uninterrupted development. In distinction from many other biological characteristics of Trichoptera, most of the modifications imposed on this generalized condition are a feature of the species rather than the genus. A life cycle of six larval instars was recorded for a European species of *Sericostoma* (Elliott 1969) and of seven for a European *Agapetus* (Nielsen 1942).

Some modifications to the generalized condition arise through intervention of diapause – a suspension of normal development at some stage in the life cycle until initiated again in response to an external environmental stimulus. Diapause is distinguished from simple quiescence in which development or activity merely slows during adverse, usually cold, periods to be resumed when conditions again become favourable. For Trichoptera, diapause functions as for other insects by suspending development until conditions in a habitat are most favourable, and also by synchronizing adult emergence after periods of dissimilar larval development.

Diapause was demonstrated in the last larval instar of the European species *Anabolia furcata* (Novák 1960); larvae were fully grown in June, fastened their cases to the substrate, and ceased feeding. Experiments revealed that diapause was terminated by short daily photoperiods, similar to those occurring naturally in late summer. Pupation occurred in nature in September, adult emergence commencing toward the end of that month. Some influence was also attributed to temperature in termination of the diapause, temperatures lower than 20°C hastening termination; adults were found to emerge earlier at higher elevations than those of the same species at lower elevations. Novák suggested that diapause in the last larval instar was a general condition for univoltine autumnal Trichoptera. This may well be so because one, and perhaps the principal, advantage of autumnal reproduction is that larvae feeding on fallen leaves of deciduous trees have a particularly rich food resource in autumn and early winter; larval development completed in winter or early spring must then be suspended until autumn approaches again. The same principle probably holds for the Nearctic genus *Pycnopsyche*; although diapause has not been demonstrated, inactivity of final-instar larvae for periods up to 6 months, followed by initiation of metamorphosis during the decreasing daily photoperiods of late summer (Cummins 1964; Mackay 1972), suggests that it is operating as in *Anabolia furcata*.

Similarly, larval feeding and growth in most species of *Neophylax* are completed in spring or early summer, whereupon larvae fasten their cases firmly to the substrate and seal off the entrance. Larvae remain within the case for several weeks, and metamorphosis occurs in late summer followed soon after by adult emergence. Most species of *Dicosmoecus* show a similar pattern. Synchronization of adult emergence may also be an advantage in these instances. It is interesting to note that species of both genera are autumnal for the most part and, perhaps as a consequence of lower temperatures at night, the adults are largely diurnal in activity. Daytime activity of caddisfly adults is found in relatively few species of Trichoptera over all, and there seems a possibility that synchronized emergence of adults might help to compensate for the greater hazards of diurnal reproductive activity. For some *Neophylax* that live in streams of temporary flow, larval diapause serves to postpone oviposition until after summer drought (Wiggins 1973a).

Diapause in the adult stage is advantageous for species of the Limnephilidae and Phryganeidae inhabiting temporary pools (Novák and Sehnal 1963, 1965; Wiggins 1973a); sexual maturity of adults emerging in spring is delayed until late summer, when development is resumed, largely through the stimulus of short daily photoperiods. This postponement of oviposition places eggs in the pool basins in autumn when the moisture level of surface soil is being replenished, thereby avoiding the summer drought. Embryonic development proceeds and larvae break out of the egg chorions but can remain in the gelatinous egg-matrix for several months, even until spring, until flooded by surface water in the pool basin (Wiggins 1973a). In this way, diapause in the adult brings about the basic delay in the life cycle, and larval development can coincide with the occurrence of temporary autumnal pools. The additional delay required for larvae to exploit temporary vernal pools appears to be provided largely by the more stable nature of the gelatinous egg-matrix and by the tendency of the larvae to remain within it until the stimulus of surface water is received. There is some evidence that small amounts of surface water can provide sufficient stimulus for larvae to leave the matrix and build cases.

Final-instar larvae of some species of *Ironoquia* aestivate in unsealed cases around the edge of temporary pools and streams during the summer periods of declining water levels, with metamorphosis taking place in late summer (Flint 1958; Williams and Williams 1975). Although photoperiod control has not been demonstrated, it seems likely that this situation differs from the preceding one only in the imposition of diapause on the last larval instar; the advantage in avoiding the summer drought period of temporary waters is the same.

Diapause of 8 to 9 months was reported in the egg of *Agapetus bifidus* in Oregon (Anderson and Bourne 1974). This may be the mechanism by which other species of *Agapetus* are able to populate temporary streams, as reported in Illinois by Ross (1944).

Life-cycle modification leading to accommodation of two congeneric species in the same habitat appears to be operating in *Neophylax*. Most species of this genus have autumnal emergence, but adults of the eastern *N. ornatus* emerge in spring. Life history data compiled by Mackay (1969) reveal that larval feeding of this species occurs from summer through to December; larvae of *N. nacatus*, an autumnal species in the same habitat, feed during winter and spring. Coexistence of species of *Pycnopsyche* in the same general habitat is accommodated through differences in their life history, behaviour, and feeding (Cummins 1964; Mackay 1972).

Although the univoltine life cycles described above are the usual condition in Trichoptera, some species are known to be bivoltine. For example, two overlapping generations per year were reported for *Glossosoma penitum* in Oregon (Anderson and Bourne 1974). Larval populations of *Ceraclea transversa* (syn. *Athripsodes angusta*) were found in two cohorts in Kentucky (Resh 1976); one overwinters in the last larval instar with cases sealed in preparation for metamorphosis and emerges in spring, and the other overwinters in the third or fourth instar and emerges in summer. By contrast, larvae of some species in the Odontoceridae, Calamoceratidae, Beraeidae, and Limnephilidae evidently require more than one year to complete a cycle; this is a difficult point to interpret from collections for in addition to collection of specimens representing a broad span of developmental stages at the same time, one also needs to have some information about the length of the adult emergence period.

These are some examples of deviations from the generalized life cycle. All of them have come to light in recent years, and it is clear that we are just beginning to appreciate how very diverse the life cycles of caddisflies really are.

The final point to be discussed under life cycles concerns the term *prepupa*. This has been widely used by workers on Trichoptera for the interval from the time when the pupal case is fastened to the substrate and the anterior and posterior openings are sealed off, to the time when larval-pupal ecdysis occurs, i.e. the point at which the external form of the insect changes from larva to pupa. But used in this sense, as Hinton (1971) has pointed out, the term prepupa covers several biological stages and events: (a) the larva sealed within the pupal case in a resting condition for intervals from several days to several months in genera such as *Pycnopsyche, Neophylax*, and *Dicosmoecus*; (b) larval-pupal apolysis in which the larval cuticle is separated from the pupal epidermis; (c) the pupa when histological reorganization of metamorphosis is taking place within but not connected to the larval cuticle. These events are terminated by larval-pupal ecdysis when the larval exuviae are cast off and the newly formed cuticle beneath gives the insect the typical pupal form.

As knowledge of the biology of Trichoptera increases, especially such aspects as the operation of diapause, it becomes necessary to identify more precisely the stages at which certain events occur. Thus, in the present work, the term *prepupa* is restricted to the period (a) above, when the larva is in a resting condition, because it establishes that the functional larva is closed off within the pupal case, and that feeding has terminated.

Larval-pupal apolysis, (b) above, terminates the prepupal stage and marks a significant change because functionally the insect is no longer a larva but a pupa. Larval-pupal apolysis in Trichoptera can probably be detected well enough for practical purposes by noting certain external changes: in the majority of families the middle and hind legs change from the normal active position in which they are partially bent and directed anteriorly to a more or less distorted position in which they become straightened and directed towards the posterior end of the insect; and in all families the eyes and the muscles of the legs and other parts no longer coincide with their position in the overlying larval cuticle.

The term *pharate pupa* (Gr. *pharos*, 'a garment') is used to designate the period, (c) above, when the developing pupa is enclosed within the separated larval cuticle (Hinton 1971). This phase of development is terminated by larval-pupal ecdysis.

When a caddisfly leaves the pupal case to come to the surface for eclosion or emergence from the pupal cuticle, it is functionally an adult, and from the time of pupal-adult apolysis it is termed a *pharate adult* (Hinton 1971). The mandibles of caddisfly pupae are, in fact, moved by muscles of the adult, and the oar-shaped legs of the pupal cuticle are powered by the slender legs of the adult which lie beneath (Hinton 1949).

Morphology

Morphological terms of a general nature can be found in an entomological text or glossary. Those that have particular application to the Trichoptera are illustrated in Figures III–VI, and some additional explanation follows.

A distinction between spines and setae should be made at the outset. *Spines* are extensions or processes of the cuticle; they may be short and pointed, longer and blade-like, comb-shaped, or of some other form. Spines are one type of surface sculpturing, and being an integral part of the cuticle, they are retained on the exuvial sclerites after ecdysis; it is likely that spines are largely protective in function. *Setae* are essentially hairs arising in alveoli or pits in the cuticle; they range widely in size to include stout bristles, flattened scale-hairs, and spurs. Setae are articulated cuticular appendages, sensory in function, and usually become separated from the exuviae at ecdysis leaving the pit to mark their position, although some pits hold sensory receptors which lack setae.

HEAD

Sclerites comprising the head capsule include a *parietal* on each side of the head (Fig. IIIE), the two parietals in contact along a dorsomedian *coronal suture* (Fig. IIIA); between the parietals dorsally is the *frontoclypeal apotome*, separated from them by the *frontoclypeal sutures* (Fig. IIIA). The coronal and frontoclypeal sutures together form the Y-shaped *dorsal ecdysial lines* along which the head sclerites separate at ecdysis (Fig. IIIA). Ventrally the genal areas or *genae* of the parietals are separated along the median line by the *ventral apotome* (gular sclerite) which may separate the genae completely (Fig. IIID)

III Morphology of caddisfly larvae. A, head and thorax of a limnephilid, dorsal, *sa*1 = setal area 1, etc.; B, labrum of a limnephilid, dorsal, primary setae numbered; C, head of a limnephilid, dorsal, primary setae numbered; D, head of a limnephilid, ventral, primary setae numbered; E, head of a limnephilid, lateral; F, gular area of head of a diplectronine hydropsychid, ventral

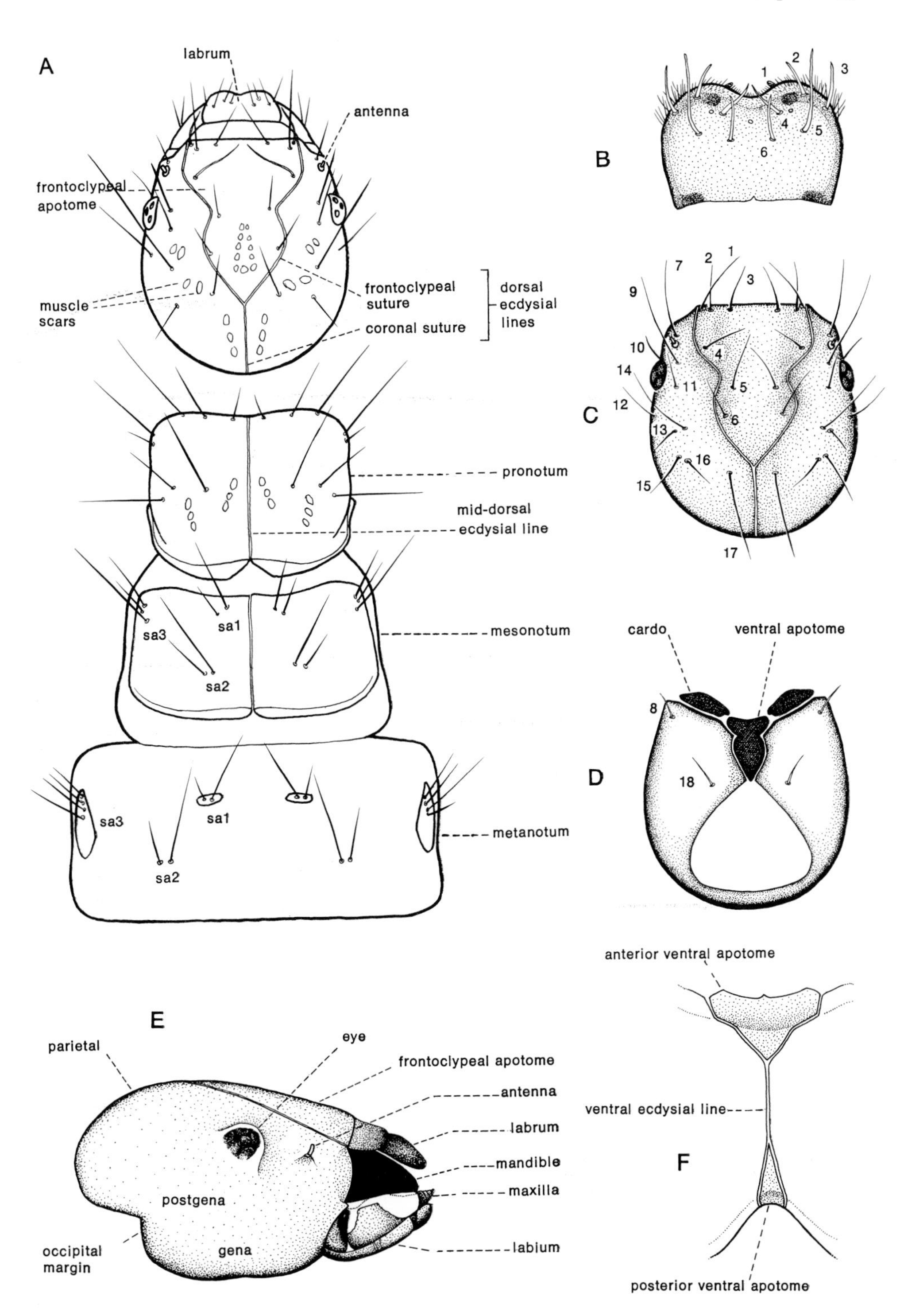
A
labrum
antenna
frontoclypeal
apotome
muscle
scars
frontoclypeal
suture
coronal suture
dorsal
ecdysial
lines
pronotum
mid-dorsal
ecdysial line
sa1
sa3
sa2
mesonotum
sa3
sa1
sa2
metanotum
B
1
2
3
4
5
6
C
1
2
3
4
5
6
7
9
10
11
12
13
14
15
16
17
D
cardo
ventral apotome
8
18
E
parietal
eye
frontoclypeal apotome
antenna
labrum
mandible
maxilla
labium
postgena
gena
occipital
margin
F
anterior ventral apotome
ventral ecdysial line
posterior ventral apotome

or only partially either as a single sclerite or as two - the *anterior* and *posterior ventral apotomes* (Fig. IIIF). At ecdysis the parietals separate ventrally along the *ventral ecydsial line* (Fig. IIIF). The exterior surface of the head capsule is often roughened by spines, ridges, or various types of sculpturing; on the interior, but often visible from the exterior, are round spots or *muscle scars* at attachment points for muscles of the head. The *labrum* is hinged to the anterior edge of the frontoclypeal apotome, and frequently bears a membranous anterolateral fringe (Fig. 10.4B,E). The *eyes* are clusters of ocelli (Fig. IIIE). Antennae in the Limnephiloidea, Glossosomatidae, and Hydroptilidae are rod-like and usually short (Fig. IIIE), but much longer in most Leptoceridae and some Hydroptilidae; in the Rhyacophilidae and Hydropsychoidea antennae are small and not clearly differentiated (e.g. Fig. 13.3B). Primary setae of the head and labrum are usually stable in position and are numbered here (Fig. IIIB,C) in accordance with the system proposed by Nielsen (1942); setae 8 and 18 arise on the ventral surface of the head (Fig. IIID). Secondary setae sometimes occur on the dorsum of the head as in Figure 10.4F.

Mouthparts (Fig. VIB) are identified by standard terms. The silk produced by caddisfly larvae is emitted through a small orifice at the tip of the labium; *labial palpi* are absent in some families, *submental sclerites* differ widely in shape and sometimes are fused. On the *maxilla* at each side of the central labium, the *galea* bears sensillae and is either the rounded lobe illustrated here (Fig. VIB) or a slender finger-like process (Fig. 17.3B). Structures identified in Figure VIB as *cardo, stipes, palpifer, palpiger,* and *mentum* are actually the sclerites borne by these parts. *Mandibles* have cutting edges of two basic types evidently correlated with feeding behaviour - a series of separate points or teeth (Fig. 10.34D), or an entire, scraper-like edge (Fig. 10.33D) - and are articulated with the head capsule by means of a knob-like ventral condyle and a dorsal cavity.

THORAX

The *pronotum* is always covered by two heavily sclerotized plates closely appressed along the *mid-dorsal ecdysial line* (Fig. IIIA); the prosternum frequently bears sclerites. In several of the case-making families there is a membranous, finger-like *prosternal horn* (Fig. VIA). The *trochantin* (Fig. VIA), a derivative of the prothoracic pleuron, is shaped characteristically in different genera.

The *mesonotum* may be largely sclerotized, the plate entire (Fig. 7.14A) or variously subdivided by median or transverse ecdysial lines; or the mesonotum may be membranous and with or without small sclerites. In most of these conditions, mesonotal setae arise in three primary locations - *setal area* 1, *sa*2, and *sa*3 (Fig. IIIA); setae are variously modified in different genera and so numerous in some that the primary setal areas cannot be distinguished (Fig. 2.5B). Arrangement of the sclerites and setae is of taxonomic significance.

IV Morphology of caddisfly larvae, illustrated by generalized limnephilids. A, entire larva, lateral, abdominal segments numbered I-IX; B, abdominal segment I, dorsal; C, abdominal segment I, ventral; D, abdominal segment bearing branched gills, lateral; E, lateral tubercles enlarged in face view and in profile, approx. x470

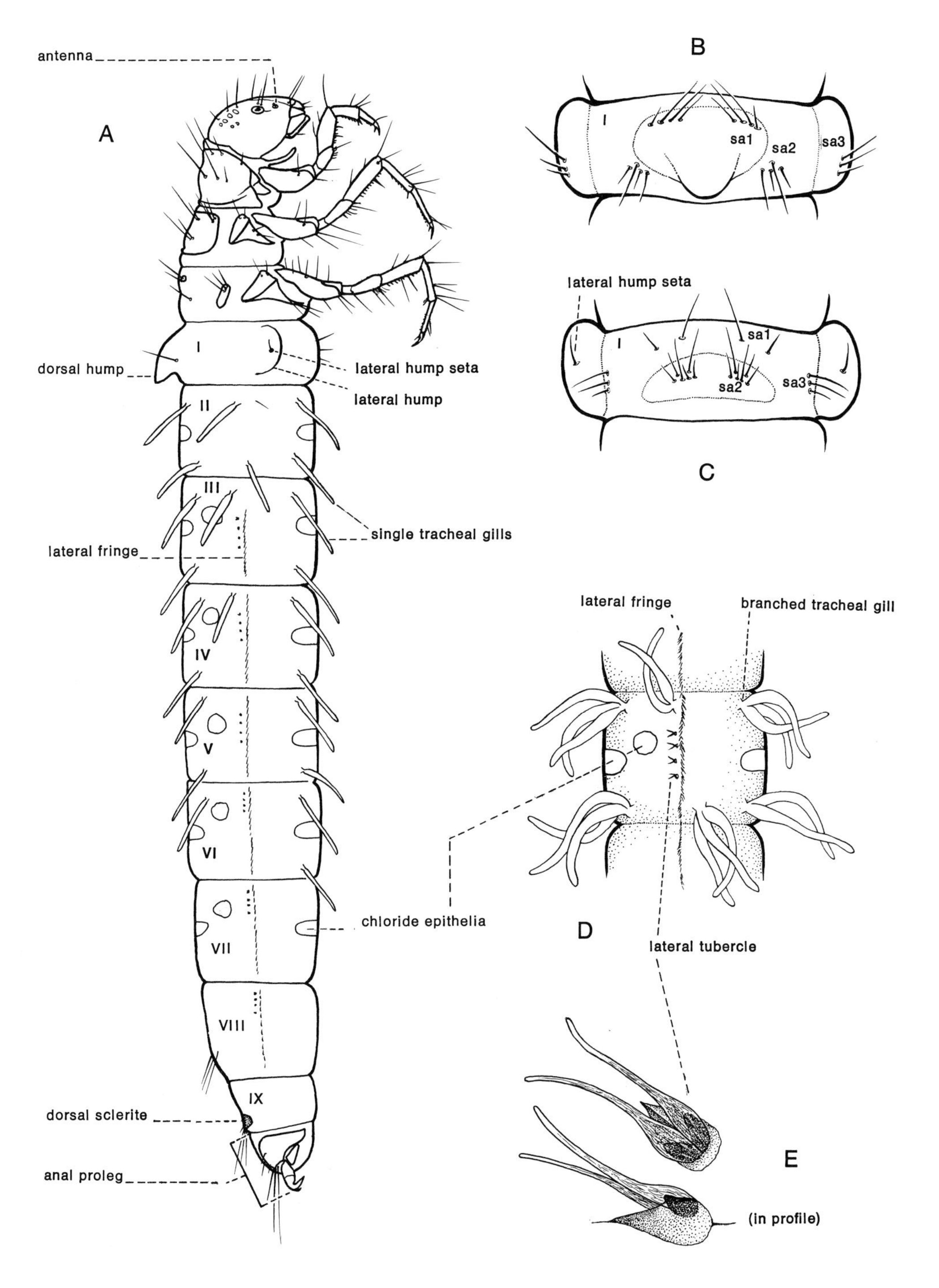
antenna
A
dorsal hump
I
II
III
IV
V
VI
VII
VIII
IX
lateral hump seta
lateral hump
single tracheal gills
lateral fringe
chloride epithelia
dorsal sclerite
anal proleg
B
sa1
sa2
sa3
lateral hump seta
C
lateral fringe
branched tracheal gill
D
lateral tubercle
E
(in profile)

Sclerotization of the *metanotum* is variable; usually the sclerites are smaller than on the mesonotum, but the setal areas are of the same primary arrangement (Fig. IIIA).

The *pleuron* of both the meso- and meta-thorax comprises an anterior *episternum* and a posterior *epimeron*, separated by a darkened depressed line - the *pleural suture* (Fig. VIA). The mesepisternum is extended anteriorly in the Goerinae (Fig. 10.31B).

The thoracic legs in some genera are all approximately the same size, but in most families the fore legs are shorter and their segments stouter. Legs (Fig. VIA) are variously armed with spines, combs, and setae; *spurs* are very stout setae, usually at the distal end of the tibia, and often paired. The *basal seta* of the tarsal claw is often enlarged to spur-like proportions; the tibia and femur are secondarily subdivided into two parts in some genera. The distinction between *major and minor femoral setae* is based on over-all length and thickness. The trochanter is usually divided into two parts; a tuft of hairs often present on the distal part is known as the *trochanteral brush.*

ABDOMEN

Segment I in most families of the Limnephiloidea bears a median *dorsal hump* and at each side a *lateral hump* (Fig. IV A); the humps are retractile and in preserved specimens are often indistinct. On both dorsum and venter of segment I (Fig. IV B,C) setal areas corresponding to *sa*1, *sa*2, and *sa*3 of the thoracic nota can be recognized. This is essentially the arrangement on segment I in such families as the Phryganeidae and Glossosomatidae, and is well developed in the Limnephilidae where the diverse arrangements are of considerable taxonomic value. In some genera of the Limnephilidae it is not possible to distinguish between these setal areas (e.g. *sa*1 and *sa*2 in Fig. 10.12D), a condition which for descriptive purposes I have usually assumed to be a result of amalgamation. The boundary between dorsal and ventral sets is usually a single seta (Fig. IV A,C), sometimes two (Fig. 10.44A), on the ventral part of the lateral hump, here designated the *lateral hump seta(e).* The lateral humps of segment I in several groups bear lightly pigmented sclerites which are important taxonomically; since these sclerites are often best distinguished by their more rigid and shiny surface, careful examination under good illumination is necessary to distinguish them.

Segment IX bears a *dorsal sclerite* (Fig. IV A) in some families, and the arrangement of setae on it is of taxonomic significance. Occasionally when the sclerite is not pigmented, it can be detected by the firm, shiny surface.

The *anal prolegs* exhibit significant structural diversity, and homologies designated here for component parts are largely those proposed by Ross (1964). The condition in Figure VA, from the Hydropsychoidea, is considered to represent the primitive type for the Trichoptera; in these retreat-making families and also in the free-living Rhyacophilidae, the anal prolegs are elongate, separate, and very mobile. Bridging the flexible membranous connection between the *lateral sclerite* and the *anal claw* is a slender *dorsal plate* and *ventral sole plate.* In the derivative condition seen in the Limnephiloidea (Fig. VB), it is believed that the prolegs have become short and thick, their bases swollen to appear almost as an additional abdominal segment (Fig. VD); short, stout prolegs seem appropriate to enable the larva to grip the interior of its close-fitting tubular case with the anal claws, usually armed with a stout *accessory hook.* In most genera of the Limnephiloidea

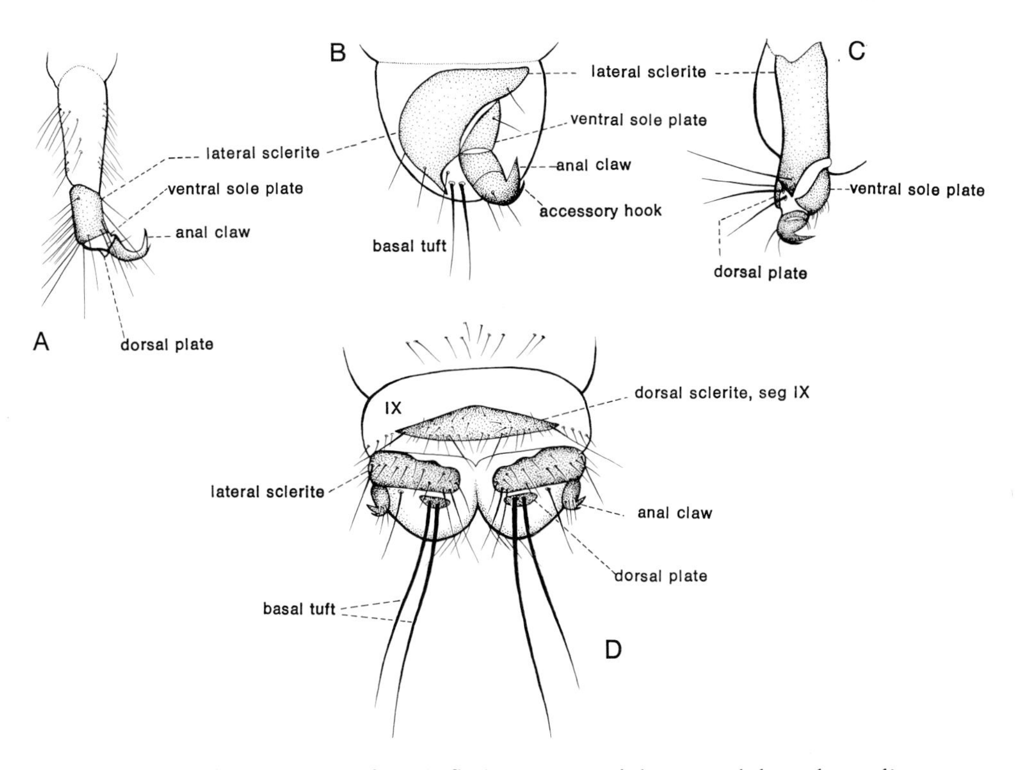

V Morphology of anal prolegs of caddisfly larvae. A, a philopotamid, lateral; B, a limnephilid, lateral; C, a glossosomatid, lateral; D, a goerine limnephilid, dorsal, including segments VIII and IX

there is little evidence of a *dorsal plate* (Fig. VB), but a small scerite incorporating the bases of the stout, terminal setae termed the *basal tuft* does occur in a few genera (Fig. VD), such as *Goereilla* (Fig. 10.20F), *Goeracea* (Fig. 10.19G), and *Goerita* (Fig. 10.21D); in *Lepania* (Fig. 10.31G) a sclerite without setae occurs ventrolaterad of the basal tuft. Both types are interpreted here as dorsal plates. The Glossosomatidae of the Rhyacophiloidea (Fig. VC) are considered to represent an intermediate condition in which some shortening of the proleg and reduction of the size of the anal claw are evident relative to the condition of these structures in the Hydropsychoidea.

The *lateral fringe* (Fig. IVA,D) is a line of fine setae usually extending along each side of most segments in genera of the Limnephiloidea. In many Limnephiloidea there is also a row of tiny forked, sclerotized processes called *lateral tubercles* (Fig. IVD,E) on each side of certain abdominal segments; these have also been termed bifid processes (Flint 1960). In the Limnephilidae, Lepidostomatidae, and Brachycentridae lateral tubercles occur on most abdominal segments; in most genera of the leptocerid branch tubercles are confined to segment VIII, although to segment VII in the Sericostomatidae. Lateral tubercles are absent from the Phryganeidae.

Respiratory exchange in Trichoptera occurs chiefly through *tracheal gills* which are filamentous extensions of the body wall. Gills are usually located on abdominal segments, although on thoracic segments, too, in some families; in larvae of a few species gills are entirely lacking. Tracheal gills are single (Fig. IV A), or branched basally (Fig. IV D), or they occur as lateral filaments from a central stalk as in the Hydropsychidae. These gills are arranged in definite positions in three horizontal series - dorsal, lateral, and ventral - on each side of most abdominal segments, with an anterior and posterior gill position in each series; a segment with a complete complement would have six gills on each side. Thus the precise position of a gill on any given segment can be designated as anterodorsal, anterolateral, posteroventral, etc. Arrangement of gills is significant taxonomically because some are often absent from particular positions, especially on the more posterior segments, although characters of gill arrangement are frequently variable and are also subject to change from one instar to another. Diagnostic gill characters offered in this study are based on final instars. Beneath the cuticle of each tracheal gill, fine tracheoles lie in the respiratory epithelium. Wichard (1973) showed in larvae of the Limnephilini that an optimum system for respiratory exchange was achieved by the arrangement of these tracheoles parallel to the long axis of the gill, with uniform intervals between tracheoles.

Osmotic regulation in at least some families of Trichoptera is mediated through rectal papillae, as is general for insects (Schmitz and Wichard, in press); this appears to represent the primitive condition for the Trichoptera, and in the Phryganeidae larvae swallow water to provide an additional source of the chloride ions absorbed by the rectal papillae (Schmitz and Wichard, in press). In addition, however, ionic absorption for osmoregulation in several families of Trichoptera is also achieved through the *anal papillae* (Nüske and Wichard 1971). These are elongate lobes arising from within the anal opening (Fig. 13.3A); anal papillae can be everted by pressure of the haemolymph and retracted by muscular action (Wichard 1976). Anal papillae occur in at least some genera of all families of the Hydropsychoidea and Rhyacophiloidea; these structures are best seen in living specimens, e.g. *Beraea* (Wiggins 1954). Nüske and Wichard (1972) found tracheoles in the anal papillae of glossosomatid larvae, and concluded that in that family at least, the structures serve as both respiratory and ion transporting epithelia. Anal papillae in Trichoptera have often been termed blood gills in the literature.

Other structures involved in ion absorption for osmoregulation are the *chloride epithelia* of the Limnephilidae (sens. lat. incl. Goerinae), Hydroptilidae, and Molannidae (Wichard 1976). In larvae of the Limnephilidae, usually on the venter of abdominal segments II-VII, chloride epithelia are seen as ovoid areas of modified cuticle bordered by a thin sclerotized line; these have been termed oval sclerotized rings in the taxonomic literature. In some genera, especially of the tribe Limnephilini, similar areas are present on the dorsum of most of these same segments (Fig. IV A,D), and there may also be smaller, circular rings of the same structure on the sides of some segments dorsal to the lateral fringe. These rings enclose areas of the hypodermis specialized for ionic transport (Wichard and Komnick 1973). In living larvae, chloride ions are absorbed from water passing through the case, accumulated in the cuticle, and transmitted to the body through the chloride epithelium in osmoregulatory compensation for ions lost through renal excretion (Schmitz

---

VI Morphology of caddisfly larvae. A, thorax and metathoracic leg of a limnephilid, lateral; B, maxillae and labium of an apataniine limnephilid, ventral

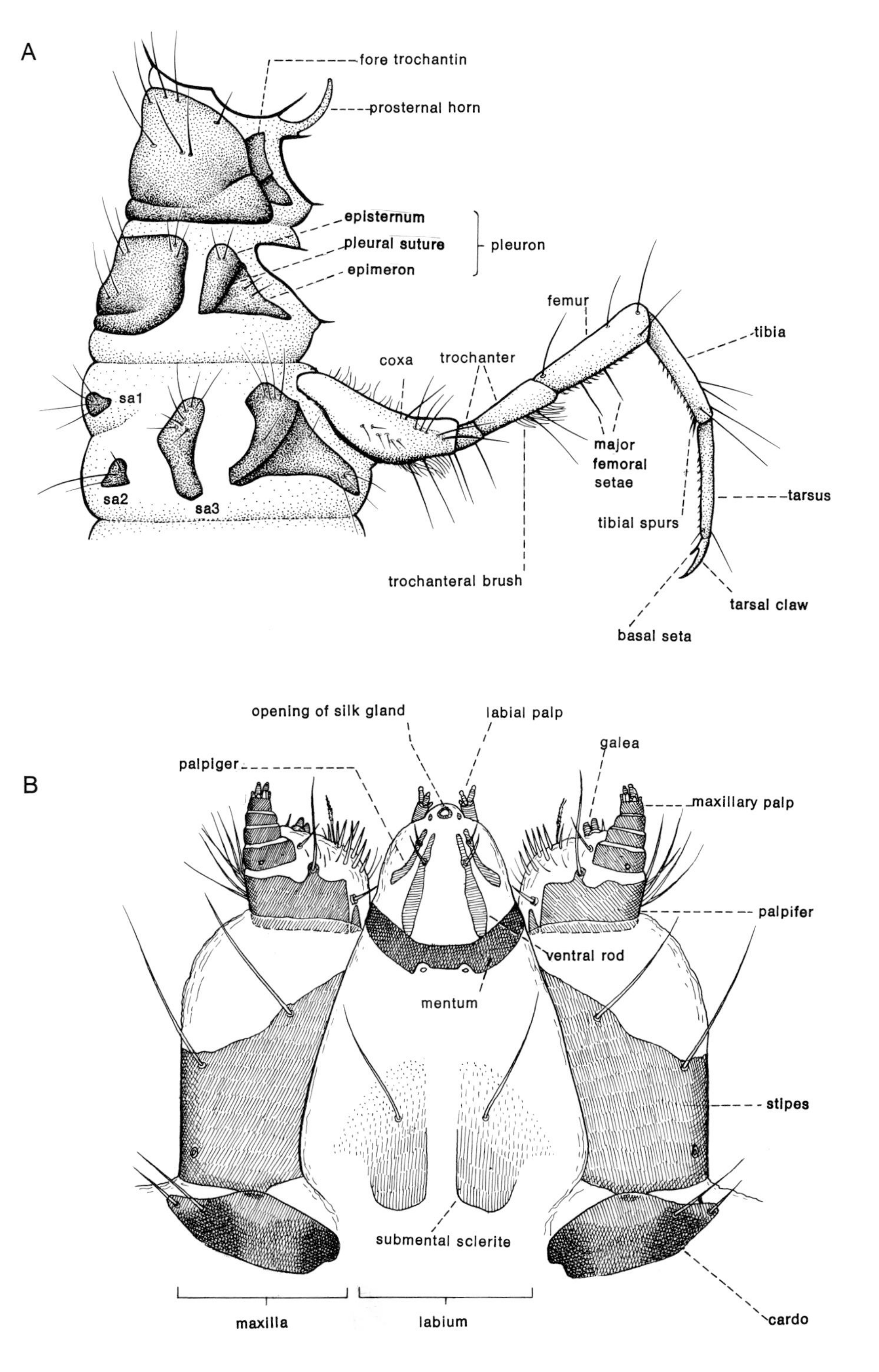
A
fore trochantin
prosternal horn
episternum
pleural suture
epimeron
pleuron
femur
tibia
coxa
trochanter
sa1
sa2
sa3
major
femoral
setae
tarsus
tibial spurs
trochanteral brush
tarsal claw
basal seta
B
opening of silk gland
labial palp
galea
palpiger
maxillary palp
palpifer
ventral rod
mentum
stipes
submental sclerite
maxilla
labium
cardo

and Wichard 1975). The peripheral sclerotized ring is a differentiation of the epicuticle along the border between chloride epithelium and normal integument. The chloride epithelial areas are sometimes difficult to distinguish, but the thin sclerous ring can usually be detected by varying the direction and brightness of the microscope illumination.

Techniques

COLLECTING

To a considerable degree, the objective in collecting will govern the method and the equipment used. The topic is first considered from the viewpoint that the aim is to discover as far as possible the full diversity of caddisfly larvae in a given site. Other objectives will be considered subsequently.

Equipment for collecting caddisfly larvae need not be elaborate. For waters that can be waded, most people find thigh-length rubber boots a useful item; the principal advantage is to have freedom to kneel down for a close look at the insects and what they are doing. A kitchen-type strainer of the finest mesh possible is useful for washing silt from samples of gravel, sand, or leaves; all-plastic ones compress easily for carrying and do not rust. Insects can then be picked from substrate materials in clean water in a small plastic basin such as a white photographic tray. A long-handled D-frame dipnet is essential, and should be provided with a bag of mesh size fine enough, e.g. 0.5 mm, to retain smaller instars. Although useful in many types of habitats, a dipnet of this type is especially valuable in recovering very small larvae from rocks and gravel in rapid currents. The method is the widely used kick-sample technique, in which the flat-side of the net is held vertically as tightly as possible against the bottom substrate. Larger rocks immediately upstream of the net are washed by hand and the current carries the dislodged insects into the net; smaller gravel is disturbed with the feet to a depth of several centimetres, exposing burrowing larvae to the force of the current. A large white tray, of the order of 50 cm x 30 cm x 5 cm, filled with water is required for sorting dipnet samples of substrate materials. Finally, stainless steel iris forceps are fine enough without being fragile, and fixed to a ring on a neck cord are almost loss-proof; a 10x loupe or hand lens fastened to the same neck cord enables one to see a surprising amount of detail in specimens in the field.

Since there is no better indication of the diversity of larval Trichoptera than the microhabitats they occupy, a collector will need to explore all of them. In a stream, the undersurfaces of large rocks in the current are usually among the first sites investigated; but then samples of the materials under the rocks should be taken, washed clean in a strainer, and inspected in a small tray. Moss growing on the tops of rocks often harbours larvae not found elsewhere. Since larvae of such genera as *Neophylax* and *Psilotreta* pupate in clusters, it is often necessary to turn over many rocks to find any of them. In shallow areas along the stream margin close visual inspection on hands and knees is informative, with substrate samples washed and sorted as described; sand and silt deposits washed in this way often yield burrowing caddisfly larvae. Accumulations of leaves and twigs are frequently productive, and samples should again be washed and sorted. Logs and branches should be inspected, and any cracks probed for pupae.

For small springs, much the same pattern can be followed as for streams, but with particular attention to washing and searching samples of the water-saturated organic muck in seepage areas at the head of the spring. Growths of liverworts and mosses in dripping areas often harbour larvae. Thin layers of water running over rocks in spring streams are sites for specialized larvae, as are spray zones above the water line.

In shallow waters along the shoreline of lakes, and in ponds and marshes where dense aquatic vegetation makes it difficult to see larvae, net samples of the plants can be turned out in a large basin for sorting. In deeper water similar samples taken from a boat in surface mats of plants such as *Ceratophyllum* often yield surprising numbers of larvae not otherwise seen. Deep bottom samples of soft materials can be taken with an Ekman grab or similar device, and the materials washed in a fine screen.

Quantitative collecting is an important objective, but one difficult to achieve. The problematic subject of quantitative benthic collecting for aquatic insect larvae is discussed at length in other references (e.g. Hynes 1970). Some relative quantitative base may be given to any of the collection methods outlined above by making as far as possible a uniform collecting effort over a given period of time in similar habitats; or substrate samples of a given area or volume can be taken and the animals removed and counted. Life history analysis is a special aspect of quantitative work, with the objective of determining the time and rate of larval development. Samples taken at regular intervals over the full life cycle are required. Since the object is to establish what proportion of the total larval population is in each instar at these intervals, it is important to use a technique that samples early instars adequately. One complication here is that caddis larvae of several groups have been found in bottom materials to depths of 20–30 cm (Williams and Hynes 1974); and in general different instars may occur in different micro-habitats. Measurement of maximum head width has proved to be the best general index for determining the instar to which a caddis larva belongs. Within a single species, series of these measurements over the period of larval development will form into clusters separated by distinct gaps, the gaps indicating the rapid increase in size of the soft cuticle at each larval ecdysis up to the head width of the succeeding instar. Examples of methods for analysing growth data of this kind in relation to time will be seen in current ecological literature (e.g. Mackay 1969, 1972).

If the objective for collecting is to obtain specimens for rearing, the possibility of physical damage in sampling must, of course, be minimized. The respiratory requirements of larvae can usually be accommodated for transit by a small amount of water in jars of appropriate size in a portable ice chest. When the air temperature is high, or when an ice chest is not available, larvae are better transported in moist leaves than submerged in water (Mickel and Milliron 1939).

ASSOCIATING LARVAL STAGES

It need hardly be emphasized that knowledge about the biology of Trichoptera depends very much on knowing what the immature stages of the species do and when they do it; that knowledge is dependent to a large extent on a sound larval taxonomy. Since only about one-third of the 1,200 or so species of Trichoptera known in Canada and the United States have been identified in the larval stage, the importance of gaining additional firm data on larval taxonomy is clear.

Of the two general approaches to the problem of establishing associations between larval and adult Trichoptera, laboratory rearing is the more precise. Usually habitat conditions can be simulated closely enough in the laboratory to sustain most larvae during development. Specifications for many types of rearing apparatus are available in the literature, and ingenuity could be taken as the guiding principle; a few examples can be found in papers by Anderson (1974b), Bjarnov and Thorup (1970), Philipson (1953a), and Wiggins (1959). These attest to the fact that simple equipment is entirely adequate for rearing; sophisticated laboratory streams are not necessary to generate this kind of information, although they are of course useful for behavioural study.

Rearing can involve a series of larvae of unknown identity collected together in the same habitat; specimens should be examined under a low-power binocular microscope to ensure that all are the same in markings, gill arrangement, and case structure, to the extent that one can discriminate. A technique by which larvae are anaesthetized with carbon dioxide for close comparative examination has been described (Bray 1966). Part of the series, say one-third to one-half, is then preserved and the remainder reared - at least five or six specimens if possible, because the method is based ultimately on an assumption that the entire initial series represents a single species. The assumption can be avoided by starting with a single egg mass, but the rearing process is much longer and feeding of early instars is often demanding. Eggs can be obtained from adults of known identity, and a procedure described by Resh (1972) provides considerable precision through stimulation of egg release from particular females as required. With caddisfly larvae obtained from either of these sources, there is a tendency to cannibalism in confined quarters, especially in omnivorous groups such as the Phryganeidae; separation of larger instars into individual containers is warranted as a routine procedure whenever the identity to be established is of critical importance. An exact record of larval structure can be obtained by removing a pharate pupa from its case; placed in an individual container, the insect sheds the larval integument in one piece and will complete development without the case (Hiley 1969).

Adults that emerge successfully should be allowed a day or two of activity to ensure that the teneral condition is past before they are preserved as study specimens; for this reason, provision should be made in the container for the adult to crawl up on some emergent object. Even if adults die during emergence, the specimens should be retained because genitalic structures are well enough formed by that time for accurate identification to species. Pupal cases vacated by pharate adults should be carefully retrieved and a small plug of cotton inserted in the anterior opening to prevent the loss of larval exuviae; this pupal case, the cast pupal skin, and the adult should each be placed in a separate shell vial labelled in some way to designate unequivocally their connection, plugged with cotton, and placed in a larger vial. All sets of these reared specimens should then be placed in a jar along with the larval specimens preserved at the outset, and fully documented to show clearly that the specimens were all collected at the same time and place and that the adults were reared. Specimens reared and documented in this way provide valuable scientific data; when time has been invested in rearing, it is certainly a loss if the results are not consolidated and documented.

The second general approach to the problem of establishing associations between larval and adult Trichoptera is a field technique known generally as the metamorphotype method. When collections are made in the same location at intervals over a year, it is often possible to acquire series of specimens representing the development of particular species from larva to the pharate adult within the pupal case. Shortly before emergence, the pharate adult has fully formed genitalic structures on which species identification can be based, especially if it is a male; and the unopened pupal case also contains the larval exuviae compressed in a mass at the posterior end. When carefully removed in a small dish of alcohol under a binocular microscope, the larval exuviae can be sorted out, and certain sclerites, particularly those of head, thorax, and legs, can be compared with final-instar larvae collected earlier to establish the association. The method is not precise but, carefully followed within a single site, it is reasonably accurate. Even when it is possible to make only a single collection at a site, long series sometimes yield final-instar larvae and pharate adults; series of this kind ought to be a primary objective in any general collecting of Trichoptera. The Leptoceridae and Molannidae present a special problem here because the larval exuviae are ejected through the posterior opening of the case shortly after ecdysis.

A collector often finds larvae and pupae but none developed to the pharate adult stage necessary for specific identification. Under this circumstance, some larvae should be preserved and a series of pupae maintained alive in a small jar on damp moss or paper towel in a portable ice chest or a refrigerator. Development is usually completed and the adults emerge, making identification possible. With a portable ice chest, this technique is especially useful for excursions of several days or weeks by automobile; we have successfully transported many specimens in this way, adults emerging continually.

Occasionally one comes upon a site where emergence is long past and vacated pupal cases are the principal evidence. By collecting as many unopened pupal cases as can be found, and opening each one carefully in a small dish of alcohol under a binocular microscope, it is sometimes possible to find a pharate male adult that died after the genitalia were sufficiently sclerotized for identification. The larval exuviae in the same case then provide some evidence for larval identification.

Specimens and exuviae supporting specific association between larval and adult stages and obtained by any of these methods should be carefully segregated in separate shell vials in a larger vial or jar and provided with a label fully documenting the circumstances of collection. It is costly in effort and money to acquire data of this kind; workers should be encouraged to treat the specimens with the care given to type specimens, and to deposit them in museum collections where they will be available for future study by qualified students of the group.

PRESERVING, STORING, AND SHIPPING

Ethyl or isopropyl alcohol diluted with water to 80% by volume is an acceptable fixative and long-term preservative for specimens; but its use requires that the fluid be changed several times in the first few days of fixation because the amount of water introduced with caddisfly larvae and cases dilutes the small volumes normally used in field collecting jars. If this precaution is not taken, larval specimens suffer from internal decomposition, become soft, break easily, and generally are inferior study material.

A superior fixing agent used increasingly by workers on larval insects of all kinds is Kahle's fluid:

Ethyl alcohol	15 parts by volume
Formalin	6
Glacial acetic acid	1
Distilled water	30

This fluid penetrates internal tissues rapidly enough to fix tissues before decomposition begins. Although most specimens are killed quickly, larger larvae can be killed faster in boiling water and transferred to the fixative. It is not necessary to change the fluid to compensate for dilution; the light abdomens of most caddisfly larvae do not darken as they usually do in alcohol, and some indication of the green and yellow colours is retained. Caddisfly larval specimens fixed in this way have proved to be superior for study in almost every way. One disadvantage of the fluid is its penetrating odour under confined conditions, a problem minimized by transferring specimens fixed in it for two or three weeks to 80% alcohol for long-term storage and study. Kahle's fluid is poured off, a water rinse added and poured off, and finally 80% alcohol is added for storage. Transfer to alcohol removes the abdominal colours retained in Kahle's fluid, but otherwise the specimens are very good. Neoprene stoppers exposed to Kahle's fluid will quickly deteriorate. Jars of 120 cc (4 oz) to 250 cc (8 oz) capacity are satisfactory as field containers; plastic tops are satisfactory if not cracked, and metal tops can be lined with a polyethylene gasket. Kahle's fluid tends to seep out under polyethylene snap-caps.

Storage of specimens by any of the standard museum procedures is satisfactory, and should preferably be in closed cabinets to exclude light. Small vials sealed with corks or screw-caps are not safe against evaporation of alcohol over the long term; neoprene stoppers provide much better protection against evaporation but should be of sulphur-free neoprene to avoid discoloration of the alcohol and possible damage to specimens over long periods. Small jars of 120 to 500 cc capacity, the lids lined with a polyethylene gasket, provide good protection against evaporation; individual lots of specimens can be kept inside the jars in shell vials plugged tightly with cotton, with each vial fully labelled.

Since larval specimens are frequently sent from one worker to another, or materials are borrowed from a museum collection, it is important that the time and scientific data they represent are not jeopardized by inadequate attention to what might seem like trivial details. Specimens preserved in fluid are relatively heavy and require a stout container and somewhat more substantial internal padding than dried insect specimens. Jars should not be allowed to come in contact with each other in transit. Any space between vials within a jar should be packed with cotton to prevent their knocking together. Vials in a jar should be padded with cotton to fill any space beneath the lid. No air bubbles should be left in the same container as specimens because the specimens are subjected to jostling and erosion of setae every time the bubble moves. This is an important advantage of using shell vials plugged with cotton within large vials or jars because air bubbles are easily avoided by inserting the cotton plug after the vial has been totally submerged in a larger vial or jar of alcohol.

SYSTEMATIC SECTION

Key to Larvae of North American Families of Trichoptera*

1 Larva with portable case of sand grains resembling snail shell (Fig. 5.1C); anal claw with comb of teeth, not hook-shaped (Fig. 5.1D). Widespread in rivers, streams, and wave-washed shores of lakes **5 Helicopsychidae**, p. 89

Larva with case not resembling snail shell, or larva not constructing portable case; anal claw with apex forming stout hook (Fig. V A-D) 2

2 (1) Dorsum of each thoracic segment largely covered by sclerotized plate (Figs. 6.6A, 7.9A) 3

Metanotum, and sometimes mesonotum, entirely membranous or largely so with several pairs of smaller sclerites or hairs (Figs. 10.1B, 13.1B) 4

3 (2) Abdomen with ventrolateral rows of branched gills, and with prominent brush of long hairs at base of anal claw (Fig. 6.6A); posterior margin of meso- and meta-notal plates lobate (Fig. 6.1A); larvae construct fixed retreats (Fig. 6.6G). Widespread in rivers and streams, occasionally on the rocky shores of lakes **6 Hydropsychidae**, p. 92

Abdomen without ventrolateral gills, and with only 2 or 3 hairs at base of anal claw (Fig. 7.10A); posterior margin of meso- and meta-notal plates straight (Fig. 7.9A,G); minute forms usually less than 6 mm long, often with purse- or barrel-shaped cases (Figs. 7.1B, 7.7B). Widespread in rivers, streams, and lakes **7 Hydroptilidae**, p. 120

4 (2) Antennae very long and prominent, at least 6 times as long as wide (Fig. 9.1B); and/or sclerotized plates on mesonotum lightly pigmented except for pair of dark curved lines on posterior half (Fig. 9.1B); larvae construct portable cases of various materials. Widespread in lakes and rivers **9 Leptoceridae**, p. 161

*Figures III to VI will be found in the Morphology section, all others under the family designated by the number preceding the period.

Antennae of normal length, no more than 3 times as long as wide (Fig. IIIE), or not apparent; mesonotum never with pair of dark curved lines as above **5**

5 (4) Mesonotum largely covered by sclerotized plates, variously subdivided and usually pigmented, although sometimes lightly (Figs. 10.1B, 18.3B); pronotum sometimes with prominent anterolateral lobe (Fig. 1.1A,B) **11**

Mesonotum usually without sclerotized plates (Fig. 13.1B), occasionally with small sclerites covering not more than half of notum (Figs. 14.5B, 14.9B); pronotum never with anterolateral lobe (Fib. 14.6A) **6**

6 (5) Abdominal segment IX with sclerotized plate on dorsum (Fig. 4.4A) **7**

Abdominal segment IX with dorsum entirely membranous (Fig. 13.1E) **9**

7 (6) Metanotal *sa*3 usually consisting of cluster of setae arising from small rounded sclerite (Fig. 14.5B); prosternal horn present (Fig. 14.5A); larvae construct tubular portable cases, mainly of plant materials. Widespread in lentic and lotic waters **14 Phryganeidae**, p. 318

Metanotal *sa*3 consisting of single seta not arising from sclerite (Fig. 4.4E); prosternal horn absent (Fig. 4.4A); larvae either without case or with tortoise-like case of stones **8**

8 (7) Basal half of anal proleg broadly joined with segment IX (Fig. 4.4F); and claw with at least one dorsal accessory hook (Fig. 4.4A); larvae construct tortoise-like portable case of small stones (Fig. 4.4B). Widespread in rivers and streams **4 Glossosomatidae**, p. 72

Most of anal proleg free from segment IX (Fig. 17.3D); anal claw without dorsal accessory hooks, although secondary lateral claw may be present (Fig. 17.3A); larvae free-living without case or fixed retreat. Widespread in rivers and streams **17 Rhyacophilidae**, p. 366

9 (6) Labrum membranous and T-shaped (Fig. 13.1D), often withdrawn from view in preserved specimens; larvae construct fixed sack-shaped nets of silk (Fig. 13.2F). Widespread in rivers and streams **13 Philopotamidae**, p. 310

Labrum sclerotized, rounded and articulated in normal way (Fig. 15.1B) **10**

10 (9) Trochantin of prothoracic leg with apex acute, fused completely with episternum without separating suture (Fig. 15.2A); larvae construct exposed funnel-shaped capture nets (Fig. 15.2D), flattened retreats (Fig. 15.3E), or tubes in substrate (Fig. 15.4D). Widespread in most types of aquatic habitats **15 Polycentropodidae**, p. 341

Trochantin of prothoracic leg broad and hatchet-shaped, separated from episternum by dark suture line (Fig. 16.3C); larvae construct tubular retreats on rocks and logs (Fig. 16.1E). Widespread in running waters **16 Psychomyiidae**, p. 356

11 (5) Abdominal segment I lacking both dorsal and lateral humps (Fig. 2.2A), each metanotal *sa*1 lacking entirely (Fig. 2.2B) or represented only by single seta without sclerite (Fig. 2.4B); larvae construct portable cases of various materials and arrangements. Widespread in running waters **2 Brachycentridae**, p. 50

Abdominal segment I always with lateral hump on each side although not always prominent, and usually with median dorsal hump (Fig. 10.3A); metanotal *sa*1 always present, usually represented by sclerite bearing several setae (Fig. 10.1B) but with at least single seta (Fig. 3.1B); larvae construct portable cases of various materials and arrangements **12**

12 (11) Tarsal claw of hind leg modified to form short setose stub (Fig. 11.1A) or slender filament (Fig. 11.2A); case of sand grains with lateral flanges (Fig. 11.1C). Transcontinental through Canada to Alaska, and eastern, in lakes and large rivers, more rarely in spring streams **11 Molannidae**, p. 288

Tarsal claws of hind legs no different in structure from those of other legs (Fig. 15.1A) **13**

13 (12) Labrum with transverse row of approximately 16 long setae across central part (Fig. 3.1D); case of plant materials variously arranged. Eastern and western, in streams **3 Calamoceratidae**, p. 64

Labrum with only 6 long setae across central part (Fig. 10.26D) **14**

14 (13) Anal proleg with lateral sclerite much reduced in size and produced posteriorly as lobe around base of stout apical seta (Fig. 1.1A); base of anal claw with mesial membranous surface giving rise to prominent brush of 25–30 fine setae (Fig. 1.1D); case of sand grains (Fig. 1.1C). Eastern part of continent, exceedingly local, in wet soil of spring seepage areas **1 Beraeidae**, p. 47

Anal proleg with lateral sclerite not produced posteriorly as lobe around base of apical seta (Fig. 10.45D), base of anal claw with mesial face largely sclerotized and lacking prominent brush of fine setae (Fig. 10.45E) **15**

15 (14) Antenna situated very close to anterior margin of eye (Fig. 8.1D); median dorsal hump of segment I lacking (Fig. 8.1A); case of various materials and arrangements. Widespread in lotic and lentic habitats **8 Lepidostomatidae**, p. 154

Antenna situated at least as close to anterior margin of head capsule as to eye (Fig. IIIE), or closer (Fig. 18.2B); median dorsal hump of segment I almost always present (Fig. IV A) **16**

16 (15) Antenna situated approximately mid-way between anterior margin of head capsule and eye (Fig. IIIA,E); prosternal horn present (Fig. VIA) although sometimes short; chloride epithelia usually present on some abdominal segments (Fig. IV A,D); case of wide range of materials and construction. Widespread in most aquatic habitats **10 Limnephilidae**, p. 179

Antenna situated at or close to anterior margin of head capsule (Fig. 18.2B); prosternal horn and chloride epithelia never present (Fig. 18.2A); cases only of mineral materials **17**

17 (16) Dorsum of anal proleg with cluster of approximately 30 or more setae posteromesiad of lateral sclerite (Fig. 18.1D); fore trochantin relatively large, the apex hook-shaped (Fig. 18.2D). Widespread in running waters and along lake shores **18 Sericostomatidae**, p. 374

Dorsum of anal proleg with approximately 5 setae posteromesiad of lateral sclerite, sometimes with short spines (Fig. 12.5F); fore trochantin small, the apex not hook-shaped (Fig. 12.1A). Widespread in running waters **12 Odontoceridae**, p. 295

1
Family Beraeidae

The Beraeidae are a small family sparsely distributed over the Holarctic region; genera from Australia, New Zealand, and South Africa have been assigned to the group, but their status is subject to confirmation. Only the single genus *Beraea* occurs in North America, and it is confined to the eastern part of the continent, where colonies are exceedingly local.

Larvae of the Holarctic genera are somewhat different structurally (Lepneva 1966), but apparently all are consistent in diagnostic characters of the anal proleg (Fig. 1.1A,D): the lateral sclerite is reduced more than in other families and extended posteriorly as a lobe from which arises a single stout seta; the mesial surface of the base of the anal claw is membranous and bulbous, giving rise to a characteristic prominent brush of 25-30 slender setae. Beraeid larvae also have a ridge along each side of the head, the antenna arising near the end on the anterior margin of the head. The lateral fringe of the abdomen is either absent (Lepneva 1966) or, as in *Beraea*, modified into a line of fewer, larger, feathery setae; tiny lateral tubercles are present on segment VIII in *Beraea*, at least. Segment IX is without a dorsal sclerite. A few single gill filaments occur in some genera. Larvae of North American representatives of the Beraeidae are small, not exceeding 6 mm in length, although those of some European species are up to 9 mm long. Larval cases, constructed of sand grains, are curved and tapered.

Beraeids live in small streams and springs where they are detritivorous; some European species live in wet leaves and moss above the water line. Larvae of at least one of the Nearctic species are confined to the wet organic muck around spring seepage areas.

1.1 Genus **Beraea**

DISTRIBUTION AND SPECIES Several species of *Beraea* are known from Europe, North Africa, and Asia. In North America the genus is known only from the east, and colonies are rare: *Beraea gorteba* Ross in Georgia; *B. fontana* Wiggins in southern Ontario; *B. nigritta* Banks from New York, which is inadequately defined.

In North America larvae are known only for *B. fontana* (Wiggins 1954).

MORPHOLOGY Larvae of *Beraea* are characterized by a sharp carina extending obliquely across the pronotum, terminating in a rounded lobe at each anterolateral corner. The head bears a ridge along each side. In *B. fontana* the mesonotum bears only a lightly sclerotized and pigmented plate which lacks a mid-dorsal ecdysial line (B); on the metanotum, sclerites are lacking, there is an anterior patch of setae continuous across the median line representing *sa*1 perhaps combined with *sa*2, and *sa*3 is a small patch of setae on each side. Sclerotized parts are for the most part brownish orange in colour. Length of larva up to 6 mm.

CASE Larval cases known for species of *Beraea* are curved and tapered, the smooth exterior surface of fine sand grains. Length of larval case up to 6 mm.

BIOLOGY Observations are available only for *B. fontana* (Wiggins 1954). Larvae live in wet organic muck around spring seepage areas, but evidently not in the exposed waters of the spring run itself. The habitat is unusual for Trichoptera, but of special interest because it is also exploited by larvae of *Moselyana*, *Lepania*, and *Goereilla*, which are western North American limnephilid genera of particular phylogenetic significance. Gut contents of *Beraea* larvae (2) were almost entirely vascular plant tissue and fine organic particles. Emergence of adults is evidently confined to late spring; collection of the last three larval instars during this same period indicates that at least two years are required for completion of the life cycle.

REMARKS Taxonomic data and references for identification of adults were given by Wiggins (1954).

Beraea fontana (Ontario, Durham Co., 12 Aug. 1953, ROM)
A, larva, lateral x29, lateral setae and anal proleg enlarged; B, head and thorax, dorsal; C, case x18; D, anal prolegs, ventral

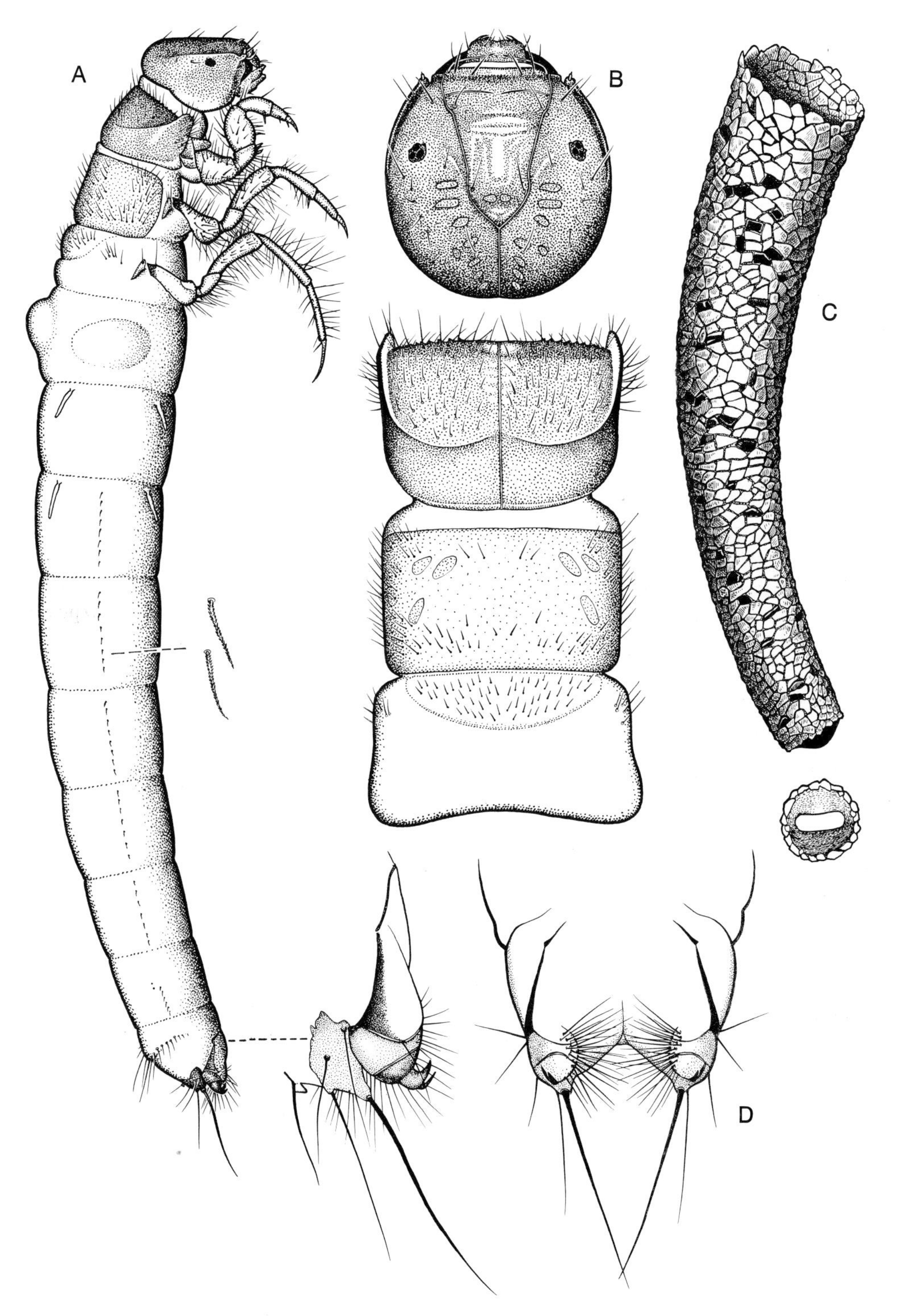
A
B
C
D

2
Family Brachycentridae

This family is widespread over much of the Holarctic region. Six genera comprising approximately 30 species are now known in the Brachycentridae in North America; representatives occur in most parts of the continent.

Brachycentrids live in running waters, ranging from cold mountain springs to the slowly flowing channels of marshy rivers. Their portable cases are usually concealed in moss in small streams, but in genera inhabiting larger rivers the larvae fix their cases to exposed substrates in the current. Species of these large rivers are often locally abundant.

Larvae of this family are unusual among the portable-case makers in having no dorsal or lateral humps on the first abdominal segment. The head frequently bears a dorsolateral ridge along each side; the ventral apotome is more or less quadrate, although of differing proportions in various genera, and always separates the genae completely. The mandibles have tooth-like points. The pronotum usually has a central, transverse ridge, or occasionally a sulcus; a short prosternal horn is present in some genera (Fig. 2.4D). Each mesonotal plate is frequently subdivided longitudinally by narrow sutures. Metanotal *sa*1 is represented only by a single seta or none at all, but a sclerite is not present. Setae on abdominal segment I are sparse or absent. Abdominal gills are single or lacking, and the lateral fringe is much reduced or absent; small lateral tubercles occur on several segments.

Larval cases are made of plant or rock materials, although one frequently finds specimens in some genera where silken secretion alone comprises a sizable part of the case. Four-sided cases are characteristic of several genera in this family.

The absence of humps on the first abdominal segment is an intriguing feature of the Brachycentridae, for this is the only Nearctic family of the Limnephiloidea (the groups whose larvae construct portable tube cases) in which these humps are not apparent. If this should prove to be the primitive condition and not a secondary loss, the Brachycentridae would be critically important in interpreting the evolution of self-ventilating respiration by the tube-case makers and by current theory (see under Biological Considerations) the exploitation of lentic habitats by Trichoptera.

Key to Genera

1 Middle and hind legs long, their femora approximately same length as head capsule (Fig. 2.3A), their tibiae produced distally into prominent process from which stout spurs arise (Fig. 2.3E) — 2

Middle and hind legs shorter, their femora much shorter than head capsule (Fig. 2.2A), their tibiae not produced distally into prominent process, although spurs arise from about same point on unmodified tibiae (Fig. 2.2F) — 3

2 (1) Mesonotum with mesial sclerites diverging posteriorly (Fig. 2.6B); head seta no. 17 not longer and stouter than other head setae (Fig. 2.6B); case circular in cross-section, tapered and composed of small rock fragments (Fig. 2.6C). Western — **2.6 Oligoplectrum**

Mesonotum with mesial sclerites diverging little if any posteriorly (Fig. 2.3B); head seta no. 17 longer and stouter than other head setae (Fig. 2.3B); case usually square in cross-section, composed of small pieces of plant materials fastened transversely (Fig. 2.3C), although case sometimes cylindrical and largely of silken secretion, or occasionally of small rock fragments. Widespread — **2.3 Brachycentrus**

3 (1) Ventral apotome of head longer than wide, somewhat narrowed at posterior end, larva with very short prosternal horn (Fig. 2.4D); case 4-sided, composed of small lengths of plant material placed crosswise with loose ends often protruding (Fig. 2.4C) — 4

Ventral apotome of head usually wider than long (Fig. 2.2D), sometimes squarish (Fig. 2.5D); larva without prosternal horn; case cylindrical, composed of lengths of plant material wound around the circumference (Fig. 2.5C) or of silk and rock materials (Fig. 2.2C) — 5

4 (3) Each half of mesonotum largely entire, lateral quarter partially delineated by variable suture, posterior margin of mesonotum raised and coloured dark brown (Fig. 2.4B). Western — **2.4 Eobrachycentrus**

Each half of mesonotum subdivided into three separate sclerites, posterior margin not conspicuously raised or coloured (Fig. 2.1B). Eastern — **2.1 Adicrophleps**

5 (3) Mesonotal *sa*1 with only a single seta (Fig. 2.2B). Western — **2.2 Amiocentrus**

Mesonotal *sa*1 with many setae extending along anterior border of sclerite, merging with *sa*3 (Fig. 2.5B). Widespread — **2.5 Micrasema**

2.1 Genus **Adicrophleps**

DISTRIBUTION AND SPECIES The genus is known only from North America by a single species, *A. hitchcocki* Flint, described from Connecticut.

We associated larvae, hitherto unknown, with adults in Pennsylvania, and collected other larvae in Maryland.

MORPHOLOGY Among North American brachycentrids, larvae of *Adicrophleps* are similar only to those of the western *Eobrachycentrus*, the two being readily separated by characters of the thorax given in the generic key. The head of *A. hitchcocki* has light-coloured muscle scars and bears a pair of longitudinal ridges along the anterolateral margins of the frontoclypeal apotome. A short prosternal horn is present. Abdominal gills are lacking and the lateral fringe is absent. Length of larva up to 6 mm.

CASE Larval cases in *Adicrophleps* are four-sided, tapered, and constructed of pieces of moss arranged transversely; trailing ends frequently left attached to the moss pieces give the case a furry appearance. The posterior opening is circular. Length of larval case up to 7 mm.

BIOLOGY Larvae were collected in several cold, rapid streams from 1 to 10 m wide from aquatic moss (*Scapania*) in riffle areas at depths not exceeding 30 cm; they were exceedingly difficult to find in the moss. Our collections indicate a single generation per year: final-instar larvae and pupae in May; adult emergence in spring and early summer; and early-instar larvae from July to October. Gut contents from larvae (2) examined were largely vascular plant tissue with some fine organic particles.

REMARKS Taxonomic data for the adult were given by Flint (1965). Our observations on the larval association and biology of this species have been greatly aided by Janice M. Glime who first found the secretive larvae in her study of aquatic mosses as habitats for aquatic insects (Glime 1968).

Adicrophleps hitchcocki (Pennsylvania, Potter Co., 20 April 1968, ROM)
A, larva, lateral x23; B, head and thorax, dorsal; C, case x18; D, head, anterior, mouthparts shaded; E, head, ventral

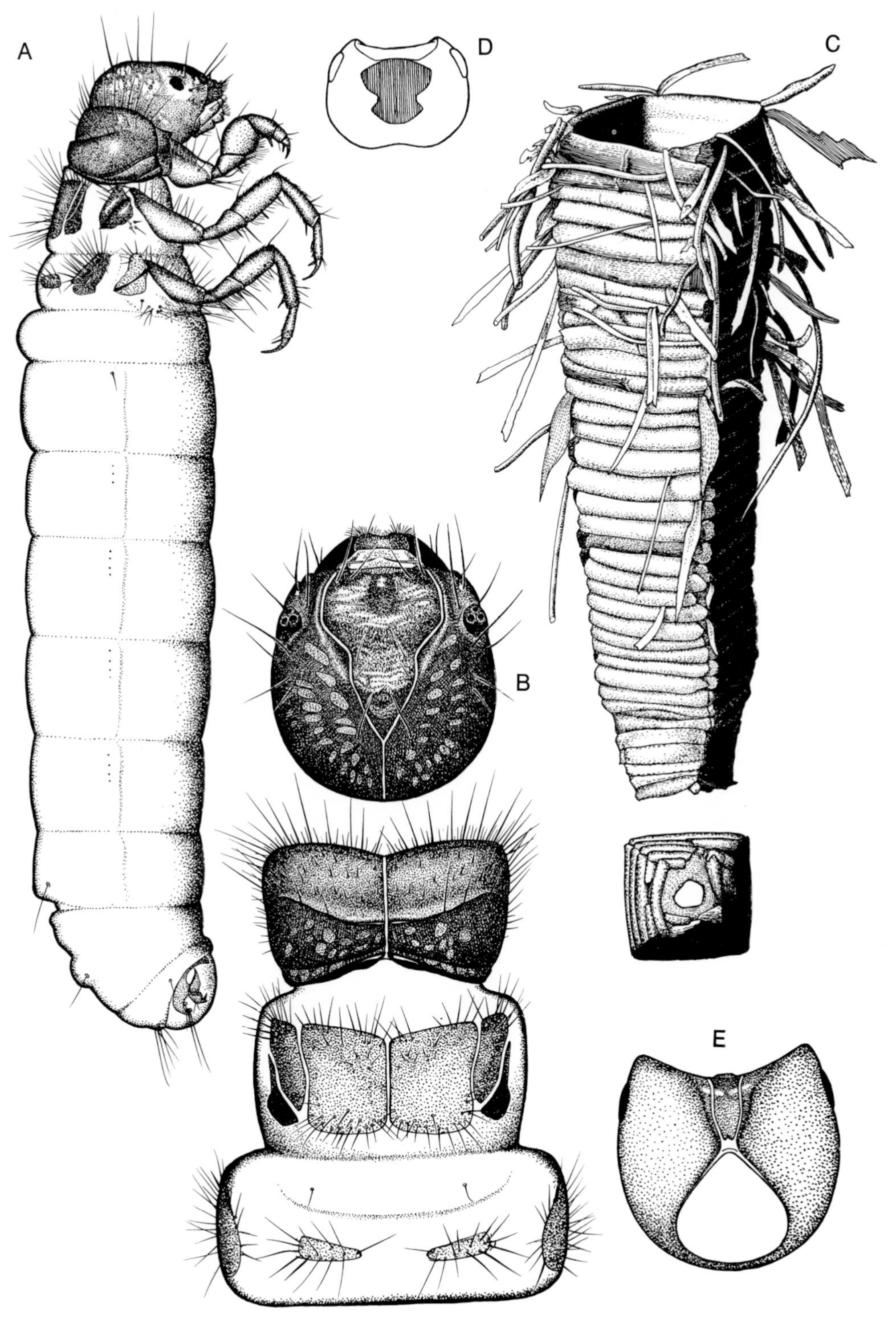
A
D
C
B
E

2.2 Genus Amiocentrus

DISTRIBUTION AND SPECIES This genus comprises the single species *A. aspilus* (Ross), widely distributed in western North America: British Columbia, California, Colorado, Idaho, Montana, Oregon, Utah, and Wyoming.

The larva was described by Wiggins (1965); we have collected larvae from several localities.

MORPHOLOGY Larvae are generally similar to those in *Micrasema*, but can be distinguished by the single seta of mesonotal *sa*1 (B). Each mesonotal sclerite is subdivided longitudinally, the two parts of each sclerite separating at ecdysis. Distal ends of the meso- and meta-tarsi are extended into angulate lobes (F). The head bears a carina along each anterolateral margin. Sclerotized parts are yellowish to light reddish brown. Single gills are present on some abdominal segments as illustrated; the lateral fringe is lacking. Length of larva up to 10 mm.

CASE Larval cases in *Amiocentrus* taper from front to rear but are straight. Thin pieces of plant materials are wound transversely around the circumference and incorporated into a layer of silk, although frequently much of the case is silk alone. Larval cases in *Amiocentrus* can be distinguished from straight cases of *Micrasema* spp. by their lack of any silken restriction around the edge of the posterior opening (C). Length of larval case up to 16 mm.

BIOLOGY The larvae are mainly characteristic of larger streams where they live in moderate currents on rooted aquatic plants and in moss on rocks. Gut contents of larvae (2) examined were largely diatoms and fine particulate matter. Anderson (1967b) studied the biology of *A. aspilus* in Oregon.

REMARKS Taxonomic discussion and references for the adult stage were given by Wiggins (1965).

Amiocentrus aspilus (Oregon, Jefferson Co., 18 June 1968, ROM)
A, larva, lateral x18; B, head and thorax, dorsal, detail of mesonotum; C, case x9; D, head, ventral; E, fore leg, lateral; F, middle leg, lateral; G, segment IX with anal prolegs, dorsal

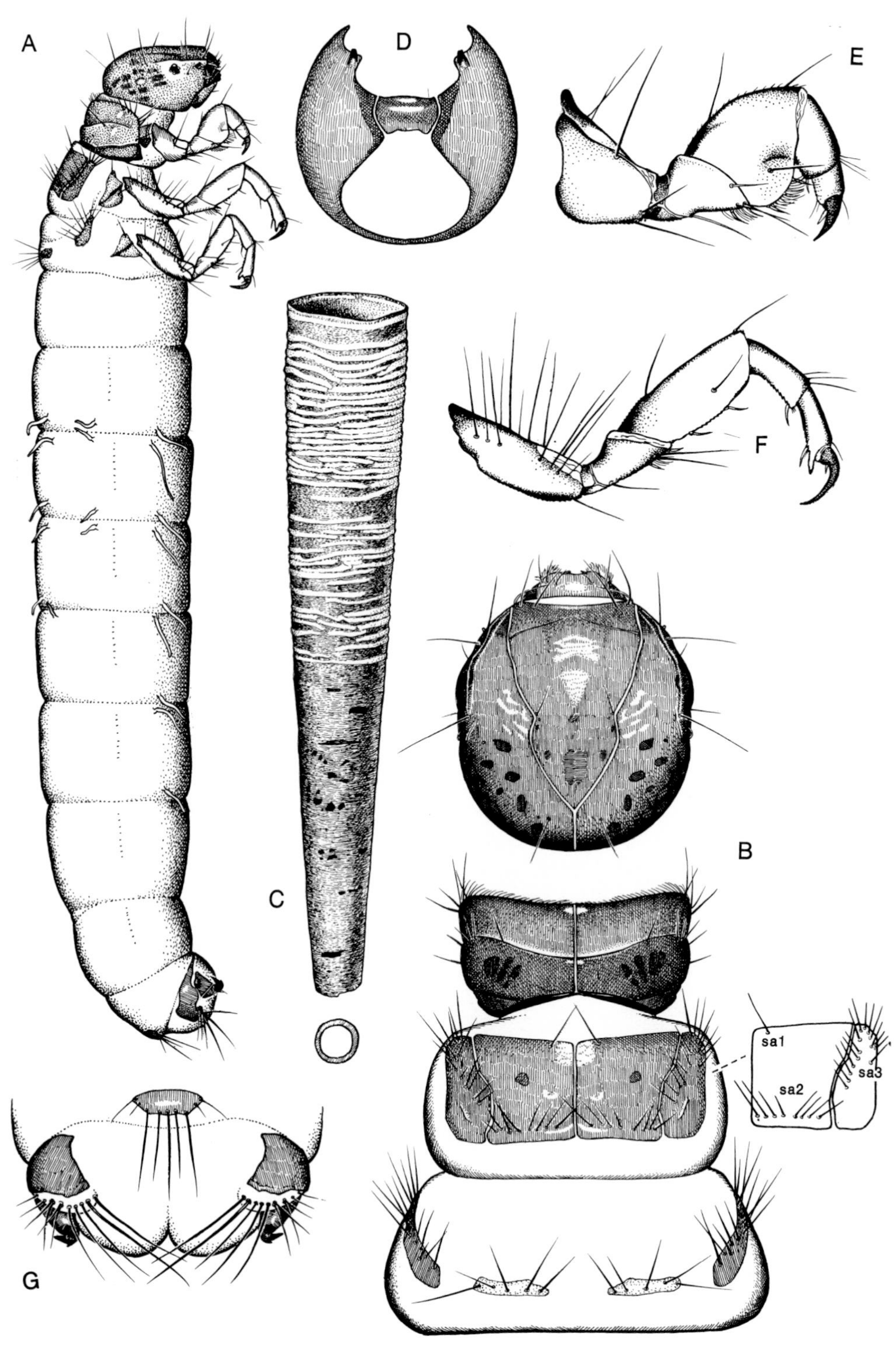
A
D
E
F
C
B
G
sa1
sa2
sa3

2.3 Genus Brachycentrus

DISTRIBUTION AND SPECIES Species of *Brachycentrus* occur in North America, Europe, and Asia; nine species are now known from this continent, where the group is common and widely distributed.

Larvae for five of the North American species have been identified in the literature: *B. americanus* (Banks) (Ross 1944); *B. etowahensis* Wallace (Wallace 1971); *B. fuliginosus* Walker (Denning 1937); *B. lateralis* (Say) (Ross 1944); *B. numerosus* (Say) (Ross 1944). We have associated larvae for *B. americanus* (Banks) and *B. occidentalis* (Banks).

MORPHOLOGY Larvae of *Brachycentrus* and *Oligoplectrum* have unusually long middle and hind legs, with the tibiae produced distally into a prominent process bearing sharp spurs (E). The head in *Brachycentrus* larvae is variously patterned or uniformly dark depending on the species, and often bears a short anterolateral carina along each side. In at least some species a short prosternal horn is present. A finger-like lobe occurs on the venter of each anal proleg (G). Length of larva up to 11 mm.

CASE Typically, larval cases of *Brachycentrus* spp. are four-sided, tapered, and constructed of narrow pieces of plant material arranged transversely (C). Occasionally, some or all of the case is of silken secretion, the silken part tending to be circular in cross-section, as are some cases constructed entirely of small rock fragments. Length of larval case up to 17 mm.

BIOLOGY Although *Brachycentrus* larvae are restricted to running waters, some do live in rather slow currents. Larvae position themselves by fastening the ventral lip of the case with silk to rock or plant substrate with the anterior opening facing into the current. Their long middle and hind legs are extended to each side of the case in what has generally been interpreted as an effective posture for filtering food particles carried by the current; but in addition to filter-feeding, larvae graze by scraping periphytic algae off the substrate (Gallepp 1974a). Studies indicate that the larvae ingest diatoms, filamentous algae, vascular plant detritus, and small animals - usually other insects (Mecom and Cummins 1964; Murphy 1919). Chironomid larvae are known to enter pupal cases of *B. occidentalis* by making openings in the silken closure membrane, causing death of the caddisfly when they pupate (Gallepp 1974b).

Brachycentrus americanus (Idaho, Bonneville Co., 9 Aug. 1961, ROM)
A, larva, lateral x13; B, head and thorax, dorsal; C, case x9; D, fore leg, lateral; E, hind leg, lateral; F, head, ventral; G, anal proleg, ventral

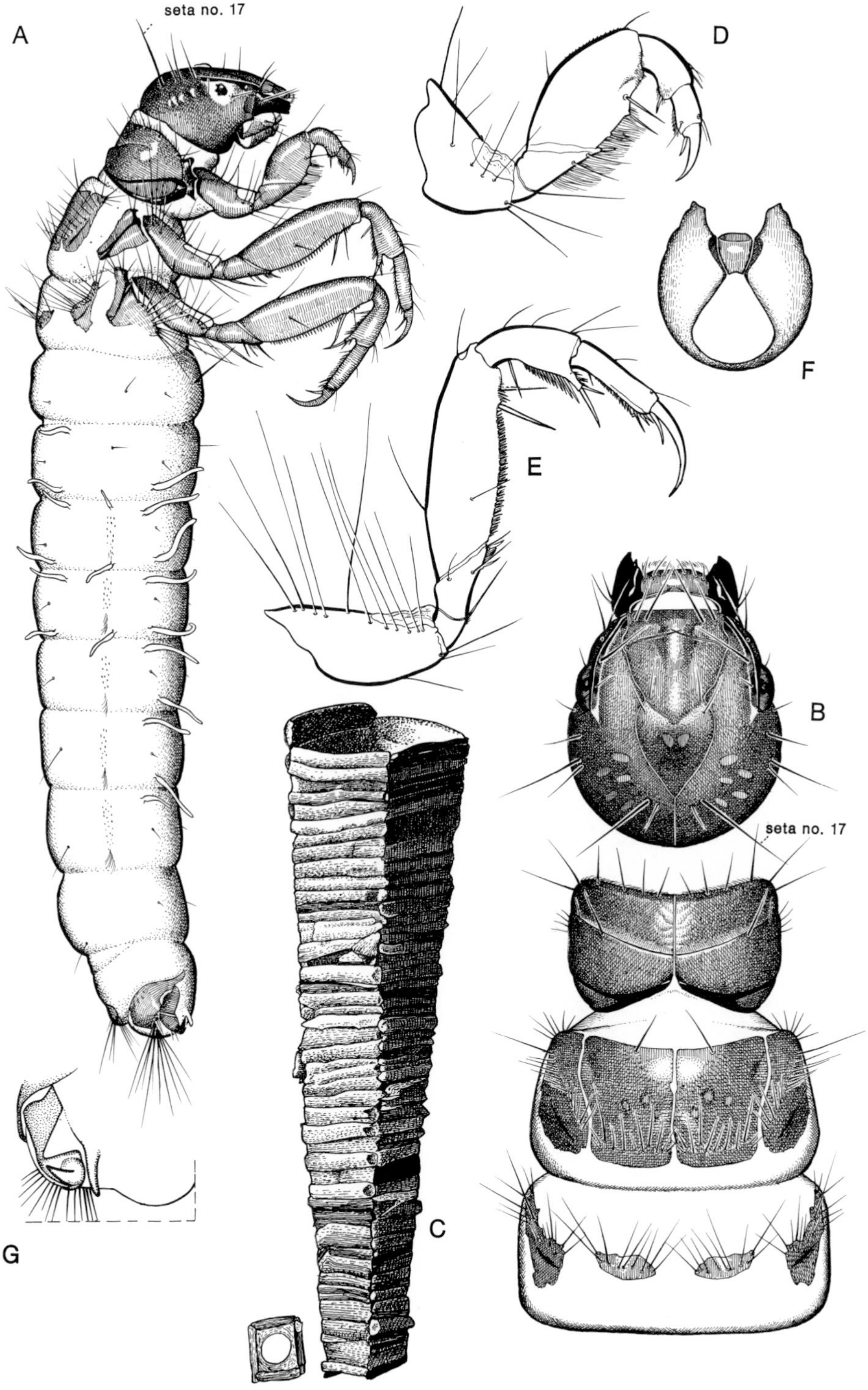

A

seta no. 17

D

F

E

B

seta no. 17

C

G

2.4 Genus **Eobrachycentrus**

DISTRIBUTION AND SPECIES This genus was created for the single species *E. gelidae* Wiggins, discovered in Oregon; we have collected additional larvae of the genus, and presumably of this species, in British Columbia and Washington. Data for larvae, pupae, and adults were outlined in the original description (Wiggins 1965).

MORPHOLOGY Larvae of *Eobrachycentrus* are readily distinguished from those of all other genera by characters provided in the key. The head is round in dorsal aspect, the dorsum flattened. A short prosternal horn is present (D). Abdominal gills are lacking, and the lateral fringe is absent. Length of larva up to 12 mm.

CASE The larval case is four-sided and evenly tapered; it is constructed of pieces of plant materials, largely moss, fastened transversely. Loose ends of the moss plants and often leaves are left projecting, giving the case a roughened appearance. The posterior opening resembles a four-leaf clover in outline. Length of larval case up to 13 mm.

BIOLOGY Larvae live in moss growing in small, very cold spring runs of mountainous terrain. Presence of the last three instars and prepupae in a June collection indicates a life cycle of at least two years. Guts of larvae (2) examined contained largely vascular plant tissue, filamentous algae, and fine particles. Adults emerge in April to crawl actively on sunny days over the still abundant snow cover.

Eobrachycentrus gelidae (Oregon, Clackamas Co., 19 April 1964, ROM)
A, larva, lateral x11; B, head and thorax, dorsal; C, case x9; D, head and part of prosternum, ventral

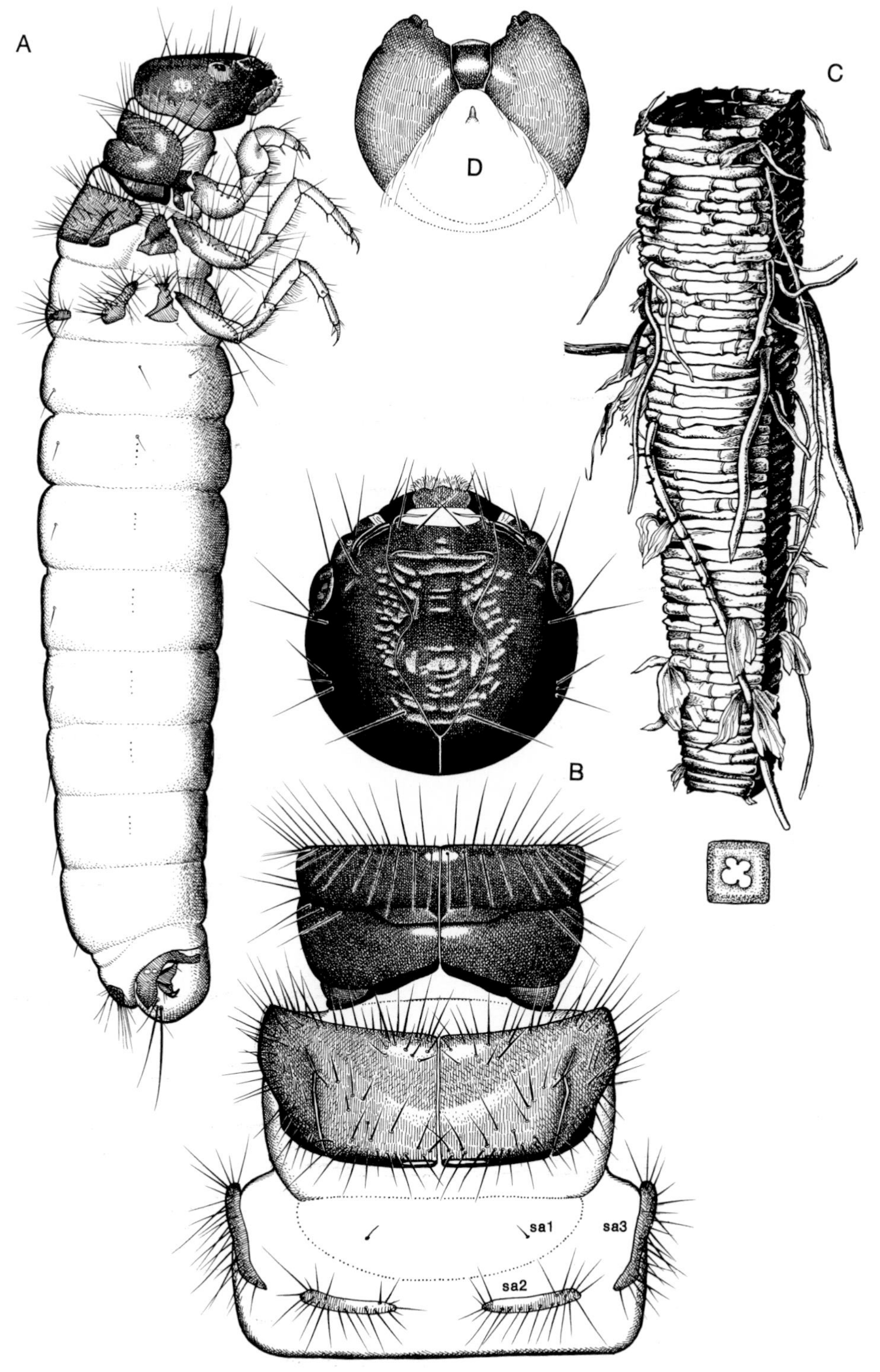
A
B
C
D
sa1
sa2
sa3

2.5 Genus Micrasema

DISTRIBUTION AND SPECIES The genus *Micrasema* is represented over a large part of the Holarctic region. Eighteen species are currently recognized in North America, and the group is represented in most parts of the continent.

Larvae have been described for *M. rusticum* (Hagen) (Ross 1944) and *M. dimicki* (Milne) (Wiggins 1965). We have associated material for *M. burksi* Ross and Unzicker, *M. scissum* McL., and *M. wataga* Ross.

MORPHOLOGY Larvae are likely to be confused only with those of *Amiocentrus*, but can be distinguished by the multiple setae of mesonotal *sa*1. The mesonotal plates are either entire (B), or in some species each one is subdivided longitudinally (Ross 1944, fig. 892) much as in *Brachycentrus* (Fig. 2.3B). The head in *Micrasema* larvae is frequently flattened dorsally and carinate, and the surface often pebbled. Sclerites of the head and thorax are usually dark reddish brown in colour. In some species of *Micrasema* the distal ends of the meso- and meta-tarsi are extended into angulate lobes, as in *Amiocentrus* (Fig. 2.2F). Abdominal gills are absent; the lateral fringe is lacking and segment VIII usually bears a protuberance on each side. Length of larva up to 8 mm.

CASE Larval cases of *Micrasema* spp. are tapered. Depending on the species involved the case is straight or curved and is constructed of sand, or of ribbon-like pieces of plant materials wound around the circumference, or largely of silk alone. Straight cases of plant materials and/or silk of *Micrasema* larvae can be distinguished from *Amiocentrus* cases of similar construction by the reduction of the posterior opening of the case with silk, the restricted opening usually three- or four-lobed (C). Length of larval case up to 10 mm.

BIOLOGY Larvae of *Micrasema* spp. live in running waters, often small, cold streams, where they are usually found on rocks in clumps of aquatic mosses. Some European species feed on periphytic algae during the first instar, and thereafter on moss (Décamps and Lafont 1974).

REMARKS Taxonomic data on adults of the North American species of *Micrasema* were summarized by Ross (1947) and Ross and Unzicker (1965).

Micrasema sp. (Oregon, Linn Co., 16 June 1968, ROM)
A, larva, lateral x19; B, head and thorax, dorsal, detail of mesonotum; C, case x13; D, head, ventral; E, segment VIII, dorsal

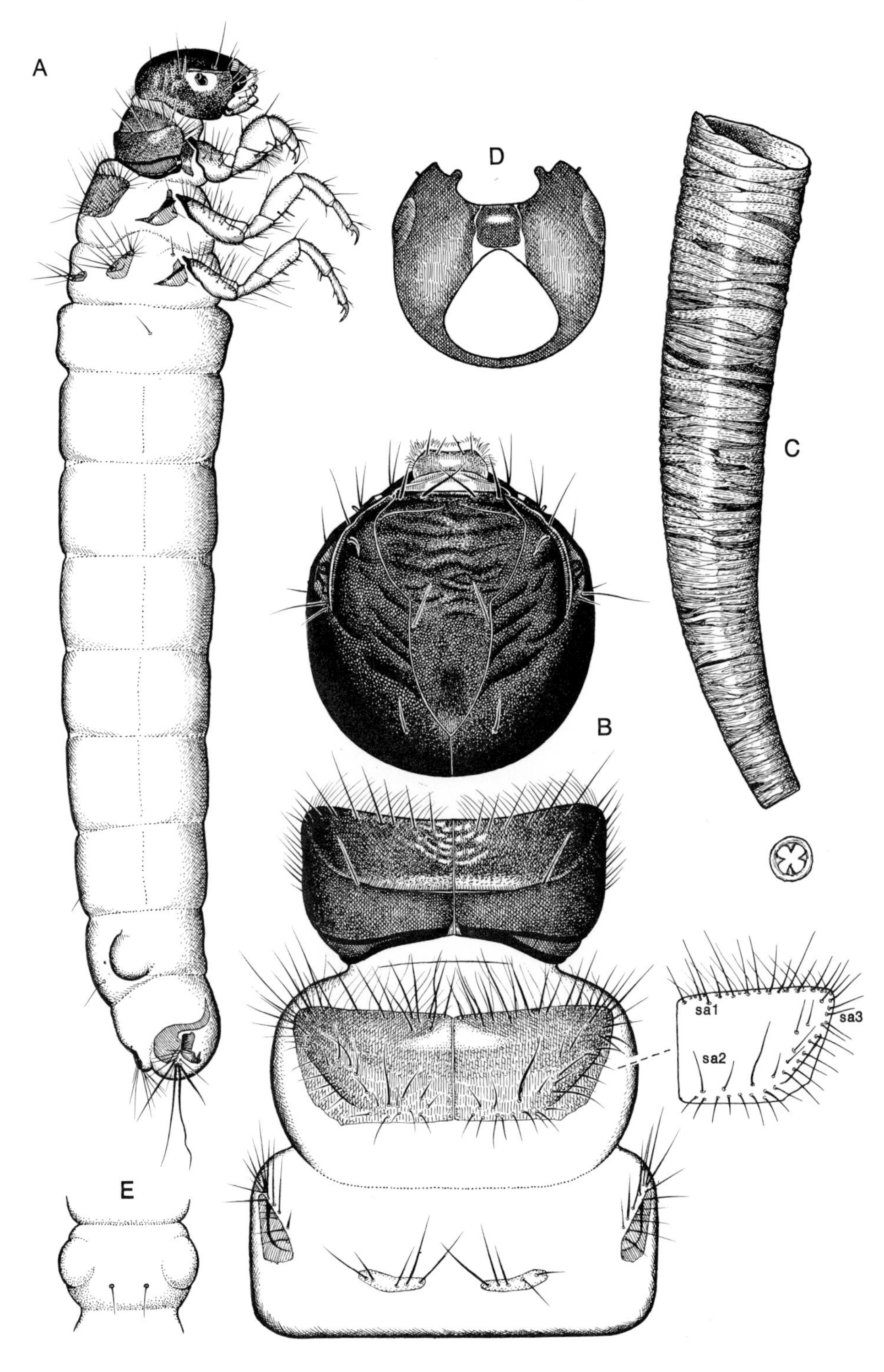
A
B
C
D
E
sa1
sa2
sa3

2.6 Genus **Oligoplectrum**

DISTRIBUTION AND SPECIES Two species are currently assigned to this genus: *O. maculatum* (Four.) in Europe, and *O. echo* Ross in California and Utah.

The larva of *O. echo* was assigned tentatively and described by Wiggins (1965); our subsequent field work has confirmed the association from collections made in Mono County, California.

MORPHOLOGY Larvae are generally similar to those in *Brachycentrus*, with long middle and hind legs. The presence of 2-5 pairs of long setae on the meso- and meta-femora, previously known only in *O. echo* (Wiggins 1965), is now known in *Brachycentrus* as well (Wallace 1971), necessitating the present generic diagnosis based on head setae and mesonotal sclerites.

The head in *O. echo* has no anterolateral carina as there is in several species of *Brachycentrus*. A short prosternal horn is present. The unsclerotized, oblique break in the lateral mesonotal sclerite (B) is variably represented and often absent. On the metathorax a line of setae extends ventromesially from the distal extremity of each metepimeron. Abdominal gills are single. A finger-like lobe is present on the venter of each anal proleg as in *Brachycentrus* (F). Length of larva up to 9 mm.

CASE The larval case of *O. echo* is straight, tapered, circular in cross-section, and constructed of small rock fragments. Thus, larvae of *Oligoplectrum* can usually be distinguished from *Brachycentrus* by their cases alone, but because cases in the latter are sometimes constructed of sand, this distinction is not always reliable. Length of larval case up to 18 mm.

BIOLOGY Larvae live in running water. Observations on populations of *O. echo* in thermal streams in Mono County, California, reveal that at a temperature of 15.6°C larvae can reach exceedingly high densities, and they persist at least to 34.4°C in waters smelling strongly of hydrogen sulphide, although at a much lower density. Gut contents of larvae (2) examined were largely diatoms and fine particles, with smaller amounts of vascular plant tissue and filamentous algae.

REMARKS Diagnostic characters of the male of *O. echo* were given by Ross (1947).

Oligoplectrum echo (California, Mono Co., 17 July 1966, ROM)
A, larva, lateral x18, abdominal spines enlarged; B, head and thorax, dorsal; C, case x11; D, fore leg, lateral; E, middle leg, lateral; F, anal proleg, ventral

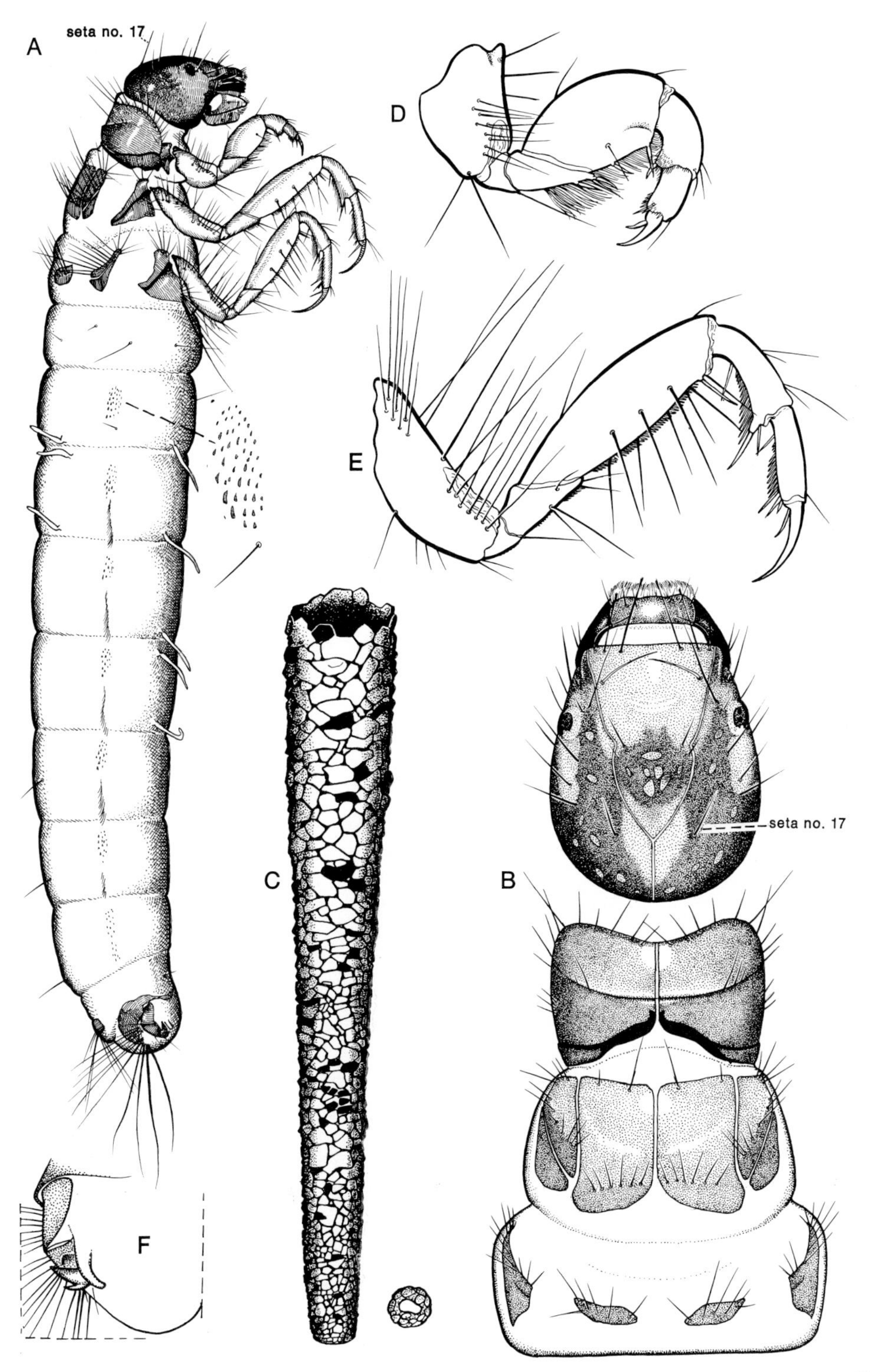
A
seta no. 17
D
E
C
B
seta no. 17
F

3
Family Calamoceratidae

This is a family of essentially tropical and subtropical distribution through all faunal regions; limited northward extensions in the Nearctic region comprise three genera with five species in eastern and western parts of the continent north of Mexico.

The North American species occur in streams where plant detritus accumulates in pools and areas of slower current. Larvae of a South American species live in water held in the leaf bases of bromeliads. Limited evidence for some species indicates that plant detritus is the major food of the larvae, although as in many detritivores it may be largely the associated microflora that is assimilated.

Calamoceratid larvae are unique in possessing a transverse row of about 16 stout setae across the central part of the labrum (Fig. 3.1D). The mandibles have tooth-like points. The pronotum is produced anteriorly in all three genera, very prominently in two of them; the fore trochantin is hooked in some genera. On the mesonotum, the *sa*3 sclerites are separate from the larger plate incorporating the other two setal areas. On the metanotum, *sa*1 and *sa*2 sclerites are reduced or absent with *sa*1 represented by a single seta, but the *sa*3 sclerite is present. Lateral humps of abdominal segment I are more ventral in position than in other families. Gills are branched or single, and the lateral fringe very dense in some genera; lateral tubercles are restricted to segment VIII.

Leaves and pieces of wood are used in several different ways to construct portable cases.

Key to Genera

1 Pronotum with anterolateral corners extended into prominent lobes (Fig. 3.1A,B); gills branched (Fig. 3.1A) ... 2

Pronotum with anterolateral corners somewhat extended but much less than above (Fig. 3.2A,B); gills single (Fig. 3.2A); case a twig with a central cavity (Fig. 3.2C). Eastern and western ... **3.2 Heteroplectron**

2 (1) Hind legs approximately same length as middle legs (Fig. 3.3A); anterolateral extensions of pronotum pointed (Fig. 3.3B); case of pieces of bark and leaves (Fig. 3.3C). Southwestern **3.3 Phylloicus**

Hind legs approximately twice as long as middle legs (Fig. 3.1A); anterolateral extension of pronotum rounded (Fig. 3.1B); case of two leaf pieces, dorsal piece overlapping ventral (Fig. 3.1C). Southeastern **3.1 Anisocentropus**

3.1 Genus **Anisocentropus**

DISTRIBUTION AND SPECIES Species assigned to this genus are recorded from all faunal regions (Fischer 1965, 1972b), but generic affinity of Nearctic and Neotropical species placed in *Anisocentropus* is in doubt, and even assignment to it of the single Nearctic species *A. pyraloides* (Walk.) is subject to verification (O.S. Flint, pers. comm.).

The larva and pupa of *A. pyraloides* were described from Georgia by Wallace and Sherberger (1970). The species was recorded from Tennessee by Edwards (1966); we collected larvae in Tennessee, and have identified larvae from Delaware.

MORPHOLOGY Larvae are readily identified by characters of the pronotum and legs given in the key. Sclerotized parts of the head and thorax are light yellowish brown in colour. The thorax and abdomen are strongly depressed; among the North American species of the Calamoceratidae, *A. pyraloides* is distinctive in this flattened form (E), and also in having hind legs twice as long as the middle legs (A). The added length is largely a result of an extremely elongate tibia that has an apparent secondary segmentation as in certain leptocerid larvae, such as *Mystacides* and *Triaenodes.* Features of larval behaviour as yet unknown may account for these distinctive morphological characters. Length of larva up to 19 mm.

CASE Larval cases of *A. pyraloides* are among the most unusual constructed by North American caddisflies. Two ovate pieces of leaves are cut and fastened together as illustrated. The larger piece forms a dorsal shield over a smaller ventral piece, and between the two is a flattened chamber for the larva. Length of larval case up to 35 mm.

BIOLOGY Observations by Wallace and Sherberger (1970) indicate that larvae live in areas of slow current of small streams in deciduous forest areas. The larvae are detrital feeders, the early instars occurring largely in accumulations of leaves and later instars on the undersides of logs and rocks. Evidence from the range of larval instars suggested a life cycle of two years.

REMARKS Diagnostic characters for adults of *A. pyraloides* were given by Betten and Mosely (1940). Specimens for illustration were loaned by J.B. Wallace, University of Georgia.

Anisocentropus pyraloides (Georgia, Heard Co., 18 March 1969, Univ. Ga. coll.)
A, larva, lateral x8, hind tibia enlarged; B, head and thorax, dorsal; C, case, ventral x3; D, labrum, dorsal; E, abdominal segments, dorsal

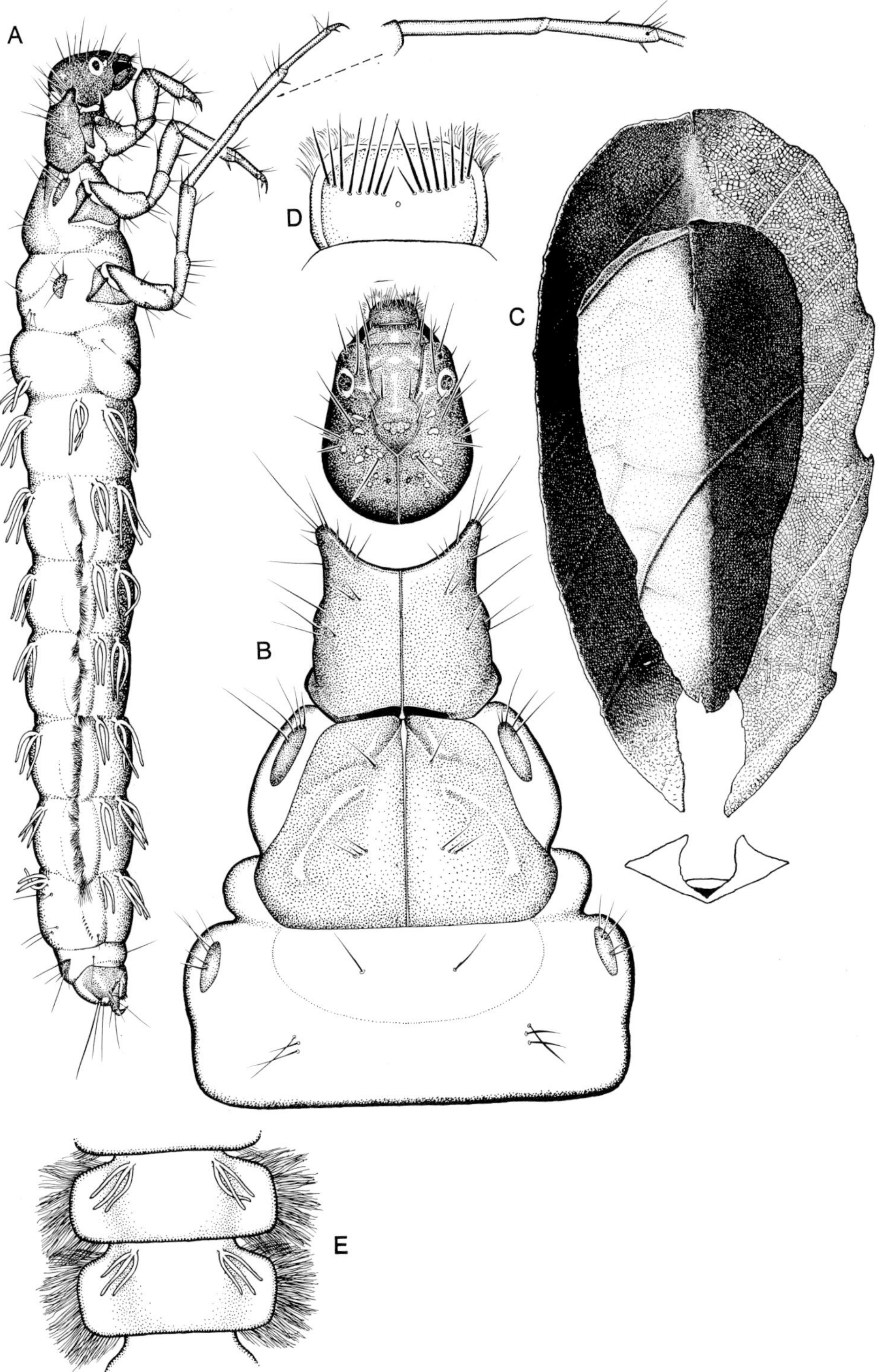
A
B
C
D
E

3.2 Genus **Heteroplectron**

DISTRIBUTION AND SPECIES The composition of *Heteroplectron* has been unclear, but the genus appears to be restricted to the Nearctic region (O.S. Flint, pers. comm.). There are two species: *H. americanum* (Walk.) (syn. *Ganonema nigrum* Lloyd) in the east from New York, New Hampshire, and Quebec southward; and *H. californicum* McL. in the west from California north to British Columbia.

The larva of the eastern species was described by Lloyd (1921); that of the western species, not previously described, is illustrated here.

MORPHOLOGY The head, pronotum, and other sclerotized parts are dark brown and shiny. The thorax and abdomen are cylindrical and not flattened as in *Anisocentropus.* Length of larva up to 25 mm.

Larvae of the two species of *Heteroplectron* can be separated by the following: in the western *H. californicum* (A) the ventral edge of the pronotum is concave to straight, forming a sharp angle where it meets the anterior margin; in the eastern *H. americanum* the ventral edge of the pronotum is convex, giving rise to a rounded anterolateral junction; and in *H. americanum* there is a raised, straight ridge on the lateral sclerite of the anal proleg, which is only weakly developed in *H. californicum.*

CASE *Heteroplectron* larvae are unusual among North American caddisflies in their use of twigs as cases: a chamber is excavated through the centre, open at both ends for circulation of water (Lloyd 1921). The chamber is lined with silk. On two occasions in California we collected larvae of *H. californicum* that were using stone cases of other caddisflies, the anterior end modified with small pieces of wood.

BIOLOGY Although characteristic of cool, running water habitats, larvae are largely confined to pools and areas of slower current where plant detritus accumulates. Observations on the life cycle of *H. californicum* were given by Winterbourn (1971a) and Anderson and Wold (1972); larvae ingested leaf fragments.

REMARKS Diagnostic characters of adults were illustrated by Kimmins and Denning (1951) for *H. californicum*, and by Betten (1934) for *H. americanum.*

Heteroplectron californicum (California, Plumas Co., 5 Oct. 1966, ROM)
A, larva, lateral x8; B, head and thorax, dorsal; C, case x4

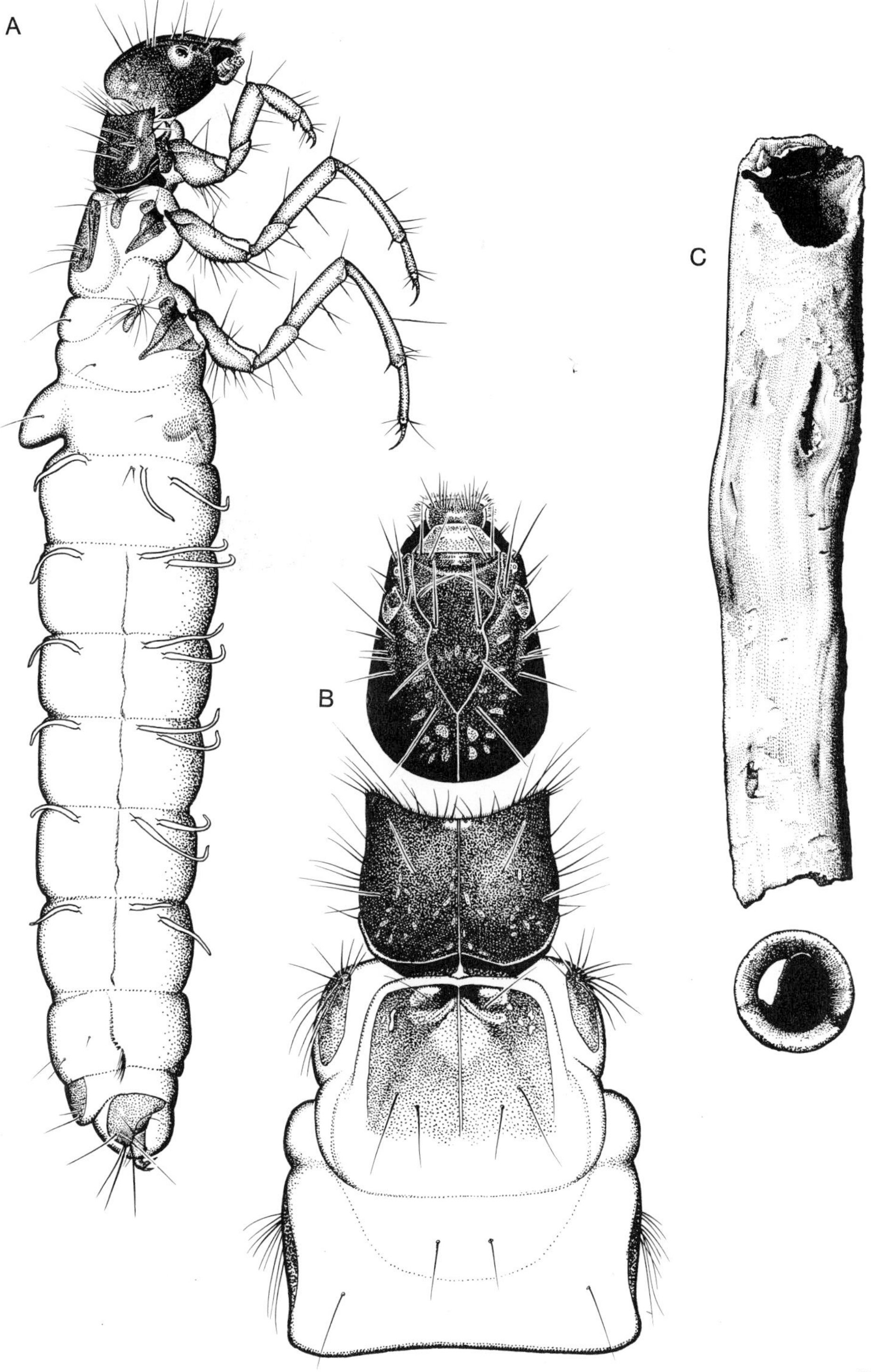
A
B
C

3.3 Genus **Phylloicus**

DISTRIBUTION AND SPECIES This is essentially a Neotropical genus, with two species recorded from the southwestern United States: *P. ornatus* (Banks) from Texas and *P. aeneus* (Banks) (syn. *Notiomyia mexicana* (Banks)) from Arizona.

A circumstantial association for larval diagnosis presumed by Ross (1959) was confirmed by Flint (1964b) for a Puerto Rican species of *Phylloicus*; we reared larvae of *P. aeneus*, illustrated here, in Arizona.

MORPHOLOGY The long, pointed processes of the pronotum (B) will readily distinguish larvae belonging to this genus. The hind legs are approximately the same length as the middle legs, and the hind tibia is not secondarily subdivided as in *Anisocentropus*. The thorax and abdomen are somewhat flattened but not as much as in *Anisocentropus*. Sclerotized parts are medium to dark brown in colour. Length of larva up to 22 mm.

CASE Larval cases are flattened and straight, consisting of dorsal and ventral halves fastened together along the edges and enclosing a central chamber. Pieces of bark, wood, or stout leaves are used as building materials, and it is not uncommon to find leaves from which circular pieces of a size appropriate for case-making have been cut. A protective hood-like piece is frequently positioned to overhang the anterior opening of the case. Length of larval case up to 40 mm.

BIOLOGY Larvae live in running water. We found *P. aeneus* in abundance in small, cool streams in arid sections of Arizona. Larvae of the last four instars, pupae, and adults from our collections made in May and June indicate a life cycle of more than one year. Gut contents from a sample of these larvae (2) were primarily filamentous algae and vascular plant tissue - probably from detritus. Prior to pupation larvae fasten their cases to the undersides of logs and rocks.

REMARKS Diagnostic characters for adults of several species of *Phylloicus*, including *P. aeneus*, were given by Flint (1967b).

Phylloicus aeneus (Arizona, Cochise Co., 23 June 1966, ROM)
A, larva, lateral x8; B, head and thorax, dorsal; C, case, ventral x4

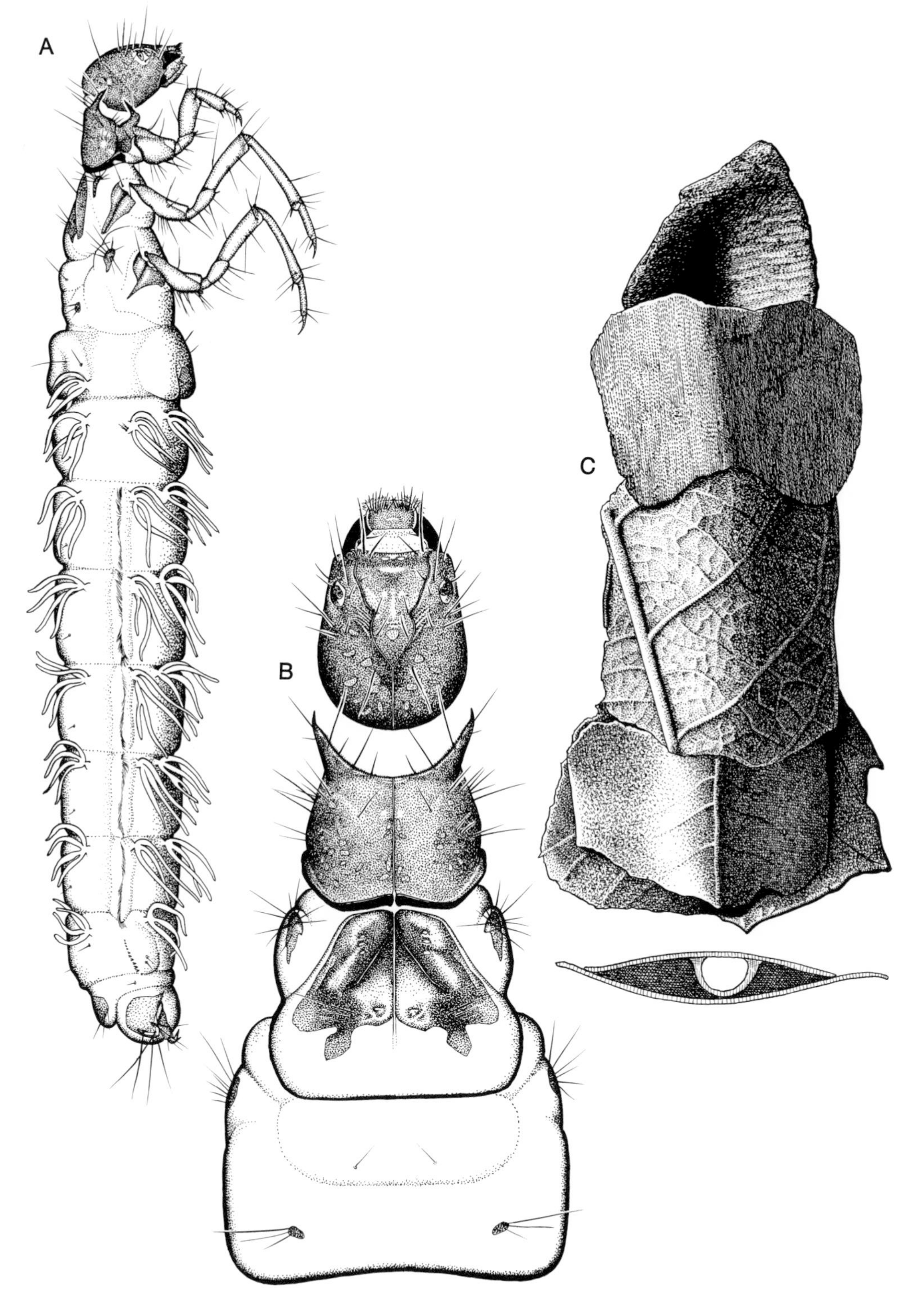
A
B
C

4
Family Glossosomatidae

Glossosomatids live chiefly in rivers and streams, and they are a common and often abundant element in these communities. A European species, *Agapetus fuscipes* Curtis, is reported from lakes (Mackereth 1956); *Glossosoma* larvae have been collected along the rocky, wave-washed shore of Lake Superior. The family is represented in all faunal regions; six genera are recognized in the Nearctic region, with approximately 80 species known in the United States and Canada.

Larvae of the family are specialized for a particular feeding site: the uppermost and generally more exposed surfaces of rocks, where they graze on diatoms, green algae, and fine organic particles (Anderson and Bourne 1974). Scraper-like mandibles lacking separate teeth and a membranous fringe on the labrum are typical adaptations for this method of feeding.

Morphologically the larvae are a homogeneous group. Only the pronotum is covered by dark sclerotized plates, the nota of the other two thoracic segments being largely membranous but with small sclerites in some genera; prosternal plates are prominent. All three pairs of legs are approximately the same size. Abdominal gills are always lacking, and there is no lateral fringe of setae; anal papillae are present. The distal half of the anal proleg in this family is free from the abdomen, a feature distinguishing it from other case-making families. One possible functional basis for this difference is that anal prolegs in those families making portable tube-cases are normally in contact only with the silken lining of the case, where the claws anchor the end of the abdomen; in the Glossosomatidae, the claws of the anal prolegs can be extended through the opening of the case to stabilize the larva on the external substrate. In terms of phylogeny, the anal prolegs of the Glossosomatidae are thought to represent a stage intermediate between those exemplified by the free-living Rhyacophilidae and tube-case makers of the Limnephiloidea (Ross 1964).

The architecture of the larval cases makes the Glossosomatidae a distinctive group. Although frequently called saddle-cases, an equally appropriate analogy for the oval, flat-bottomed domes of tiny rock fragments is with the shell of a tortoise; a broad strap across the venter leaves an opening at each end through which head and legs are extended. Certain features of these cases are characteristic for particular genera, and are described under the generic headings. All cases have some openings between stones to permit free circulation of water current over the larvae for respiration. There is a tendency in several genera for larvae to spin a flexible silken membrane around the rim of the ventral openings; in observing living larvae one can see that these membranes fold over the aperture when the larva withdraws into its case. In some exotic genera, tubular extensions are added to these openings (e.g. Flint 1963, fig. 7). The hinged stone closure of cases in a series of larvae collected in Arizona and illustrated here (Fig. 4.6G) is a precise mechanism for closing the holes, confirming that there is selective pressure for a case-closure mechanism; possibly this serves to keep predacious insects from entering the case. Glossosomatid larvae reverse position within their cases easily; thus, with no distinction between the front and rear of the case, both openings are built alike. Larvae can feed while completely beneath the case, probably an important feature on the exposed upper surfaces of rocks where diatoms, their principal food, grow. It is likely, in fact, that portable cases in the Glossosomatidae are a protective device for exploiting a food resource rather than an enhancement of respiratory efficiency as in tube-case-making families.

Another unique feature of the glossosomatid case is that it is not suited to continual enlargement as are portable cases in other families; usually each instar builds a larger case attached to one end of the existing case, finally cutting the old one away (Anderson and Bourne 1974). This behaviour results in cases being discarded several times during larval growth, and appears, superficially at least, to be an inefficient way of investing energy in the construction of a case. But evidently these insects are geared to profligacy in cases, for in a Minnesota stream caseless larvae of *Glossosoma intermedium* drifted at rates up to 350 per hour per foot of stream width (Waters 1962). Under conditions of stress such as reduced current flow or high temperature, glossosomatid larvae abandon their cases, usually building new ones when conditions return to normal (Anderson and Bourne 1974). Prior to pupation, the larva cuts out the ventral strap, fastens the outer rim of the case to a rock, and spins an oval brown cocoon of silk.

From several lines of evidence, then, case-making by glossosomatid larvae can be regarded as basically different in procedure and function from that of other families.

Three subfamilies have been recognized, and all are represented in North America.

AGAPETINAE: *Agapetus* Head with ventromesial margins of genae not thickened, mid-ventral ecdysial line approximately 1½ times longer than each divergent branch enclosing ventral apotome; submental sclerites paired and separate; pronotum with numerous setae around most of periphery; mesonotum with two sclerites, metanotum with two smaller sclerites; tarsal claws with basal seta sessile; anal claw with one or two accessory hooks; anal opening without sclerotized bar along each side. The group is represented in all faunal regions except the Neotropical, where it evidently is replaced ecologically by the Protoptilinae.

This group has been designated as a tribe of the Glossosomatinae (Ross 1956), but the subfamilial level proposed by Martynov (1924) seems more in keeping with the larval data outlined here.

GLOSSOSOMATINAE: *Anagapetus, Glossosoma* Head with ventromesial margins of genae thickened, mid-ventral ecdysial line approximately same length as each divergent branch enclosing ventral apotome; submental sclerites fused; pronotal setae sparse; meso- and meta-nota lacking sclerites; tarsal claws with basal seta on raised truncate base; anal claw with one or two accessory hooks; anal opening with sclerotized bar along each side. The group is Holarctic and Oriental in distribution.

PROTOPTILINAE: *Culoptila, Matrioptila, Protoptila* Head with ventromesial margins of genae not thickened, mid-ventral ecdysial line approximately 1½ times longer than each divergent branch enclosing ventral apotome; submental sclerites paired and separate; ventral apotome reduced to slender V-shaped sclerite; pronotal setae sparse; mesonotum bearing median sclerite with smaller sclerite at each side; metanotum with two small sclerites, tarsal claws with basal seta frequently arising from side of basal process, claws sometimes trifid; anal claw with four or more accessory hooks; anal opening lacking sclerotized bar along each side. The group is restricted to the New World; most genera occur in the Neotropical region (Flint 1963), where the Protoptilinae are the only glossosomatids present.

Another genus, *Mexitrichia*, essentially Neotropical from Mexico to South America, might occur within the southern limits of the Nearctic region; diagnostic characters of larvae were given by Flint (1963).

Key to Genera

1 Mesonotum without sclerites (Fig. 4.4E); head with ventromesial margins of genae thickened, median ventral ecdysial line approximately as long as each divergent branch enclosing ventral apotome (Fig. 4.4D); anal opening bordered on each side by dark, sclerotized line (Fig. 4.4F) (subfamily Glossosomatinae) **2**

Mesonotum with two or three sclerites (Figs. 4.1A, 4.6A); head with ventromesial margins of genae not thickened, median ventral ecdysial line approximately 1½ times longer than each divergent branch (Figs. 4.1D, 4.3D); anal opening without dark, sclerotized line on each side (Fig. 4.1G) **3**

2 (1) Pronotum in lateral view about one-third excised anterolaterally to accommodate coxa (Fig. 4.4C). Widespread **4.4 Glossosoma**

Pronotum in lateral view about two-thirds excised anterolaterally to accommodate coxa (Fig. 4.2E). Western montane regions **4.2 Anagapetus**

3 (1) Mesonotum with three sclerites (Fig. 4.6A); head with ventral apotome reduced to slender, V-shaped sclerite (Fig. 4.6D). Widespread (subfamily Protoptilinae) **4**

Mesonotum with two sclerites (Fig. 4.1A); head with anterior ventral apotome not slender as above (Fig. 4.1D). Widespread (subfamily Agapetinae) **4.1 Agapetus**

4 (3) Each tarsal claw terminating in three acute points, one representing the basal seta, but all approximately equal in length (Fig. 4.5D). Southeastern **4.5 Matrioptila**

Each tarsal claw terminating in normal single acute point, but with smaller basal seta (Figs. 4.3E, 4.6E) 5

5 (4) Tarsal claws with basal seta long and slender, arising from side of stout process at base of claw (Fig. 4.6E). Widespread **4.6 Protoptila**

Tarsal claws with basal seta stout, larger than process at base of claw (Fig. 4.3E). Widespread **4.3 Culoptila**

4.1 Genus Agapetus

DISTRIBUTION AND SPECIES *Agapetus* is represented in all faunal regions except the Neotropical. Approximately 30 species are known north of Mexico, and although the group is widespread over much of North America, the diversity of species is greater in southern and western areas.

Larvae have been associated and characterized for *A. minutus* Sibley (Sibley 1926), *A. illini* Ross (Ross 1944), *A. diacanthus* Edwards (Edwards 1956), and *A. bifidus* Denning (Anderson and Bourne 1974), but there is as yet no over-all basis for identification of larvae to species.

MORPHOLOGY Study of our larval glossosomatid collections indicates that *Agapetus* larvae are the only ones in North America with two mesonotal sclerites (A); they are also distinctive in having more setae on the pronotum than larvae of other genera, especially near the posterior margin. In general, *Agapetus* larvae are similar to those of the Protoptilinae. Length of larva up to 6 mm.

CASE Larval cases in *Agapetus* (B,C) show some general similarity to those of *Protoptila* in having a relatively large stone on each side, and in being higher in relation to their length than cases of other glossosomatids. Edges of the main openings frequently have silken flaps which are manipulated by the legs to effect almost complete closure (Anderson and Bourne 1974). Length of larval case up to 7.5 mm.

BIOLOGY Larvae of *Agapetus* live in streams of hilly terrain, generally in waters intermediate between the colder sites frequented by *Anagapetus* and *Glossosoma* and warmer sites where *Protoptila* occurs. Larvae of *A. bifidus* were, however, found in the same small Oregon stream with *Glossosoma* and *Anagapetus* larvae, but were more abundant in slower sections than either of the others (Anderson and Bourne 1974); *A. bifidus* was also found to overwinter in the egg stage in diapause for 8-9 months, with rapid development of the larvae after hatching in March. The spring occurrence of *A. illini* larvae in temporary streams in Illinois (Ross 1944) might be dependent upon a long egg-diapause. Food of *Agapetus* larvae in a Colorado stream was found to fluctuate seasonally between diatoms and detritus (Mecom 1972a).

REMARKS Phylogeny, biogeography, and taxonomy of the adults were summarized by Ross (1951b, 1956). The European species *A. fuscipes* is probably the best known member of this genus; among several aspects, case-building was studied by Hansell (1968a) and eggs by Anderson (1974b), and Nielsen (1942) found it to be one of the few Trichoptera with seven larval instars.

Agapetus sp. (Virginia, Giles Co., 18 April 1968, ROM)
A, head and thorax, dorsal x58; B, case, ventrolateral x20; C, case, dorsal; D, head, ventral; E, mandible, ventral; F, tarsal claw of middle leg, lateral; G, anal prolegs, caudal

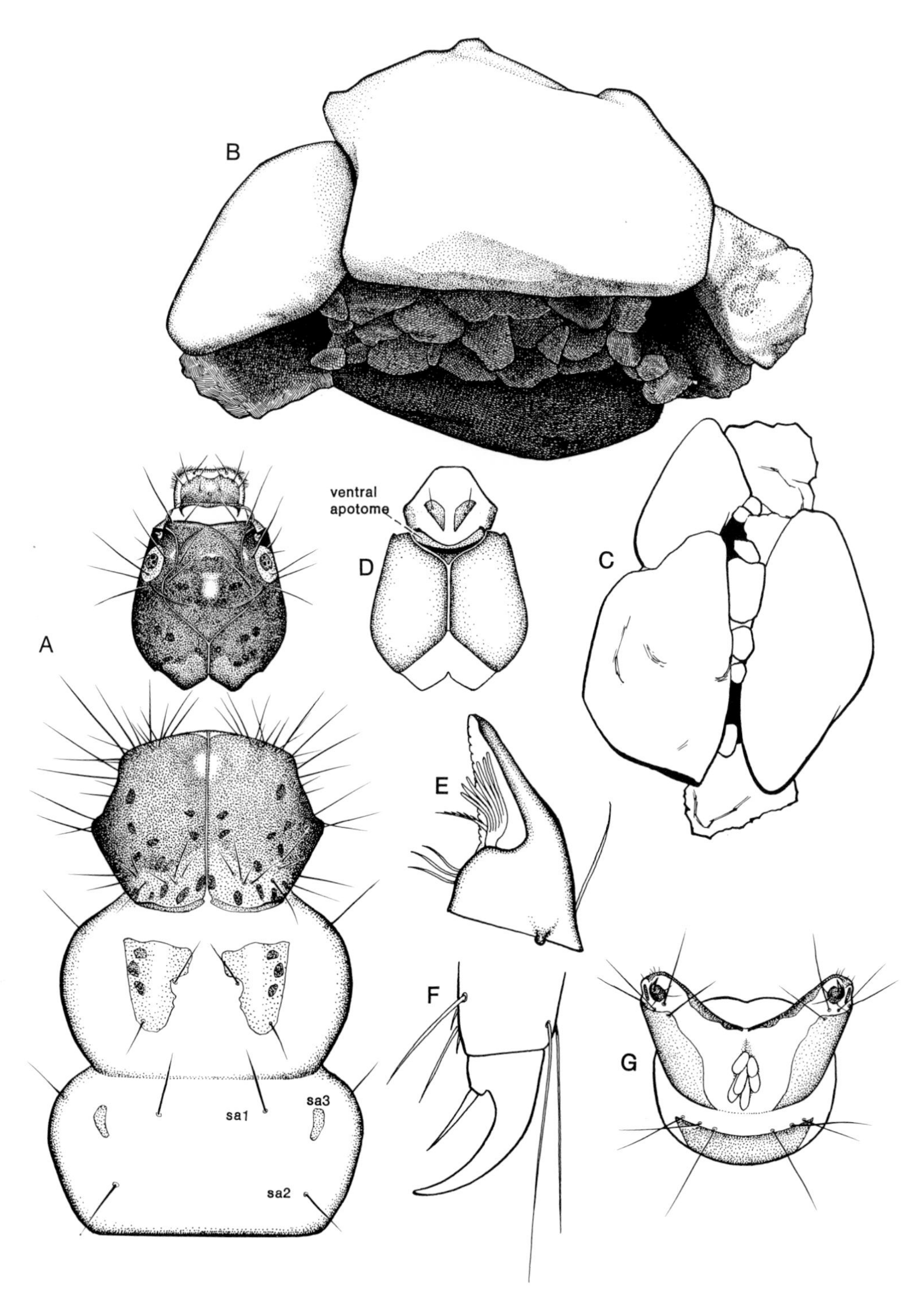
B
ventral
apotome
D
C
A
E
F
G
sa1
sa3
sa2

4.2 Genus **Anagapetus**

DISTRIBUTION AND SPECIES This group is confined to North America where six species are known, but only in western montane areas.

Diagnostic characters at the generic level given by Ross (1959) are not consistent for all species; we have associated material for four species.

MORPHOLOGY Larvae are similar to *Glossosoma*, sharing all characters listed for the Glossosomatinae. *Anagapetus* larvae can be distinguished by the more posterior origin of the fore legs, resulting in a deep lateral excision in the pronotum to accommodate the coxa (E). Length of larva up to 6.5 mm.

CASE Larval cases are similar in general shape and structure to those of *Glossosoma*, but frequently have some silk and sand grains around the ventral openings (B). Length of larval case up to 7.5 mm.

BIOLOGY *Anagapetus* larvae are most likely to be found in the cooler headwater sections of mountain streams. Larvae of *A. bernea* Ross had one generation per year but, in contrast to those of *Agapetus bifidus* that occurred in the same Oregon stream, fed during the winter months to reach the final instar by spring; diatoms, green algae, and detritus were ingested during summer months (Anderson and Bourne 1974).

REMARKS Taxonomic data for adults were summarized by Ross (1951a); phylogeny and biogeography of the species have also been studied (Ross 1956).

Anagapetus bernea (British Columbia, E.C. Manning Prov. Park, 6 July 1969, ROM 690179)
A, head and thorax, dorsal x57; B, case, ventrolateral x17; C, case, dorsal; D, head, ventral; E, prothorax, lateral, coxa outlined; F; tarsal claw of middle leg, lateral; G, anal prolegs, caudal

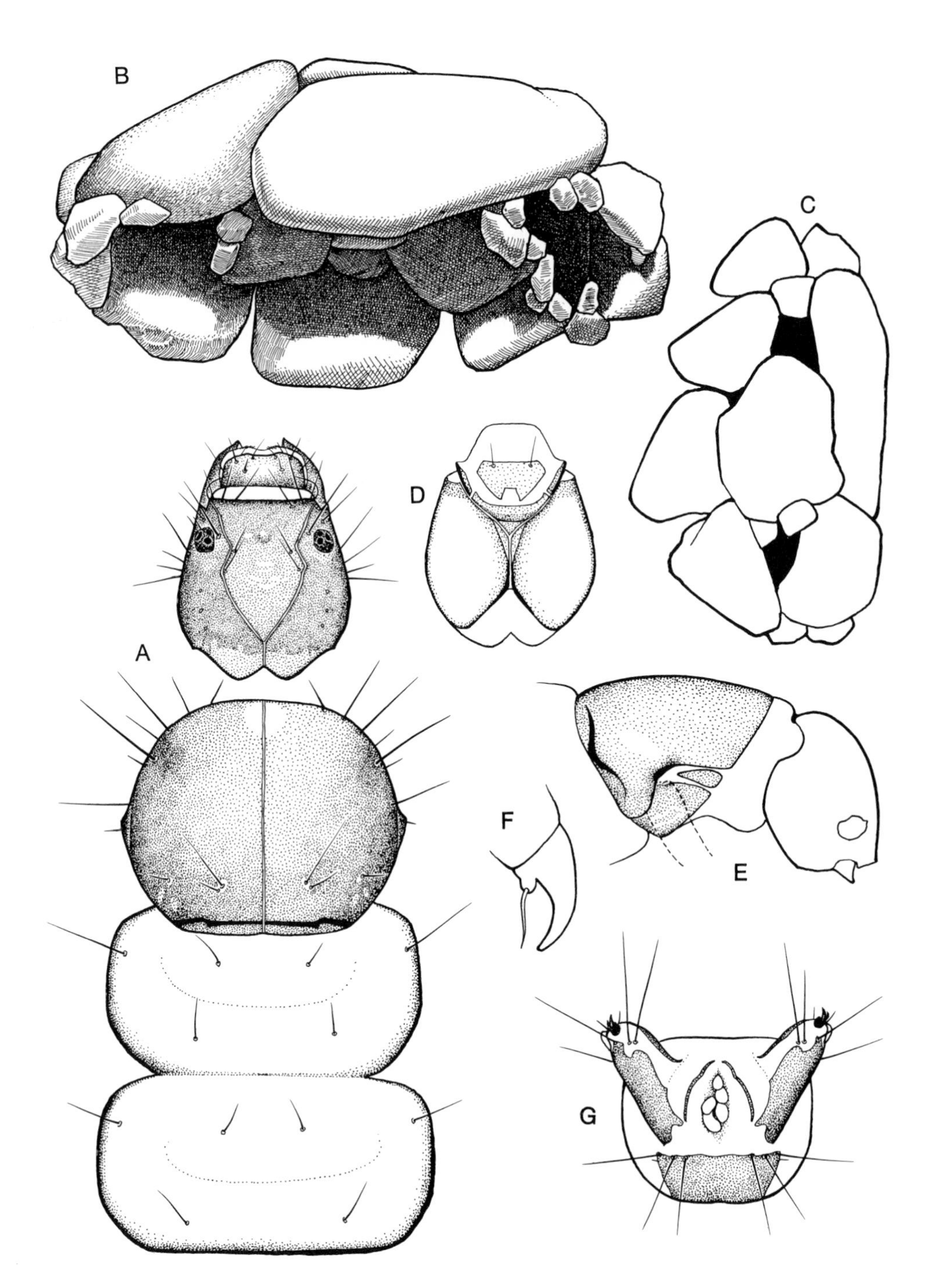
B
C
D
A
F
E
G

4.3 Genus **Culoptila**

DISTRIBUTION AND SPECIES This genus is restricted to the New World, where most species occur in Mexico and Central America. Four species are, however, now known north of Mexico, largely in the southwest. Two of these, *C. cantha* (Ross) and *C. thoracica* (Ross), both formerly placed in *Protoptila*, also extend northward to Wyoming. The same two species are recorded from the eastern part of the continent, *C. cantha* from Maine and Maryland and *C. thoracica* from North Carolina (Flint 1974b).

The larva for *Culoptila* has not been identified in the literature before, and I am indebted to O.S. Flint, Smithsonian Institution, for providing the associated series of *C. moselyi* Denning from Arizona, illustrated here. We collected another series of larvae in Utah which appear to be of the same genus.

MORPHOLOGY Available larval material in *Culoptila* is typical for the Protoptilinae, and similar to *Protoptila*; the median mesonotal sclerite (A) is rounded posteriorly rather than angulate as in *Protoptila* (Fig. 4.6A), but this may prove not to be consistent at the generic level. The only reliable basis that we have found for distinguishing larvae of *Culoptila* from *Protoptila* is the much stouter basal seta of the tarsal claw (E), which is articulated at its own base and evidently homologous with the slender basal seta of *Protoptila* (Fig. 4.6E). Examination for this character is best done with a compound microscope on a temporary slide mount of a leg. Length of larva up to 3.5 mm.

CASE Larval cases in *Culoptila* are not at all similar to those in *Protoptila*, for instead of the large single side stones of the latter genus, the sides of the case are formed of many small rock fragments of fairly uniform size. Partial collars of silk are fastened around the external periphery of each opening (B). Length of larval case at least up to 3 mm.

BIOLOGY Little is known about the biology of *Culoptila*. The Arizona larva illustrated was collected in Turkey Creek, Chiricahua Mountains; our Utah specimens were collected 29 July, in the Green River, a wide, rather warm and silt-laden river near Dinosaur National Monument. The type localities cited for *C. cantha* and *C. thoracica* (Ross 1938) suggest a large-river habitat as well.

Culoptila moselyi (Arizona, Chiricahua Mts., 9 June 1968, USNM)
A, head and thorax, dorsal x113±; B, case, ventrolateral x28; C, case, dorsal; D, head and prothorax, ventral; E, tarsal claw of middle leg, lateral

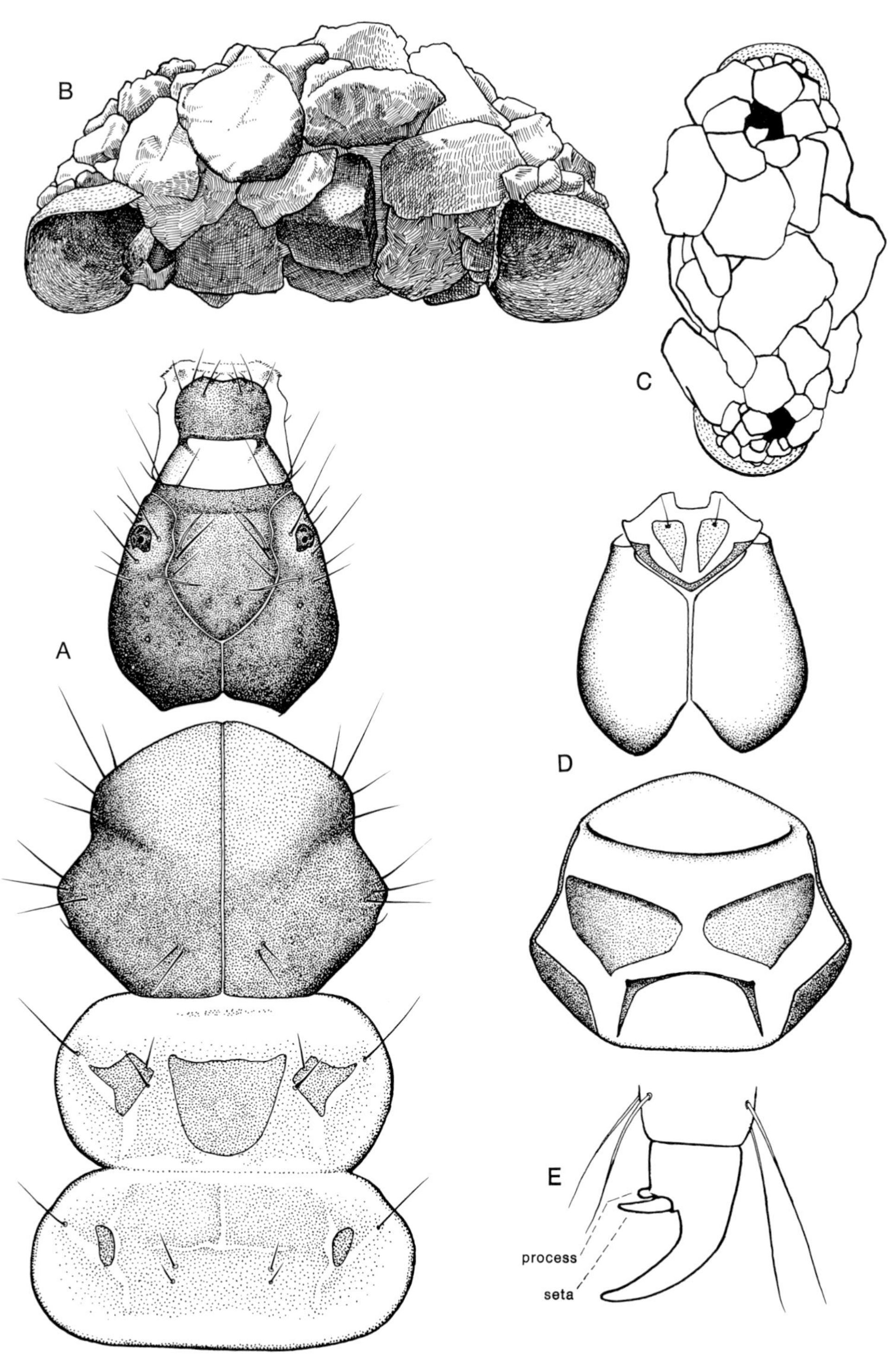

B
C
A
D
E
process
seta

4.4 Genus Glossosoma

DISTRIBUTION AND SPECIES Species of *Glossosoma* occur over much of the Nearctic, Palaearctic, and Oriental faunal regions. Approximately 25 species are known in North America, all but three of them confined to western montane areas. Two of these three, *G. nigrior* Banks and *lividum* (Hagen), are common in the northeastern part of the continent, the former extending westward to Minnesota; the remaining species, *G. intermedium* (Klapalek), is Holarctic and recorded in North America through much of the eastern and central parts of the continent to British Columbia.

Although larvae have been identified to the generic level (Ross 1944, 1959), they are a homogeneous group, and no species taxonomy has yet been developed. The larva of what is probably *G. lividum* (Ross 1956) was described as *G. americanum* by Lloyd (1921), Sibley (1926), and Betten (1934). We have associated material for 11 species.

MORPHOLOGY Larvae of *Glossosoma* share with those of *Anagapetus* the several distinctive characters of the Glossosomatinae, and since the latter is confined to western montane areas, *Glossosoma* in the rest of the continent can be recognized from the set of subfamilial characters. In the west where *Anagapetus* also occurs, the shape of the pronotum is diagnostic for *Glossosoma* as indicated in the key. Length of larva up to 9.5 mm.

CASE Larval cases of *Glossosoma* are similar to those of *Anagapetus*, but tend to be larger and usually lack membranous silk around the main openings; rock fragments of fairly uniform size are used in both genera. Length of larval case up to 12 mm.

BIOLOGY *Glossosoma* is the dominant saddle-case maker of cold, rapid streams (Ross 1951a); we have larvae from the Ontario shoreline of Lake Superior. *G. penitum*, studied in an Oregon stream, was found to have two overlapping generations per year (Anderson and Wold 1972; Anderson and Bourne 1974). In a study of the feeding habits of *G. nigrior*, Cummins (1973) found that larvae in a Pennsylvania stream fed largely on periphytic algae, but in a Michigan stream detritus was the dominant food.

REMARKS Ross (1956) discusses phylogeny and biogeography of *Glossosoma*, and summarizes taxonomic data for adults; the three species recorded from eastern North America belong to the subgenus *Eomystra* (syn. *Mystrophora*) along with one western species, most western species to the subgenus *Ripaeglossa*, but one, *G. penitum*, to *Anseriglossa*.

Glossosoma sp. (Ontario Durham Co., 6 Oct. 1951, ROM)
A, larva, lateral x21, tarsal claw of middle leg and anal proleg enlarged; B, case with larva, ventrolateral x17; C, prothorax, lateral, coxa outlined; D, head, ventral; E, head and thorax, dorsal; F, anal prolegs, caudal

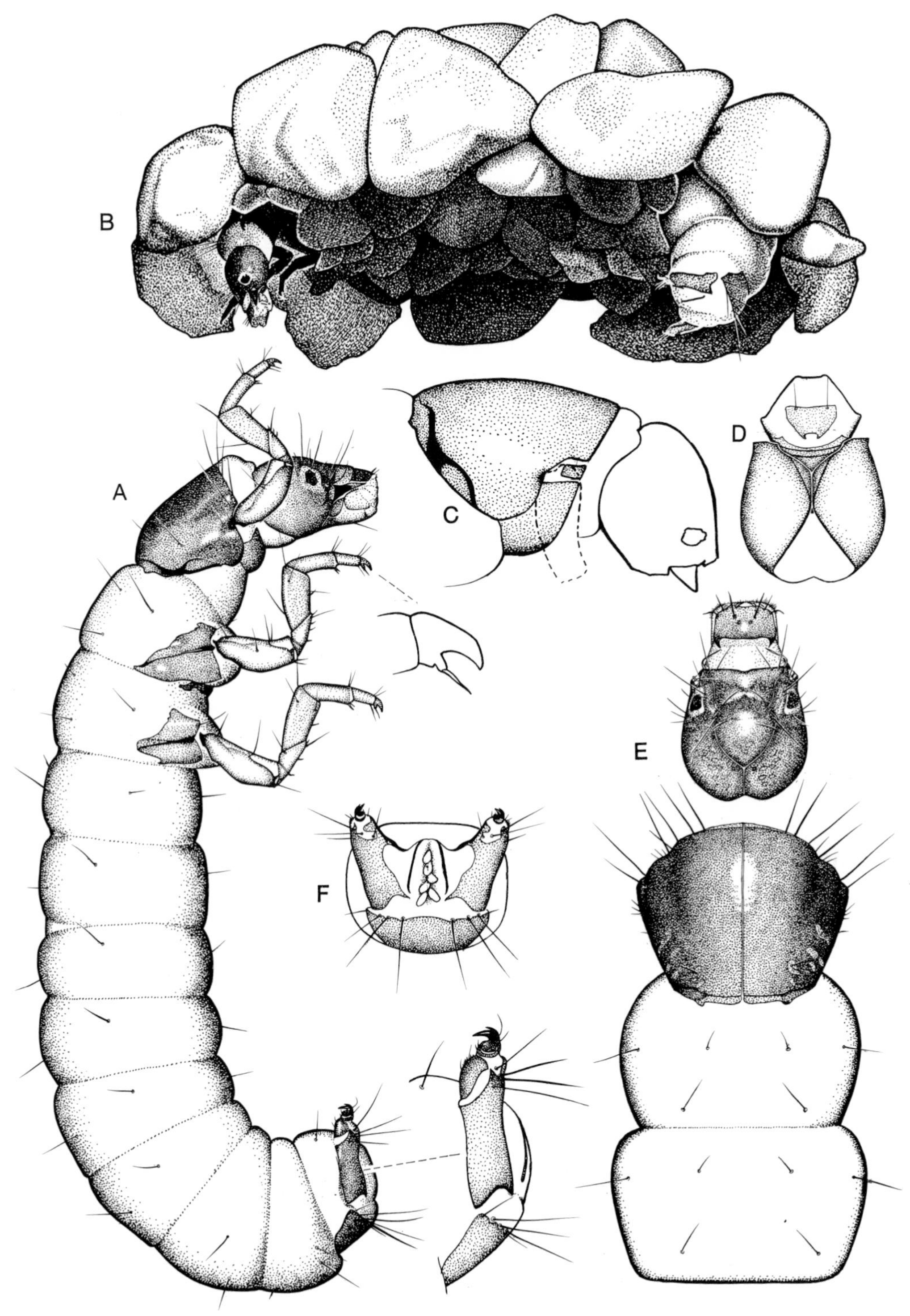
B
A
C
D
E
F

4.5 Genus **Matrioptila**

DISTRIBUTION AND SPECIES The genus *Matrioptila* consists of only a single species, *M. jeanae* (Ross), known from mountainous sections of southeastern North America: Tennessee, North and South Carolina, Georgia, and Kentucky.

Larval and pupal stages were described by Flint (1962b); we collected larvae of this genus in South Carolina and Georgia.

MORPHOLOGY Larvae are typical members of the Protoptilinae in having three mesonotal sclerites and two small metanotal sclerites (A). *Matrioptila* larvae are, however, distinctive in having each tarsal claw terminating in three points (D) of approximately the same length, but asymmetrical; these points are evidently homologous with the seta, process, and claw in other protoptiline genera (Figs. 4.3E, 4.6E). Length of larva up to 3.5 mm.

CASE Although of typical glossosomatid architecture, cases of *Matrioptila jeanae* are flatter and more depressed than others in the Nearctic fauna. Mineral particles of small and rather uniform size are used, and in series of larvae we collected many cases were constructed at least partially of transparent particles, although this could be as much a consequence of the materials available to the larvae as of their behaviour. Length of larval case up to 3 mm.

BIOLOGY Larvae were collected in cold, mountain streams. Final-instar larvae and pupae occur in May (Flint 1962b), adults in June. Gut contents of larvae (3) examined were largely fine organic particles.

REMARKS Diagnostic characters of adults of *M. jeanae* were given by Ross (1938). The genus is regarded as one of the most primitive of the Protoptilinae (Ross 1956).

Matrioptila jeanae (South Carolina, Oconee Co., 18 May 1970, ROM 700354)
A, head and thorax, dorsal x114; B, case, ventrolateral x36; C, case, dorsal; D, tarsal claw of middle leg, lateral

B

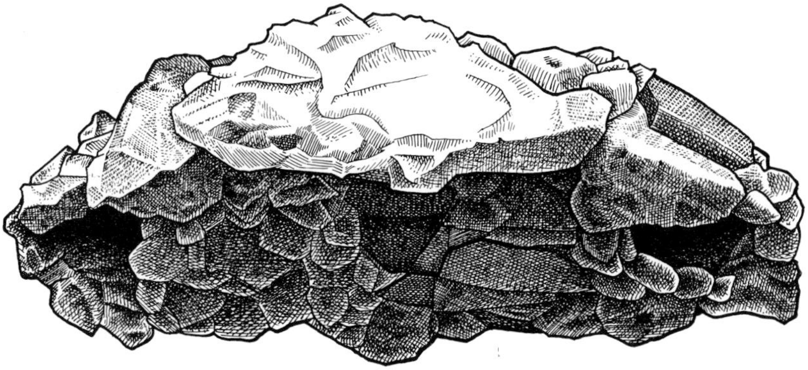

A

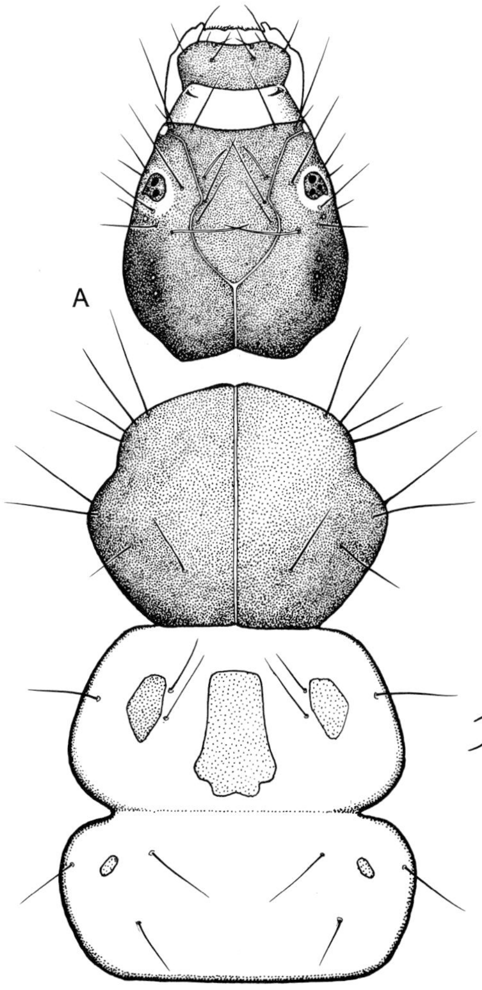

C

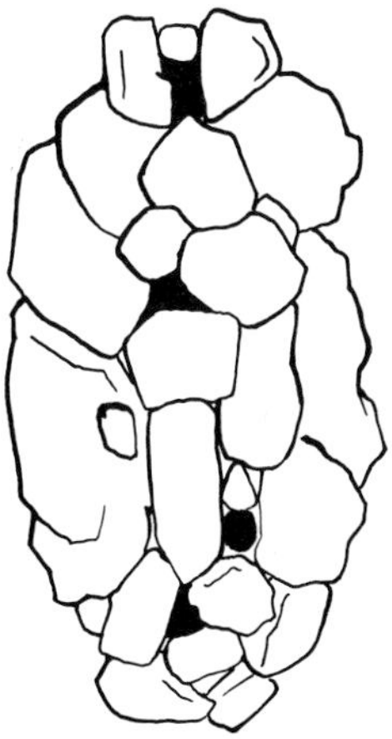

D

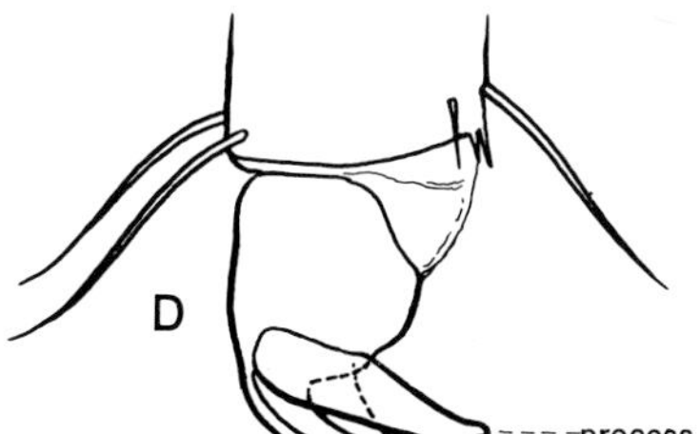

process
seta
claw

4.6 Genus Protoptila

DISTRIBUTION AND SPECIES *Protoptila* occurs only in the New World, but is widely distributed in both the Neotropical and Nearctic faunal regions. Approximately 13 species are known north of Mexico, and the group is represented in most parts of the continent, north at least as far as the central parts of Manitoba and Alberta.

The larva was identified for *P. maculata* (Hagen) by Ross (1944); we have associated material for the same species and have collected series of *Protoptila* larvae from many parts of North America.

MORPHOLOGY Larvae are typical of the Protoptilinae, similar to those of *Culoptila*, and evidently are reliably distinguished only by the slender basal seta of the tarsal claws (E) as outlined in the key. Length of larva up to 3.5 mm.

CASE Larval cases in *Protoptila* (B,C) can usually be identified by the relatively large stone incorporated into each side; the cases are higher in relation to width than those in most other genera although similar to *Agapetus* in this respect. Length of typical larval cases up to 4 mm.

In a series we collected in Arizona larvae and cases are similar to those of *Protoptila*, but each opening of the case is fitted with a stone hinged with silk along the outer edge (G); living larvae examined under a binocular microscope pushed aside the hinged stone when they extended their head and thorax, and the stone closed back into place when the larva withdrew within its case. Because no other larvae that we have examined, including all others assigned to *Protoptila*, have a closure mechanism of this type, there is a possibility that these larvae belong to some other genus. One possibility is *Mexitrichia*, but larval cases described for that genus are different (Flint 1963). Length of larval cases in this series up to 5 mm.

BIOLOGY Because *Protoptila* larvae live in somewhat warmer streams, often larger and slower flowing, than other members of the family, they are the glossosomatids of drier and more evenly contoured central parts of the continent. Gut contents of larvae (3) examined were largely fine organic particles with some diatoms.

Protoptila sp. (Ontario, Durham Co., 25 May 1953, ROM)
A, head and thorax, dorsal x83; B, case, ventrolateral x25; C, case, dorsal; D, head, ventral; E, tarsal claw of middle leg, lateral; F, anal claw with four accessory hooks
G, ? *Protoptila* sp. (Arizona, Yavapai Co., 4-5 July 1966, ROM), ventral portion of larval case showing hinged closure stones

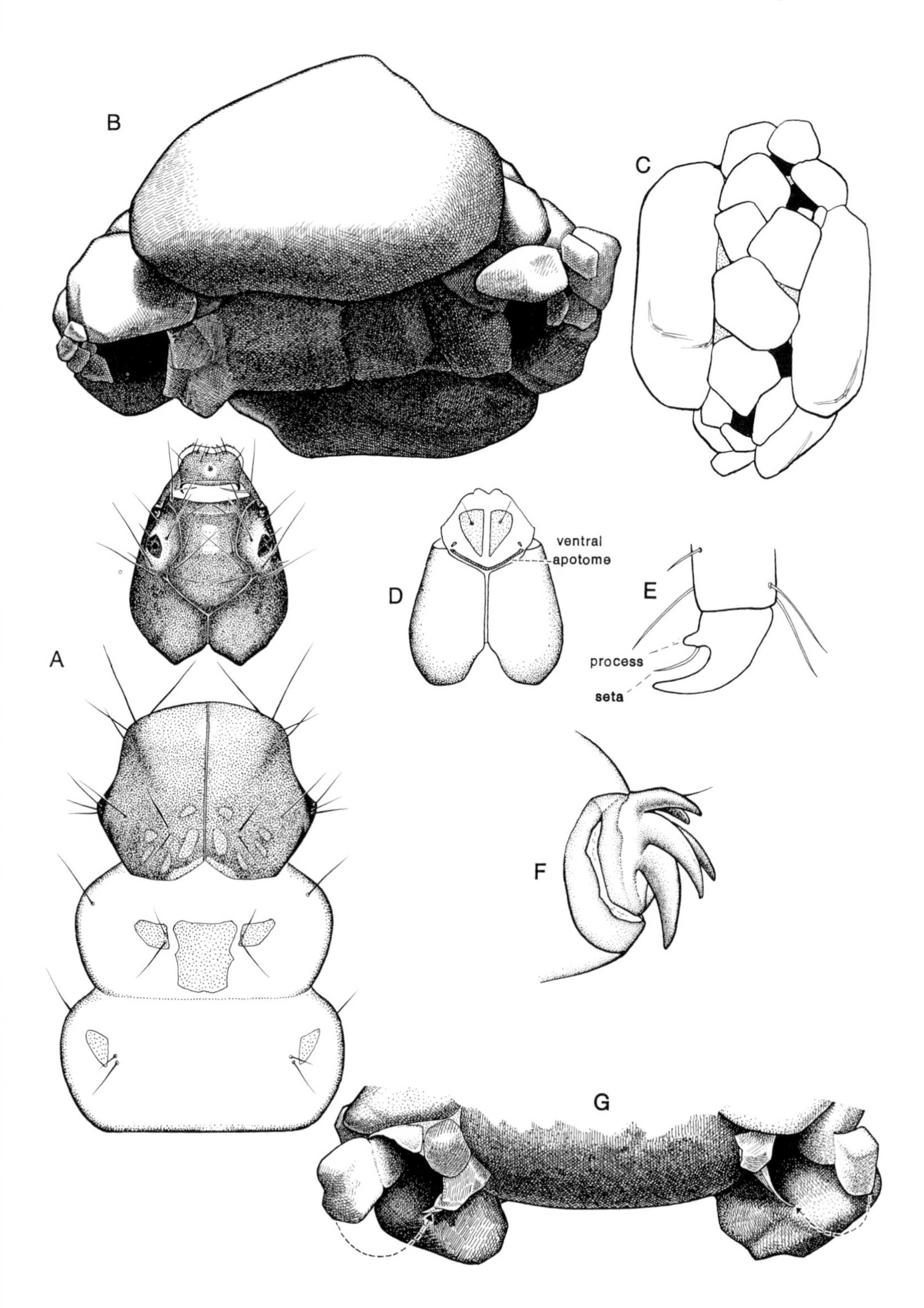
B
C
A
D
ventral
apotome
E
process
seta
F
G

5
Family Helicopsychidae

So abundant and widespread are these caddisflies, it is easy to forget that the larvae are among the most remarkable of all insects. Their helical cases of closely fitted rock fragments are an outstanding example of the elegance and precision of insect behaviour. The family is relatively small, but representatives are widely distributed over most faunal regions. A single genus, *Helicopsyche*, occurs in North America. Aspects of the morphology and biology of a European species were outlined by Botoşăneanu (1956).

Larval cases of most species resemble tightly coiled snail shells although a Cuban species has an open-coiled case suggestive of *Baikalia*, an unusual endemic snail genus of Lake Baikal (Botoşăneanu and Sýkora 1973). The case of the most common North American species, *Helicopsyche borealis*, was originally described as a snail having 'the remarkable property of strengthening its whirls by agglutinations of particles of sand, by which it is entirely covered' (Lea 1834). It is likely that the coiled-case-making behaviour of the Helicopsychidae was derived from the more usual tube-case-making activity; perhaps there is some advantage for the larvae in consolidating the mass of the case. Helical cases, for example, seem well suited to interstitial habitats; larvae of *H. borealis* were found to depths of 30 cm below a stream bed (Williams and Hynes 1974). Within their cases, larvae lie on one side, the abdomen coiled more dorsoventrally than laterally. An opening on the spire of cases of *H. borealis* corresponds to the posterior opening of tube-cases, and it may facilitate water circulation through the case for respiration. Helical cases are not totally confined to the Helicopsychidae, but are also constructed by larvae in the South African leptocerid genus *Leptecho* (Scott 1961).

5.1 Genus **Helicopsyche**

DISTRIBUTION AND SPECIES The genus is represented in most faunal regions. In North America a dozen or so species are known from Mexico, but only four north of the Rio Grande: *H. piroa* Ross in Texas; *H. mexicana* Banks in Arizona and Texas; *H. limnella* Ross in Arkansas and Oklahoma; and *H. borealis* (Hagen) widespread and common over much of the continent. The northern limit of *H. borealis* is uncertain, but we have records to 55°N lat. from Saskatchewan.

Of these four species, only the larva of *H. borealis* has been described, by Vorhies (1909) and Elkins (1936) among others.

MORPHOLOGY Structural features are evident in the illustration. The fore trochantin is unusually long (A), and the comb-like structure of the anal claw (D) is unique among North American larvae. Lateral tubercles (E) occur on segment VIII.

CASE The snail-like cases made of sand grains provide an unmistakable diagnosis for larvae of *Helicopsyche*. The dorsal lip of the anterior opening is extended as a hood, covering the larva as it grazes on rock surfaces. Diameter of case up to 7 mm.

BIOLOGY Larvae of *Helicopsyche* spp. are normally associated with running water, but those of *H. borealis* are also common in the littoral zone of lakes. Vorhies (1909) found larvae to depths of 8–10 feet in Wisconsin lakes. *H. borealis* larvae have an exceptionally broad temperature tolerance; we collected larvae of this species in thermal streams of Yellowstone National Park, Wyoming, where temperatures ranged up to 34°C and no other caddis larvae were found. In other streams with thermal affluents in California and Montana, *Helicopsyche* larvae were always among the few Trichoptera present. Food of *H. borealis* larvae, analysed by Coffman et al. (1971) and Mecom (1972a), consists of algal, detrital, and animal materials. There appears to be a continual emergence of adults from spring to early autumn (Ross 1944), followed by a long egg-diapause of 5-6 months (Williams and Hynes 1974).

Helicopsyche borealis (Ohio, Ashland Co., 2 Aug. 1968, ROM)
A, larva, lateral x21; B, head and thorax, dorsal; C, case, lateral x16; D, anal claw; E, lateral tubercles on segment VIII, x580; F, case, dorsal

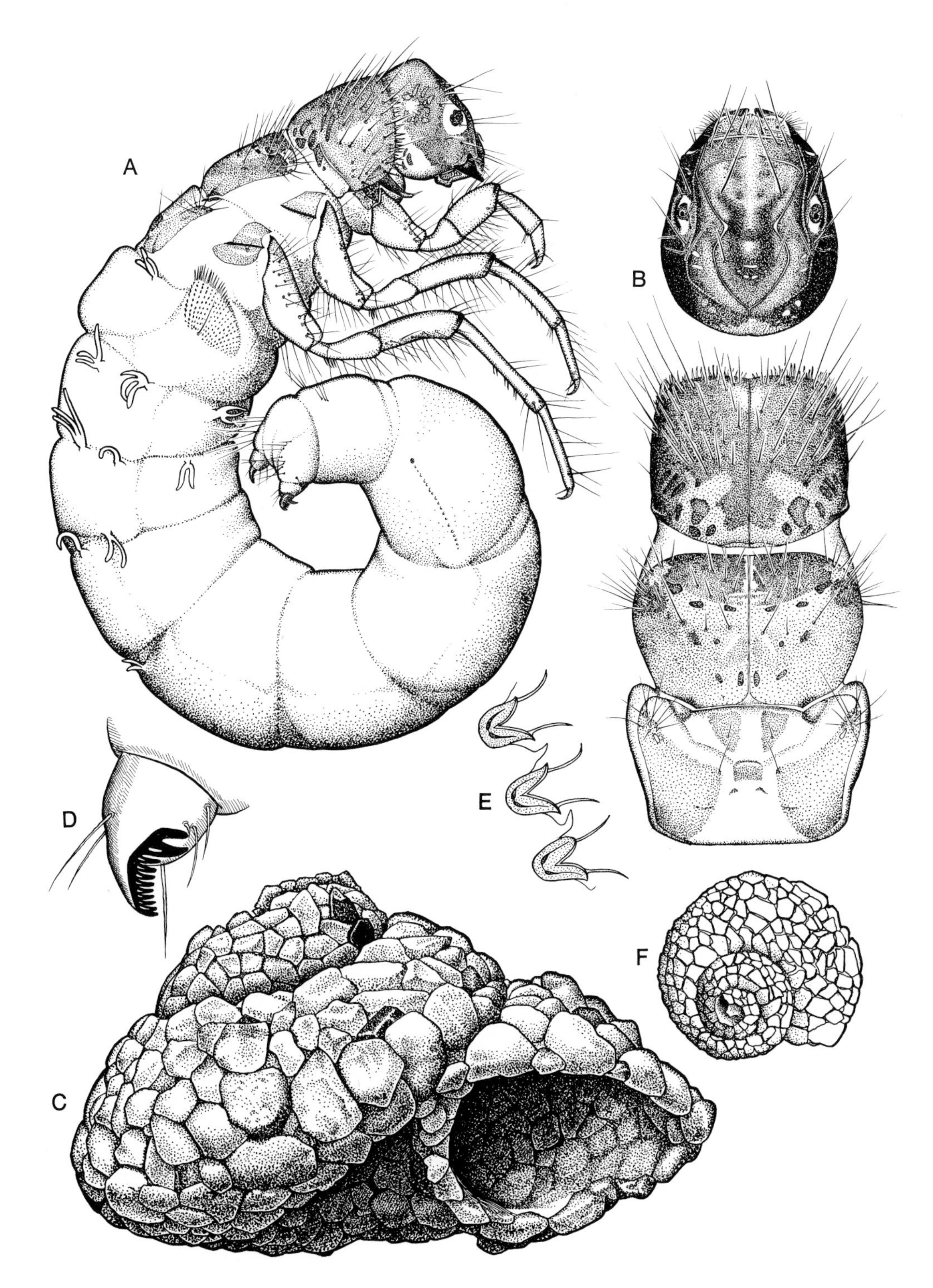
A
B
C
D
E
F

6
Family Hydropsychidae

The Hydropsychidae are a large and dominant family of caddisflies living in running waters over much of the world; a few species live along wave-washed shorelines of lakes. Eleven genera are known in the Nearctic region, and usually several of them are represented in most running waters throughout the continent. Approximately 145 species are recorded north of Mexico.

Hydropsychid larvae are distinguished from all others by extensive sclerotization on the dorsum of each thoracic segment, by branched gills on the ventral surface of the abdomen and last two thoracic segments, and usually by a tuft of long hairs near the apex of each anal proleg. In some genera, and perhaps all, first instars are without gills, and might then be confused with the Hydroptilidae; but first instars of Hydropsychidae can usually be distinguished by the lobate hind margin of the mesonotal plate (Fig. 6.1A). Characters of the gills and other structures diagnostic for hydropsychid genera are often not fully developed in early instars. Larvae have dense brushes of setae at each side of the labrum which probably function in feeding. The meso- and meta-notal plates do not have a median dorsal ecdysial line, as does the pronotal plate, but in some subfamilies these two sclerites do have a transverse ecdysial suture. On the venter of segments VIII and IX in all genera there are sclerites with stout, backward-directed setae that probably serve in locomotion.

Hydropsychid larvae are best known for the elegant silken capture nets they spin to strain from the current much of their food (Figs. 6.6G, 6.8F). Studies in Europe (Sattler 1958, 1963b) and North America (Wallace 1975a,b) have revealed intricate details of the net-spinning behaviour and the tendency for larvae in different genera to weave meshes within certain dimensions, thereby specializing their filtering for food particles of a particular size range. Food studies on larvae in several genera indicate that algae, fine organic particles, and small aquatic invertebrates are ingested.

Larvae of this family are able to produce sound by moving their femora across transverse ridges on the ventral surface of the head (Johnstone 1964). The femur bears a dorsomesial prominence (Fig. 6.11D) that functions as a scraper. The ventral aspect of the head illustrated in following pages for each hydropsychid genus shows that the structure of the transverse ridges is different for each one, although considerable general similarity exists within subfamilies. Whether these ridges produce different sounds has not been demonstrated. From his observations, Johnstone suggested that a larva might produce sound to repel potential predators, or to discourage other hydropsychid larvae from constructing shelters within its filtering territory.

All four subfamilies of the Hydropsychidae are represented on this continent; following are the Nearctic genera in each and a summary of larval characters.

ARCTOPSYCHINAE: *Arctopsyche, Parapsyche* Head tending to be quadrate in dorsal aspect; frontoclypeal sutures sigmoid, short cuticular inflection on frontoclypeal apotome near eyes; mandibles with teeth grouped near apex; ventral apotome not subdivided, thereby separating genae through their entire length; ventral, transverse ridges of head broken by longitudinal gaps, the roughened area ovoid in outline; meso- and meta-notal plates subdivided by transverse ecdysial line; abdominal and thoracic gills with all branches arising from apex of central stalk; lateral gills similar to ventral gills; sclerites on venter VIII short and wide.

This group is ranked as a separate family Arctopsychidae by some authors (e.g. Lepneva 1964; Schmid 1968a), but I share the view taken by most North American workers (Ross 1967; Flint 1974a; Wallace 1975a) that subfamilial status is more appropriate to the differences involved. The group is considered to be the most primitive in the Hydropsychidae.

DIPLECTRONINAE: *Aphropsyche, Diplectrona, Homoplectra, Oropsyche* Head tending to be globose; frontoclypeal sutures sigmoid or nearly so; mandibles with teeth distributed along mesial edges; ventral apotome of head in two parts, both parts approximately the same length; ventral ridges of head usually broken by longitudinal gaps, the roughened area broadly tapered posteriorly in outline; meso- and meta-notal plates subdivided by transverse ecdysial line; abdominal and thoracic gills rather sparsely branched, some branches arising along central stalk, most arising from apex of stalk; lateral gills, when present, reduced to short lobes; sclerites on venter VIII ovoid.

HYDROPSYCHINAE: *Cheumatopsyche, Hydropsyche, Potamyia, Smicridea* Head tending to be quadrate in dorsal aspect, the dorsum flattened; frontoclypeal sutures generally more nearly straight than sigmoid; mandibles with teeth distributed along mesial edges; ventral apotome in two parts, the posterior part minute; ventral ridges of head not broken, the entire roughened area generally rectangular; fore trochantin frequently forked; meso- and meta-notal plates not subdivided transversely; abdominal and thoracic gills with lateral branches numerous; lateral gills when present reduced to short lobes; sclerites on venter VIII triangular and elongate, sometimes fused.

Another genus, *Plectropsyche* allied to *Cheumatopsyche*, is known from Mexico (Ross 1947), evidently outside the Nearctic region.

MACRONEMATINAE: *Leptonema, Macronema* Mandibles with teeth distributed along mesial edges; ventral apotome in two parts, posterior part minute or lacking; meso- and meta-notal plates not subdivided transversely; abdominal and thoracic gills feather-like, apical and lateral branches equally dense, lateral gills similar to ventral gills.

6 Family **Hydropsychidae**

Key to Genera

1 Ventral surface of head with genae entirely separated by ventral apotome (Fig. 6.2C) (subfamily Arctopsychinae) **4**

Ventral surface of head with genae not entirely separated by ventral apotome (Fig. 6.11G) 2

2 (1) Ventral apotome of head in two parts, anterior and posterior, posterior ventral apotome at least half as long as median ecdysial line separating the two parts (Fig. 6.1E) (subfamily Diplectroninae) **5**

Ventral apotome of head usually in two parts, as above, but posterior ventral apotome reduced and much less than half as long as median ecdysial line (Fig. 6.11G), or absent (Fig. 6.8E) 3

3 (2) Abdominal gills consisting of elongate central stalk with numerous filaments arising more or less uniformly along entire length (Fig. 6.7A); fore trochantin never forked (Fig. 6.8B) (subfamily Macronematinae) **11**

Abdominal gills with central stalk, but filaments fewer and not arising uniformly, frequently in apical tuft with fewer gill filaments arising from basal part of central stalk (Fig. 6.3A); fore trochantin usually forked (Fig. 6.6B) (subfamily Hydropsychinae) **8**

4 (1) Dorsum of most abdominal segments with tuft of several long setae and/or scale-hairs in *sa*2 and *sa*3 positions (Fig. 6.10B); ventral surface of head usually but not always with ventral apotome roughly rectangular (Fig. 6.10D). Widespread **6.10 Parapsyche**

Dorsum of most abdominal segments with only one long seta in *sa*2 and *sa*3 positions, frequently with one or two shorter ones as well, but not a tuft as above (Fig. 6.2E); ventral surface of head with ventral apotome narrowed posteriorly (Fig. 6.2C). Widespread **6.2 Arctopsyche**

5 (2) Pronotum with transverse sulcus behind which the posterior one-third or so of pronotum is constricted from anterior two-thirds (Fig. 6.1B) 6

Pronotum lacking transverse sulcus as above, constricted only slightly at posterior border (Fig. 6.4B). Eastern and western **6.4 Diplectrona**

6 (5) Mesial surface of middle and hind femora with some stout setae subdivided into a cluster of several flattened lobes (Fig. 6.5B). Western **6.5 Homoplectra**

Mesial surface of middle and hind femora with setae not as above (Fig. 6.1C) 7

7 (6) Anterior margin of frontoclypeal apotome convex and symmetrical (Fig. 6.9C). Southeastern **6.9 Oropsyche?**

Anterior margin of frontoclypeal apotome asymmetrical, broadly notched on left side (Fig. 6.1D). Southeastern **6.1 Aphropsyche?**

8 (3) Venter VIII with single median sclerite bearing stout setae (Fig. 6.12E); anterior border of submentum entire (Fig. 6.12C). Southwestern **6.12 Smicridea**

Venter VIII with pair of sclerites bearing stout setae (Fig. 6.6D); anterior border of submentum with median division (Fig. 6.6F) **9**

9 (8) Venter of prothorax with pair of enlarged sclerites in intersegmental fold posterior to prosternal plate (Fig. 6.6C); anterior border of frontoclypeal apotome never with small, median notch (Fig. 6.6E). Widespread **6.6 Hydropsyche**

Venter of prothorax with two, usually tiny, sclerites in intersegmental fold posterior to prosternal plate (Fig. 6.3B); these sclerites may be larger than illustrated, but anterior border of frontoclypeal apotome usually bears small, median notch (Fig. 6.3D) **10**

10 (9) Anterior ventral apotome of head with prominent anteromedian protuberance (Fig. 6.11G); venter IX with posterior margin of setate sclerites entire (Fig. 6.11H); lateral border of mandibles flanged (Fig. 6.11E); fore trochantin forked or unforked (Fig. 6.11B,C). Central and eastern **6.11 Potamyia**

Anterior ventral apotome of head lacking prominent anteromedian protuberance (Fig. 6.3E); venter IX with posterior margin of setate sclerites notched (Fig. 6.3F); lateral border of mandibles not flanged (Fig. 6.3C), fore trochantin forked (Fig. 6.3A). Widespread **6.3 Cheumatopsyche**

11 (3) Tibia and tarsus of fore leg with dense setal fringe (Fig. 6.8B); dorsum of head flattened and with sharp carina (Fig. 6.8D). Eastern **6.8 Macronema**

Tibia and tarsus of fore leg lacking dense setal fringe; dorsum of head without sharp carina (Fig. 6.7E). Mexico and southeastern Texas **6.7 Leptonema**

6.1 Genus Aphropsyche

DISTRIBUTION AND SPECIES The genus *Aphropsyche* is known only from eastern North America where there are two species: *A. doringa* (Milne) (syn. *A. aprilis* Ross), recorded from Tennessee, North Carolina, and New Hampshire; and *A. monticola* Flint from Virginia.

Larvae have not been identified in the literature, and the tentative association offered here is based on circumstantial evidence. The larva suggested as a possible candidate for this genus (Ross 1944:83, Hydropsychid Genus A) is now known to be the aberrant larva of a species of *Diplectrona* (Ross 1970). Series of diplectronine larvae that I have studied from four localities (Arkansas, Indiana, and Tennessee) appear morphologically identical, yet can be distinguished from *Diplectrona.* A reasonable possibility is that these larvae belong to either *Aphropsyche* or *Oropsyche.* Because of the more extended distribution known for *Aphropsyche* species, I am tentatively assigning these larvae to that genus.

MORPHOLOGY Larvae here assigned tentatively to *Aphropsyche* are typical members of the Diplectroninae, but have a pronounced indentation on the left side of the anterior margin of the frontoclypeal apotome. On the basis of present knowledge, these larvae are likely to be confused only with *Diplectrona metaqui* Ross, a form strangely discordant with other species in that genus; but in *D. metaqui* (Fig. 6.4F,G) the frontoclypeal notch is more nearly median and the left mandible bears a large dorsal protuberance. Furthermore, the pronotal sulcus and waist-like constriction of *Aphropsyche*(?) larvae (B) are not present in any *Diplectrona* species now known. Length of larva up to 16 mm.

RETREAT Unknown.

BIOLOGY The four larval collections available to me were all made from small streams in mountainous terrain.

REMARKS Taxonomic data for adults of *A. doringa* were given by Ross (1944), and for *A. monticola* by Flint (1965). I am indebted to O.S. Flint, Smithsonian Institution, and H.H. Ross, University of Georgia, for loaning some of the larvae studied.

Aphropsyche ? sp. (Arkansas, Logan Co., 16 May 1958, USNM)
A, larva, lateral; B, pronotum, dorsal; C, middle leg, mesial, one seta enlarged; D, head, dorsal; E, head, ventral

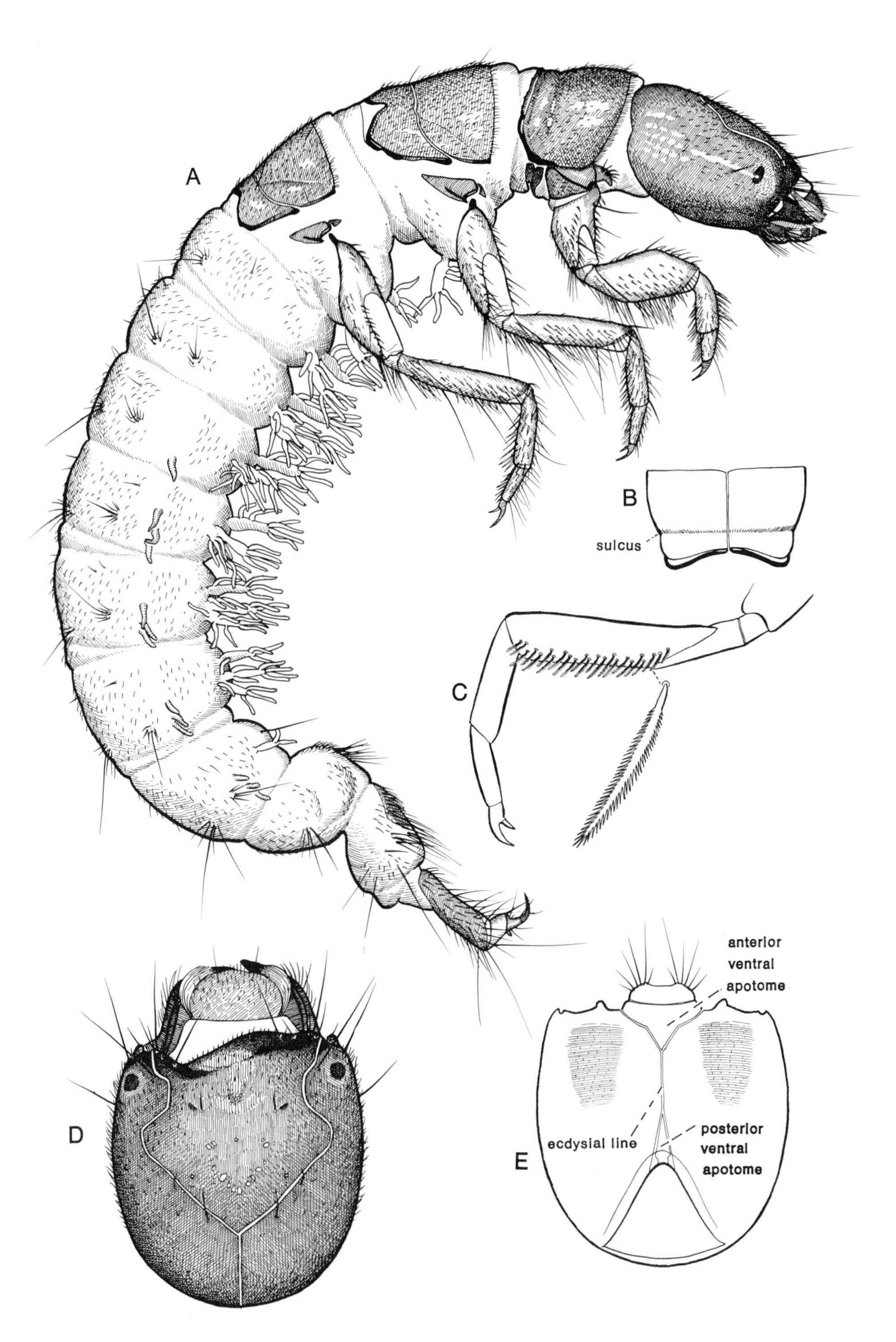
A
B
sulcus
C
D
anterior
ventral
apotome
ecdysial line
posterior
ventral
apotome
E

6.2 Genus **Arctopsyche**

DISTRIBUTION AND SPECIES *Arctopsyche* is a genus of some 17 species widely distributed in boreal and montane sections of the Oriental, Palaearctic, and Nearctic regions. Four species are known in North America: *A. californica* Ling from California; *A. grandis* (Banks), widespread and common in the west from New Mexico and California to Alberta and British Columbia, and recorded also from northern Quebec (Schmid 1968a); *A. irrorata* Banks from the southeast; *A. ladogensis* (Kolenati), Holarctic through northern Europe, Siberia, and northern Canada to Maine and New Hampshire.

Larvae have been described for *A. grandis* (Smith 1968a) and *A. irrorata* and *A. ladogensis* (Flint 1961a).

MORPHOLOGY Larvae of the subfamily Arctopsychinae, with the two genera *Arctopsyche* and *Parapsyche*, are readily separated from all others because the ventral apotome separates the genae entirely; other characters of the two genera are outlined under the subfamily. In larvae of *Arctopsyche* the ventral apotome is narrowed posteriorly (C), but the same is true in at least one species of *Parapsyche* (q.v.). Therefore, it is necessary to make a further distinction for larvae of *Parapsyche* based on the tufts of several long setae or scale-hairs in the *sa*2 and *sa*3 positions of most abdominal segments. *Arctopsyche* larvae now known have only a single long seta in this position, although there may be one or two shorter ones as well (E). Most larvae of *Arctopsyche* have brownish yellow sclerites and a pale median dorsal band on the head and thorax. Length of larva up to 30 mm.

RETREAT Observations on the larval retreat and capture net of *A. irrorata* and *A. grandis* (Wallace 1975a) indicate that these structures are essentially similar to those of *Hydropsyche* and *Diplectrona*; mesh sizes of the nets of the two *Arctopsyche* species are, however, the largest known in the family.

BIOLOGY Larvae inhabit cold, running waters, where retreats are located in strong currents. Information on life cycle was given by Flint (1961a) and Smith (1968a). Available data indicate that *Arctopsyche* larvae feed mainly on other aquatic insects (Mecom 1972a,b; Wallace 1975a); evidently capture nets are maintained throughout the winter by *A. irrorata* in Georgia (Wallace 1975a).

REMARKS Taxonomy of adults has been summarized by Schmid (1968a).

Arctopsyche grandis (Wyoming, Yellowstone Nat. Park, 12 July 1961, ROM)
A, larva, lateral x13; B, head, dorsal; C, head, ventral; D, segments VIII and IX with bases of anal prolegs, ventral; E, abdominal segment, lateral

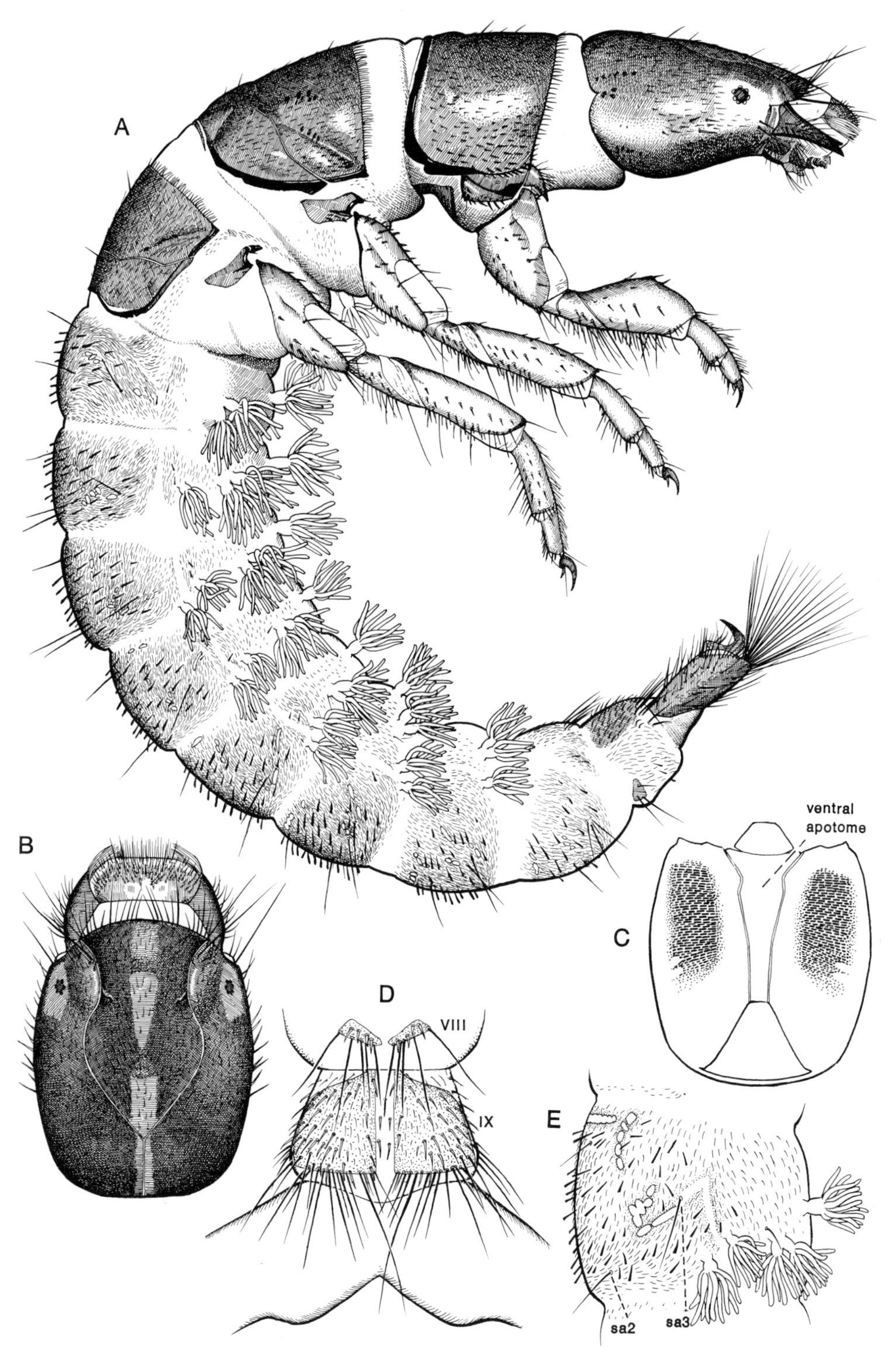
A
B
C
D
E
ventral apotome
VIII
IX
sa2
sa3

6.3 Genus Cheumatopsyche

DISTRIBUTION AND SPECIES *Cheumatopsyche* is a large genus represented in all faunal regions except the Neotropical. Approximately 39 species are now known in Canada and the United States, and the group is common almost everywhere.

Larval stages have been associated for several species, but diagnostic characters for their separation have been difficult to find (Ross 1944). We have associated material for eight species.

MORPHOLOGY North American larvae of *Cheumatopsyche* tend to be uniformly dark on the head, without the diverse colour patterns of *Hydropsyche* larvae. In the larva of at least one species, *C. etrona* Ross, the sclerites posterior to the prosternal sclerite are unusually large (J.B. Wallace, pers. comm.), leading to possible confusion with *Hydropsyche*. Most *Cheumatopsyche*, including *C. etrona*, are distinctive, however, in having a median notch on the anterior margin of the frontoclypeal apotome (D), although the notch is absent in some species. Length of larva up to 13 mm.

RETREAT Fremling (1960) described the capture net of *C. campyla* Ross as more voluminous and flimsy than that of *Hydropsyche orris*, and therefore suited for relatively weaker currents.

BIOLOGY According to Ross, larvae of *Cheumatopsyche* tend to be more dominant in warmer streams than *Hydropsyche* (1959), and to be successful in streams too polluted for most other caddisflies (1944). *C. campyla* larvae were found to be more abundant in slower currents than *H. orris* Ross, and conversely in stronger currents (Freming 1960). *Cheumatopsyche* larvae have also been found to depths of 20 cm or more in the interstitial habitat of stream beds (Williams and Hynes 1974). We collected larvae along the shore of Lake Winnipegosis, Man., together with *Hydropsyche recurvata*. Food of *Cheumatopsyche* larvae in a small Pennsylvania stream was found to be largely algae and animals, with a small detrital component (Coffman et al. 1971). *C. lasia* Ross and *C. campyla* were reported to have two generations per year in a Texas river (Cloud and Stewart 1974).

REMARKS Taxonomic data for adults of the North American species were given by Gordon (1974).

Cheumatopsyche pettiti (Ohio, Knox Co., 2 Aug. 1968, ROM)
A, larva, lateral x26; B, prothorax, ventral; C, right mandible, dorsal; D, head, dorsal; E, head, ventral; F, segments VIII and IX, ventral

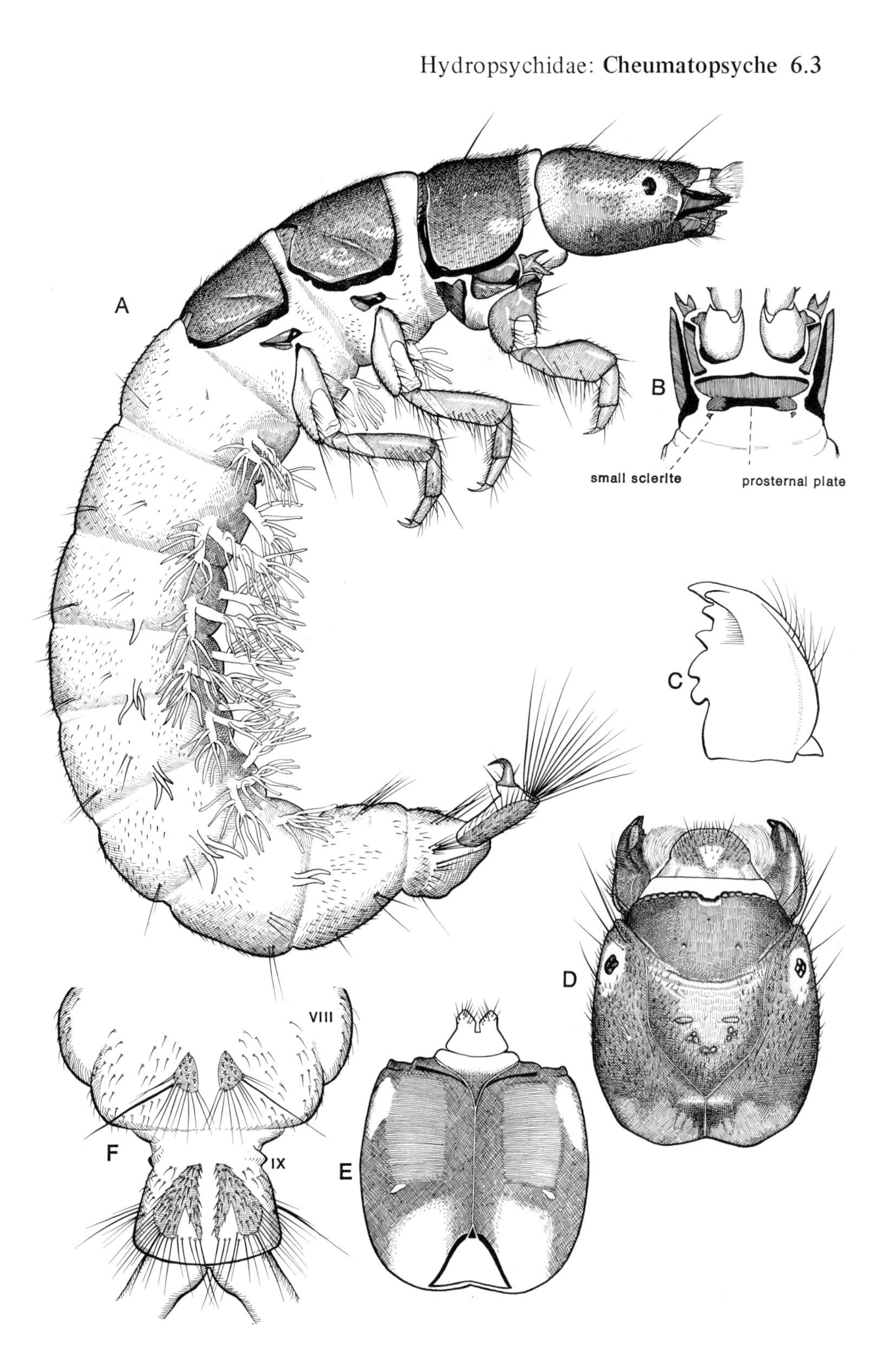
A
B
small sclerite
prosternal plate
C
D
VIII
F
IX
E

6.4 Genus Diplectrona

DISTRIBUTION AND SPECIES The genus is represented in all faunal regions except the Ethiopian and Neotropical. Three species are known in America north of Mexico: *D. modesta* Banks, widespread throughout the east to South Dakota and Oklahoma; *D. metaqui* Ross in Georgia, Tennessee, Kentucky, Indiana, and Illinois (Ross 1970); and *D. californica* Banks from California (Flint 1966).

Diagnostic data are available for larvae of *D. modesta* (Ross 1944, 1959) and for *D. metaqui* (Ross 1944 and 1959 as Hydropsychid Genus A; Ross 1970). We have larvae of both.

MORPHOLOGY The larvae known for two North American species are typical members of the Diplectroninae in having the posterior ventral apotome of the head well developed (C). But the larva of *D. metaqui* is so grossly different from any other of our hydropsychids in the large thumb-like process on the left mandible (F,G) that it was assigned tentatively to an unknown genus, Hydropsychid Genus A (Ross 1944, 1959), until finally shown to belong to a new species *Diplectrona metaqui* (Ross 1970). *D. metaqui* is also different from *D. modesta* in having a prominent notch on the anterior border of the frontoclypeal apotome (F). Both species are, however, consistent in lacking a transverse sulcus on the pronotum, and thus can be distinguished from the other diplectronine genera in which the sulcus is present. Sclerites of the head and thorax are reddish brown in colour. Length of larva up to 15 mm.

RETREAT According to Sattler (1963a) the retreat and capture net of the European *Diplectrona felix* McL. are constructed essentially as in *Hydropsyche*, although the retreat is almost entirely of plant materials and without the considerable proportion of sand utilized by *Hydropsyche* larvae.

BIOLOGY Larvae live in rapid portions of small, cool streams. Mackay (1968) found them in moss on submerged rocks and in leaf packs; the species (prob. *D. modesta*) was univoltine, eggs hatching in late summer and autumn, larvae overwintering to mature in early summer.

Diplectrona spp.

A-E, *Diplectrona* prob. *modesta* (Pennsylvania, Lycoming Co., 7 May 1968, ROM)

A, larva, lateral x15; B, pronotum, dorsal; C, head, ventral; D, head, dorsal; E, segments VIII and IX, ventral

F, G, *D. metaqui* (Tennessee, Franklin Co., 14-15 May, ROM 700337)

F, head, dorsal; G, left mandible, lateral

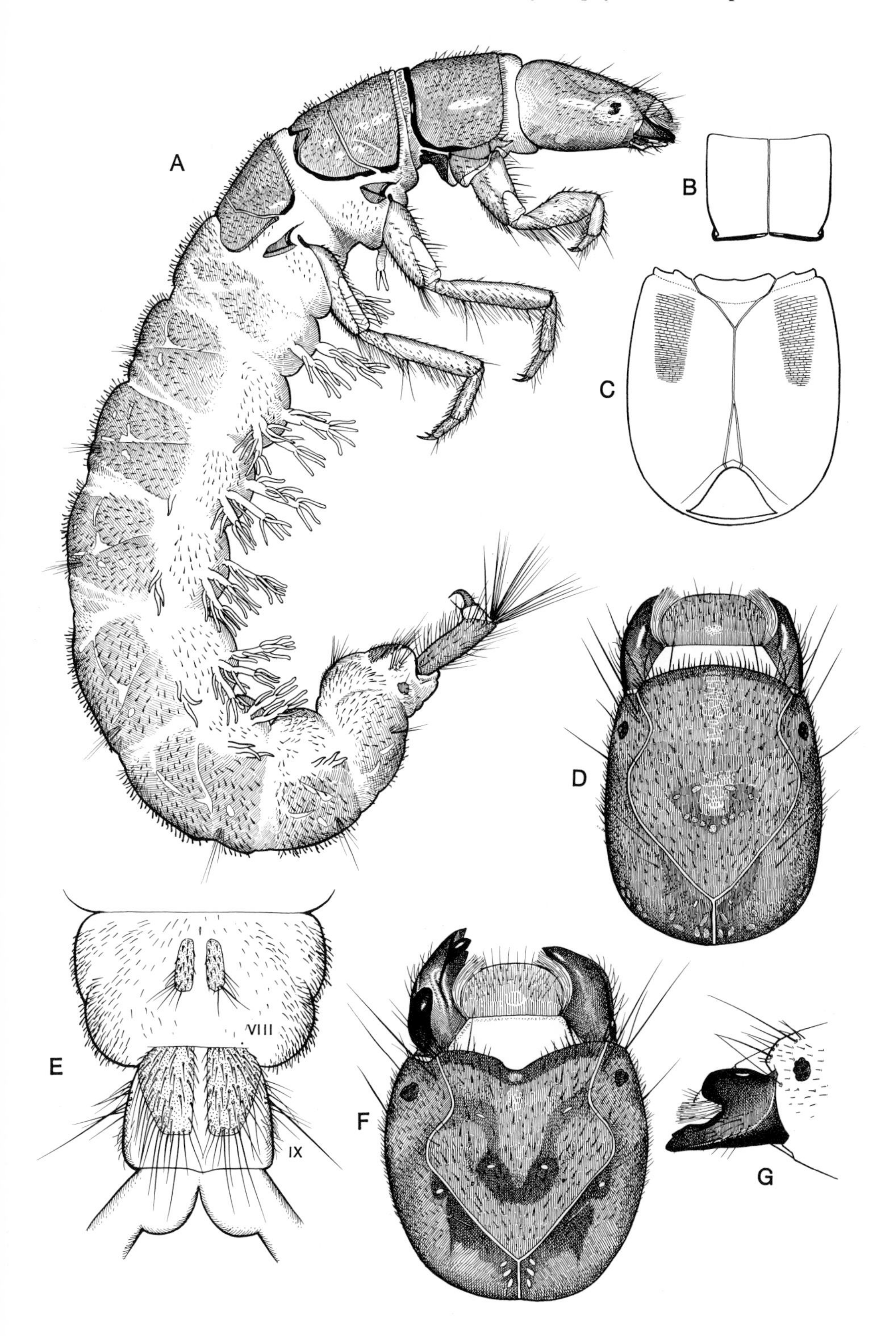
A
B
C
D
E
VIII
IX
F
G

6.5 Genus **Homoplectra**

DISTRIBUTION AND SPECIES This genus is entirely North American and the eight species now known are recorded only from the Coast and Cascade Mountains of Oregon and California. The species appear to be local in occurrence.

No larva was previously known for any species in this genus. In Linn Co., Oregon, we collected an associated series of larvae and adults of a species of *Homoplectra* as yet unidentified; and we have larvae from three other Oregon localities which appear to belong to this genus.

MORPHOLOGY The larva is a typical diplectronine with a globose head; the anterior margin of the frontoclypeal apotome is concave and symmetrical. Larvae of *Homoplectra* studied can be distinguished from those of all other genera by distinctive setae on the mesial surface of the middle and hind femora, each seta subdivided into a cluster of several flattened lobes (B). Sclerotized parts are brownish red in colour. Abdominal gills are more sparsely branched than in other genera. Length of larva up to 10.5 mm.

RETREAT No detailed observations.

BIOLOGY Larvae have been collected only in small, cold trickles and streams of mountainous terrain.

REMARKS Diagnostic characters for males of five species of *Homoplectra* were given by Denning (1956).

Homoplectra sp. (Oregon, Linn Co., 22 June 1968, ROM)
A, larva, lateral x19; B, middle leg, mesial, one seta enlarged; C, head, dorsal; D, head, ventral

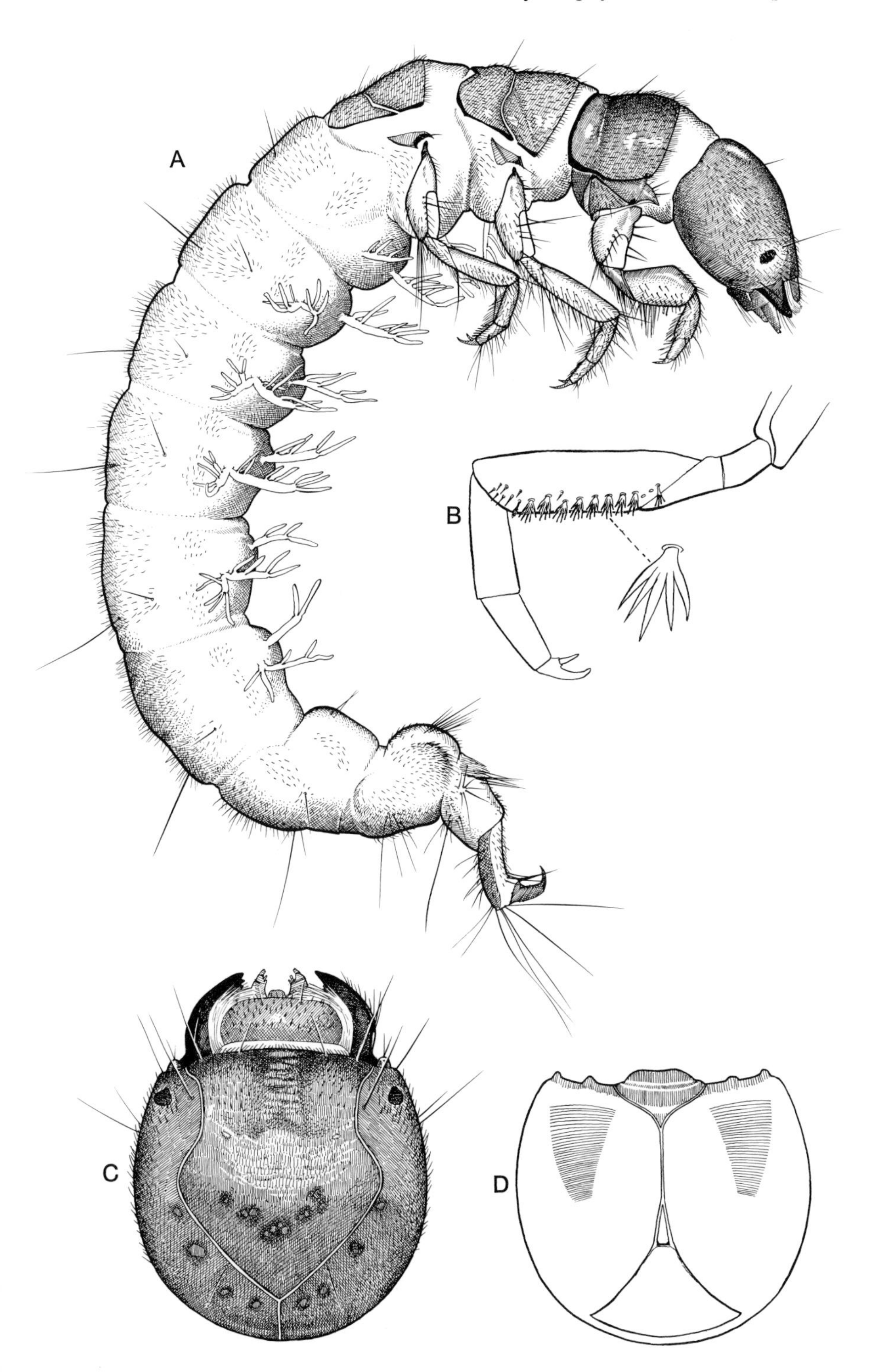
A
B
C
D

6.6 Genus **Hydropsyche**

DISTRIBUTION AND SPECIES *Hydropsyche*, the largest genus in the family, shares with *Cheumatopsyche* a broad global distribution through all faunal regions except the Neotropical. Approximately 70 species are now known in America north of Mexico.

Larval stages were associated for a number of species by Ross (1944).

MORPHOLOGY Larvae are typical members of the Hydropsychinae, and usually can be distinguished from other genera by the pair of large sclerites in the intersegmental fold posterior to the prosternal plate (C) (but see under *Cheumatopsyche*). The dorsal ramus of the fore trochantin (B) does not begin to develop until the second instar and becomes larger with successive moults (Siltala 1907a); the paired sclerites posterior to the prosternal plate are also not well developed in early instars. Colour patterns of the head exhibit a wide range of interspecific differences (Ross 1944, figs. 346–357). Length of larva up to approximately 16.5 mm.

RETREAT The capture net of *Hydropsyche* larvae is usually taut and the perimeter supported by pieces of debris that are themselves stabilized by silken guy-lines (G; Fremling 1960, fig. 8). Nets are not maintained during low winter temperatures (Malicky 1973).

BIOLOGY According to Ross (1944) the *Hydropsyche* species studied in Illinois live in most types of permanent streams in the state; some species were found in abundance in larger rivers, others were restricted to small spring streams. It is likely that most species are similarly adapted to particular habitats. Although hydropsychids are generally restricted to running waters, larvae of *H. recurvata* Banks have been collected on rocks along the edge of Lake Michigan (Ross 1944); we have collected larvae and pupae of the same species along the shores of other large lakes. Food of *Hydropsyche* larvae has been analysed many times, and algae, detritus, and animals found to be ingested in varying proportions at different seasons (Coffman et al. 1971; Mecom 1972a). A detailed account of the life history of *H. orris* was given by Fremling (1960).

Hydropsyche morosa (Ontario, Peel Co., 15 Sept. 1970, ROM)
A, larva, lateral x15; B, fore trochantin, lateral; C, prothorax, ventral; D, segments VIII and IX, ventral; E, head, dorsal; F, head, ventral; G, retreat and capture net

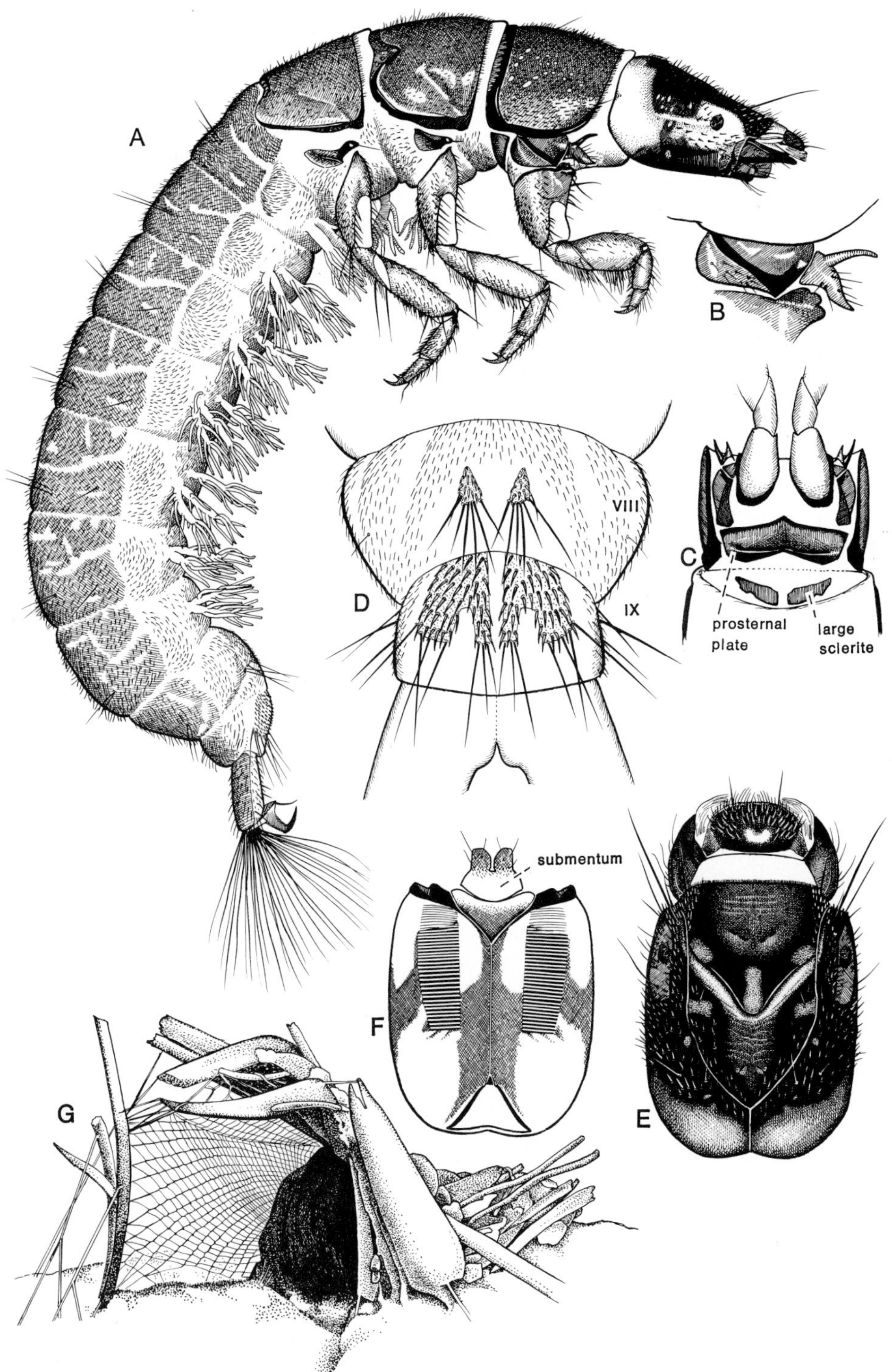
A
B
C
prosternal plate
large sclerite
D
VIII
IX
submentum
F
E
G

6.7 Genus **Leptonema**

DISTRIBUTION AND SPECIES *Leptonema*, primarily a genus of the Neotropical region with some Ethiopian elements, is widespread in Mexico and evidently reaches the Nearctic region along the southeastern border of Texas (Flint 1968a); there are adult specimens from San Antonio in the Museum of Comparative Zoology, Harvard University (O.S. Flint, pers. comm.).

Larvae of New World *Leptonema* species have been described by Flint (1964b, 1968a); larval characters for the genus were summarized by Ulmer (1957).

MORPHOLOGY Although *Leptonema* is classified in the subfamily Macronematinae, larvae show little obvious similarity to those of *Macronema*. One diagnostic larval character shared by the two genera is the structure of the abdominal gills - a central stalk with filaments radiating uniformly throughout its length, and somewhat feather-like in appearance (A). Larvae are densely covered with short, dark setae, a feature especially noticeable on the abdomen. In the few series of Mexican larvae I studied there is a single wide sclerite immediately posterior to the prosternal plate (C); a prominent mesial lobe is located on the distal end of the fore coxa, and the scraper on the mesial face of the fore femur is especially well developed (B). Length of larvae examined up to 24.5 mm.

RETREAT According to Flint (1968a), larvae of *Leptonema* construct typical hydropsychid retreats with silken capture nets; the retreat is covered with sand grains, and fastened to a rock.

BIOLOGY Larvae live in running waters, ranging in size from trickles and tumbling mountain brooks to large lowland rivers (Flint 1964b, 1968a).

Leptonema sp. (Mexico, Veracruz, 17 Dec. 1948, ROM)
A, larva, lateral x11, gill enlarged; B, fore leg, mesial; C, prothorax, ventral; D, head, ventral; E, head, dorsal

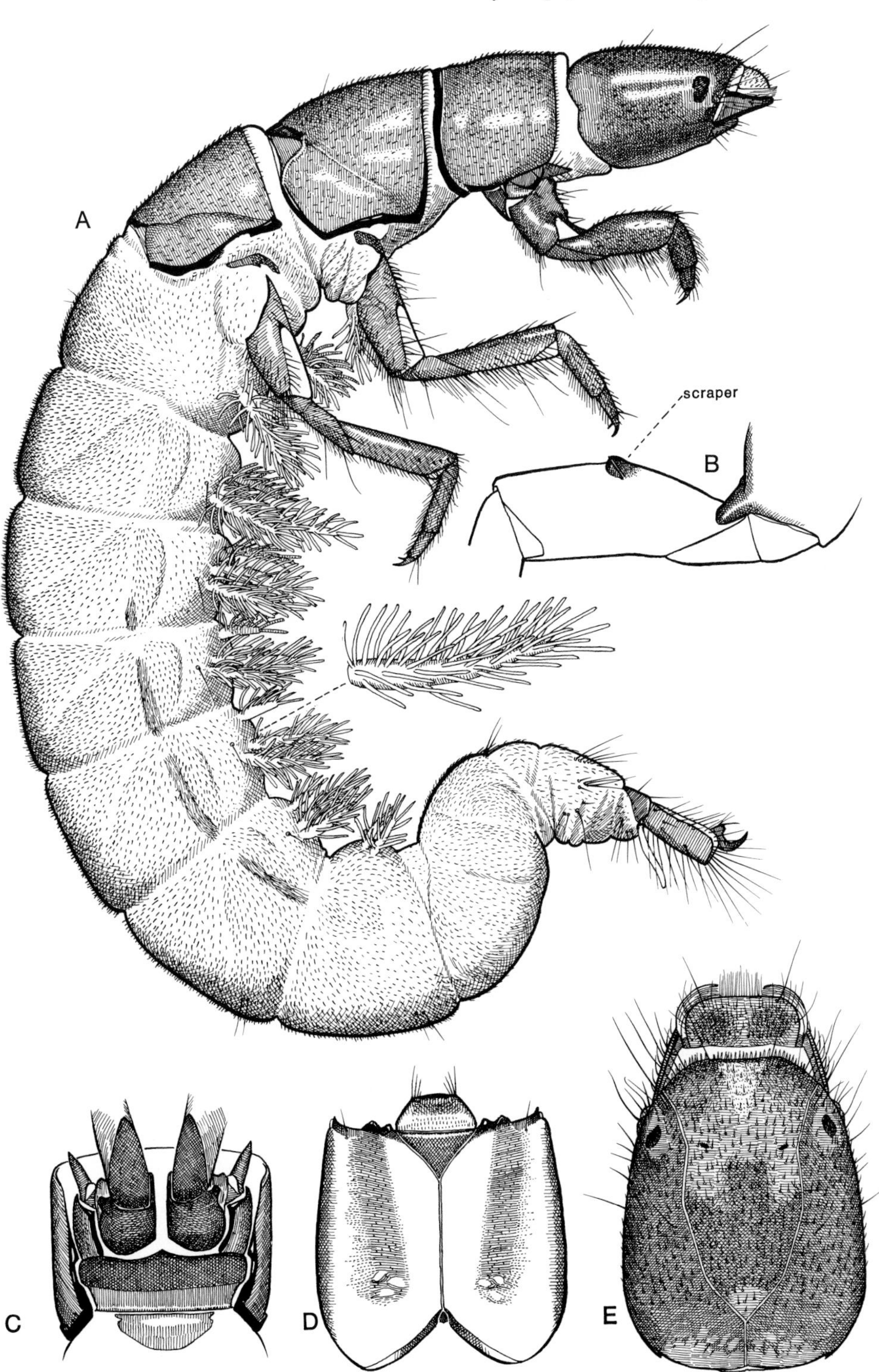
A
scraper
B
C
D
E

6.8 Genus Macronema

DISTRIBUTION AND SPECIES *Macronema* is a large genus represented in all faunal regions. North of Mexico the Nearctic fauna consists of three species: *M. carolina* (Banks) from New York and Florida west to Oklahoma; *M. transversum* (Walker) from Georgia and Indiana; and *M. zebratum* (Hagen) from Minnesota, Quebec, and New Hampshire to Illinois and Georgia, with a record from Utah.

Diagnostic characters for larvae of all three species were given by Ross (1944).

MORPHOLOGY Larvae of *Macronema* are very distinctive from all other North American genera, primarily because of the flattened head with a sharp U-shaped carina, and the pair of sclerites at the base of the labrum (D); also unusual is the dense fringe of setae on the fore tibia and tarsus (B), and the prominent process at the base of the femur (C). Length of larva up to 17.5 mm.

RETREAT Larval retreats and capture nets of *Macronema* species are among the most highly specialized in the family (Sattler 1963b, 1968; Wallace 1975b). Larvae of *M. zebratum* construct an open-ended chamber of fine gravel particles and silk (F) on rocks, situated to intercept the current. Small particles carried by the current are filtered out by a silken net spun across the chamber. The larva rests in a tubular diverticulum, its head close to the upstream surface of the net; dense setae on the fore legs and labrum are believed to function as brushes for collecting fine particles from the net. Current also flows through the larval tube, meeting respiratory requirements and carrying away faeces. The flattened head of *Macronema* larvae is thought to function in restricting openings in the chamber, thereby controlling the amount and direction of the water current passing through (Wallace and Sherberger 1974). Length of retreat illustrated approximately 15 mm.

BIOLOGY These are insects of large rivers in North America. Larvae ingest fine particulate matter, phytoplankton, and bacteria, and among the North American Hydropsychidae are the most efficient collectors of such small particles because of the very small mesh size of their nets - in *M. carolina* 5 x 40 μm, which is one-sixth the size of the finest plankton nets available commercially (Wallace and Sherberger 1974). Even so, these meshes are considerably larger than those constructed by larvae of the Philopotamidae (q.v.).

Macronema zebratum (Ontario, Haliburton Dist., 29 May 1967, ROM)
A, larva, lateral x11; B, fore leg, lateral; C, fore leg, base of femur, mesial; D, head, dorsal; E, head, ventral; F (same data as above, but Sept. 1975), larval retreat and feeding net, ventral, portion of wall removed, meshes enlarged, arrows indicating direction of water flow x3

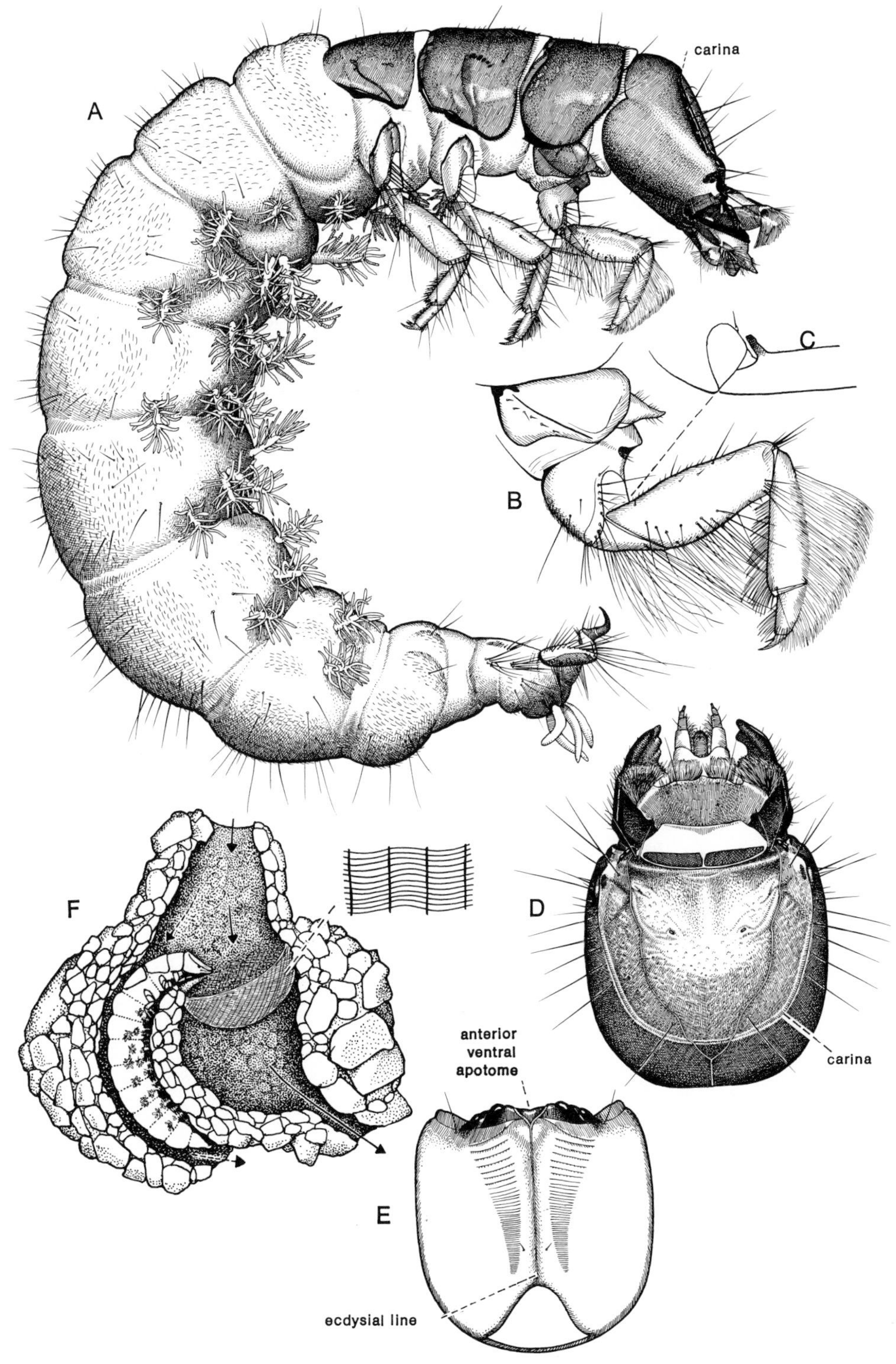
A
carina
B
C
D
carina
E
anterior
ventral
apotome
ecdysial line
F

6.9 Genus **Oropsyche**

DISTRIBUTION AND SPECIES *Oropsyche* has a single species, *O. howellae* Ross, known only from North Carolina.

Larvae have not been positively associated with this species, and the association tentatively proposed here is based on circumstantial evidence. Having made the assumption for larvae of *Aphropsyche* (q.v.), there remained one collection of unidentified diplectronine larvae which appeared to be generically distinct. Because the single species in *Oropsyche* is known only from North Carolina, in fact from Jackson Co. which is immediately adjacent to Macon Co. where the larvae were collected, I am assuming tentatively that these larvae belong to *O. howellae.*

MORPHOLOGY Larvae have globose heads and resemble those of the western genus *Homoplectra*, but their femoral setae are similar to those of *Aphropsyche* (?) and are not divided into a cluster of flattened lobes as in *Homoplectra*. The anterior margin of the frontoclypeal apotome is symmetrical as in *Homoplectra*, lacking the notch of *Aphropsyche* (?). Sclerotized parts are brownish orange. Length of larva up to 10 mm.

RETREAT Unknown.

BIOLOGY These larvae were collected in a small, cool stream part way up Wayah Bald, a mountain rising in excess of 5,300 ft (1,600m) in North Carolina.

REMARKS Diagnostic characters for the male of *O. howellae* were given by Ross (1944). The material studied and illustrated was obtained on loan from the United States National Museum through O.S. Flint.

Oropsyche ? *howellae* ? (North Carolina, Macon Co., 8 June 1961, USNM)
A, larva, lateral; B, segments VIII and IX with bases of anal prolegs, ventral; C, head, dorsal; D, head, ventral

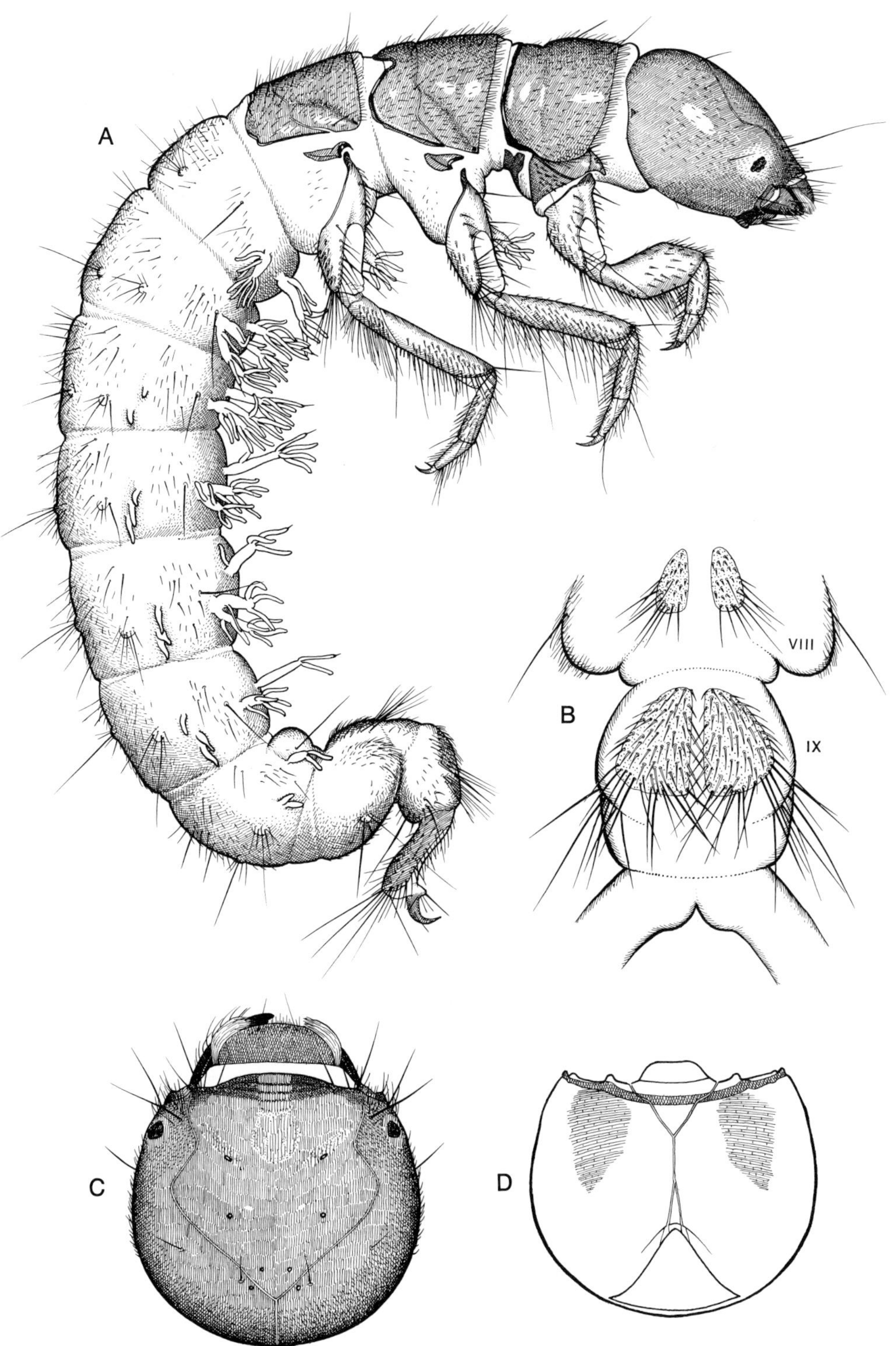
A
B
VIII
IX
C
D

6.10 Genus Parapsyche

DISTRIBUTION AND SPECIES Species of *Parapsyche*, of which more than 20 are now known, occur in Japan, China, Burma, and the Himalaya as well as North America. Seven species are known from this continent, five in the west and two in the east.

Larvae have been described for the eastern species *P. apicalis* (Banks) and *P. cardis* Ross by Flint (1961a), and for two western species, *P. almota* Ross and *P. elsis* Milne, by Smith (1968a). We have associated material for these four.

MORPHOLOGY *Parapsyche* and *Arctopsyche* comprise the subfamily Arctopsychinae, distinguished primarily by a ventral apotome that entirely separates the genae. In larvae of several species of *Parapsyche* the lateral margins of the ventral apotome are parallel for the most part (D), clearly distinct from the convergent margins of this sclerite in *Arctopsyche*. But in at least one species, *P. elsis*, the sclerite is tapered as in *Arctopsyche* (Smith 1968a). Diagnosis for larvae of *Parapsyche* must be further qualified, then, to include species with a tuft of several long setae or scale-hairs in the *sa*2 and *sa*3 positions of most abdominal segments (B). Sclerites of the head and thorax are usually brown in colour. Length of larva up to 29 mm.

RETREAT Larvae construct a typical hydropsychid retreat of small stones and pieces of detritus, and spin a silken capture net.

BIOLOGY Larvae of *Parapsyche* are characteristic of small, cold streams where they construct retreats in strong currents. Data for the life cycle of the eastern species were given by Flint (1961a) and Mackay (1969), and for two western species by Smith (1968a).

REMARKS Schmid (1968a) has reviewed taxonomy of the adults.

Parapsyche apicalis (Virginia, Madison Co., 23 May 1970, ROM 700372)
A, larva, lateral x14; B, abdominal segment, lateral; C, head, dorsal; D, head, ventral

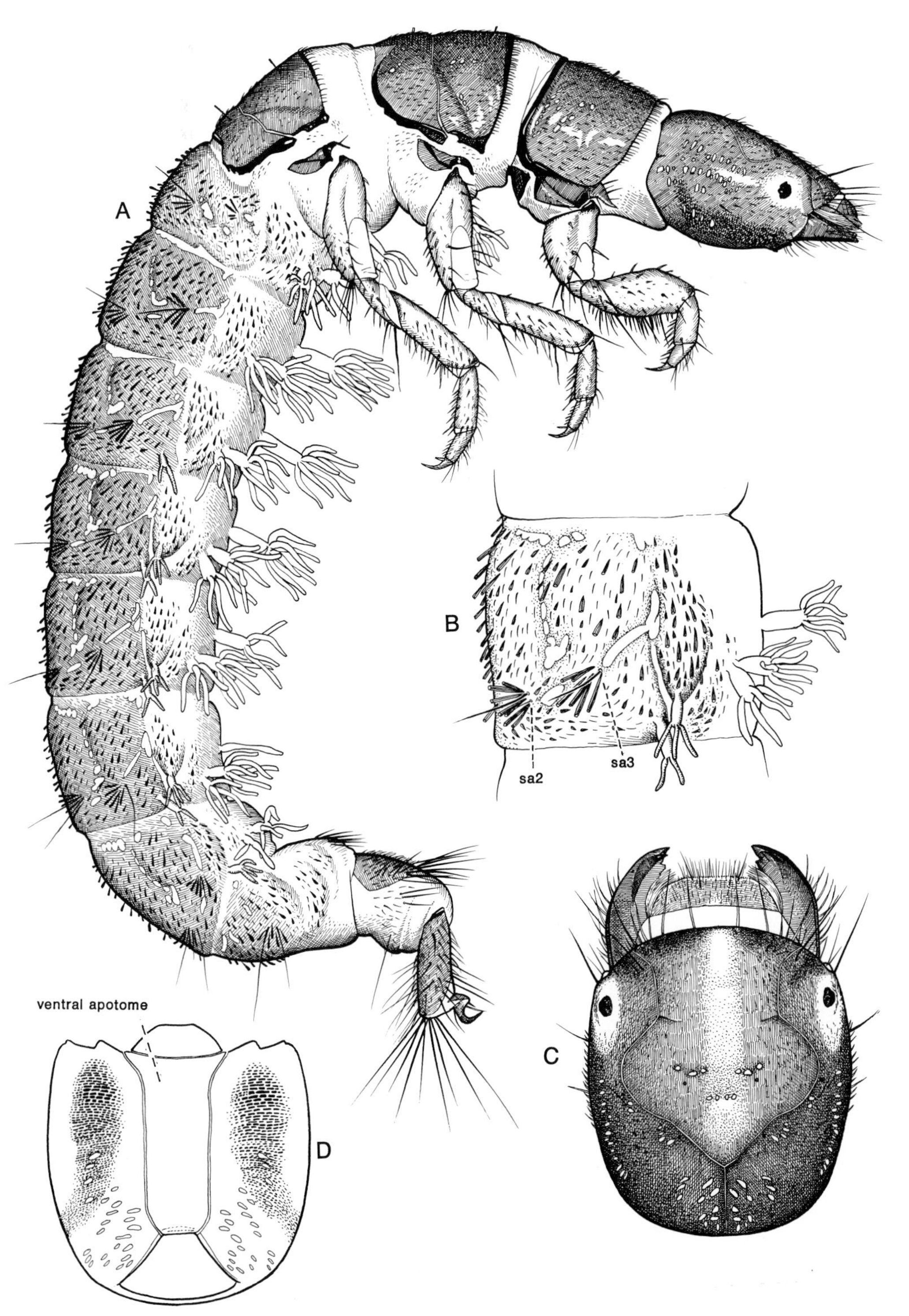
A
B
sa2
sa3
C
ventral apotome
D

6.11 Genus Potamyia

DISTRIBUTION AND SPECIES The only North American species in this genus is *Potamyia flava* (Hagen), an inhabitant of larger rivers over much of the eastern half of the United States and through Minnesota and South Dakota to Montana. The status of a species of *Potamyia* from China has been confirmed (Schmid 1965). No others are known.

Diagnostic features for the larva of *P. flava* have been given previously by Ross (1944, 1959).

MORPHOLOGY Larvae of *Potamyia* are generally similar to those of other Hydropsychinae; but in collections we have examined the fore trochantin ranges from a simple, unforked condition (C) through intermediates to a forked condition (A,B). The conditions illustrated are all from a single series of 18 specimens. *Potamyia* larvae with a forked fore trochantin could, then, be confused with *Cheumatopsyche*, but the characters in the key should effectively separate the two. The head and thorax are light yellowish brown in colour, with few distinctive markings. Each mandible bears a prominent lateral flange (E,F). Length of larva up to 14.5 mm.

RETREAT Fremling (1960) described the capture net of *P. flava* as a loose, voluminous structure, similar to nets of *Cheumatopsyche campyla*; both partially collapsed when the substrate was removed from water.

BIOLOGY Larvae of *Potamyia* are characteristic of large, relatively warm rivers. In the upper Mississippi River, where Fremling (1960) studied the biology of *P. flava*, the greatest concentrations of larval nets were on rocks in sandy, silt-free bottom materials. Larvae of *P. flava* and of *Cheumatopsyche campyla* were relatively more abundant in sites with slower current than those of *Hydropsyche orris*, which were dominant in stronger currents.

Potamyia flava (South Dakota, Yankton Co., Feb.-July 1968, ROM)
A, larva, lateral x16; B, forked fore trochantin of another larva, lateral; C, fore trochantin of another larva showing vestigial dorsal ramus, lateral; D, fore leg showing scraper on femur, mesial; E, right mandible, dorsal; F, head, dorsal; G, head, ventral; H, segments VIII and IX, ventral

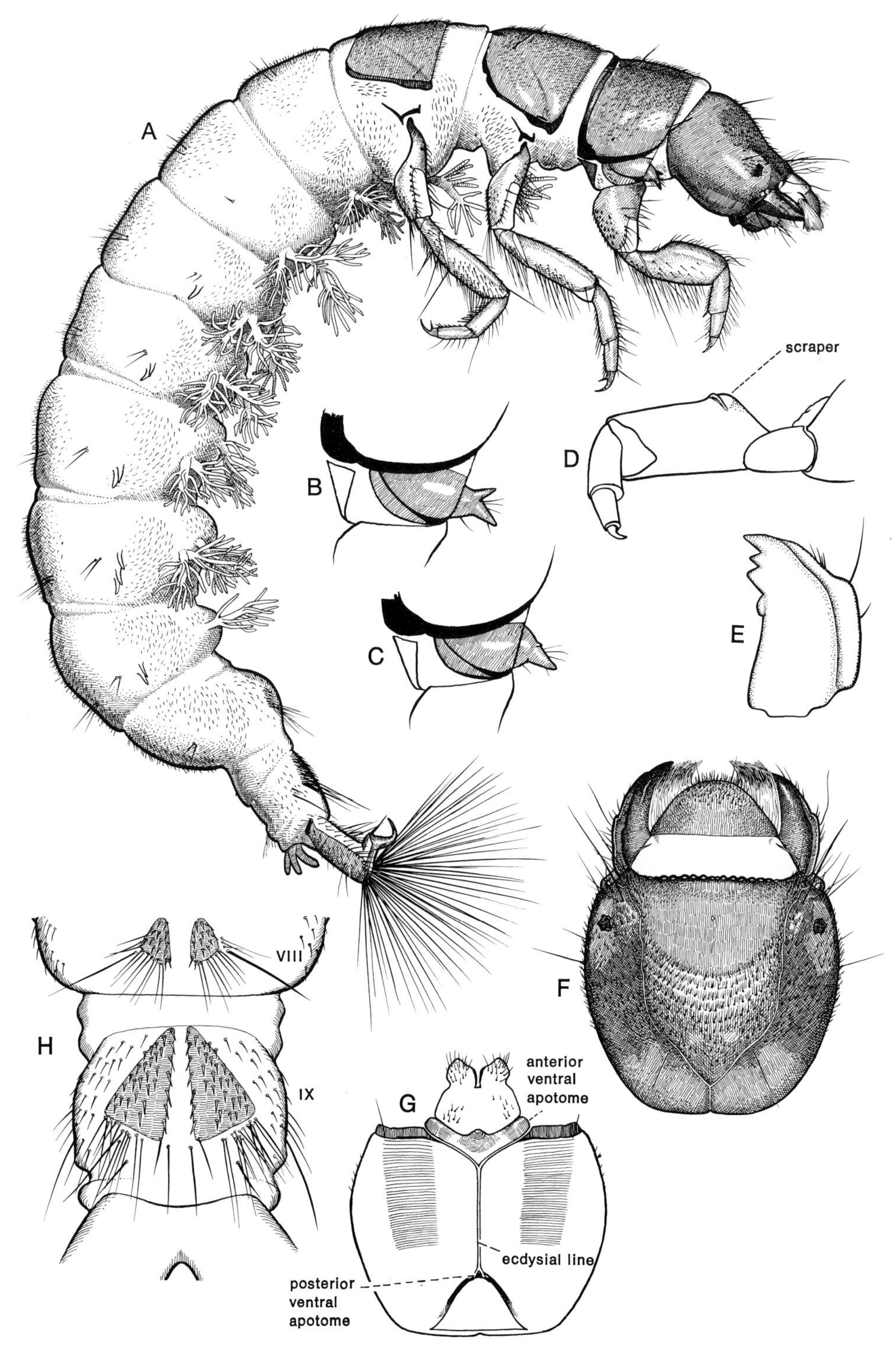
A
B
C
D
scraper
E
F
H
VIII
IX
G
anterior
ventral
apotome
ecdysial line
posterior
ventral
apotome

6.12 Genus **Smicridea**

DISTRIBUTION AND SPECIES Species of *Smicridea* occur in Australia, throughout South and Central America and the major islands of the Antilles, and into the southwestern United States. The radiation of species in the Neotropical region (Flint 1974a) indicates that *Smicridea* functions there as the ecological equivalent of *Hydropsyche* and *Cheumatopsyche*, both of which are absent from South America. North of Mexico four species are known, roughly within the area that includes and is bounded by Texas, Oklahoma, Colorado, and California. Within this area *Smicridea* species are sympatric with *Hydropsyche* and *Cheumatopsyche.*

Diagnostic characters for the larva of *S. fasciatella* McL. were given by Ross (1944); Flint (1974a) described the larva and pupa of this species and identified larvae of two other Nearctic species, *S. dispar* (Banks) and *S. signata* (Banks).

MORPHOLOGY Larvae of *Smicridea* are generally similar to those of other Hydropsychinae but are distinctive in having a single, median sclerite on the venter of segment VIII (E), and a submentum that is entire without a median cleft (C). The fore trochantin is simple and not bifurcate (B), distinguishing the larvae at once from those of *Hydropsyche* and *Cheumatopsyche*; and the sclerites posterior to the prosternal sclerite are small as in most *Cheumatopsyche.* Length of larva up to 9.5 mm.

RETREAT According to Flint (1974a) larvae of *Smicridea* construct a typical hydropsychid retreat with a capture net extending across the current.

BIOLOGY Larvae occur in running waters of a wide range, and are often abundant in streams of the southwest (Flint 1974a). Observations on *Smicridea* larvae in a Cuban seepage stream were given by Botoşăneanu and Sýkora (1973).

REMARKS Taxonomic data for Nearctic species of *Smicridea* were summarized by Flint (1974a); the two subgenera *Rhyacophylax* and *Smicridea* are distinct in characters of adults, but there appear to be no corresponding differences in larvae. *Smicridea* is sufficiently discordant from other genera of the Hydropsychinae to elicit views that it should not be assigned to that subfamily (Ulmer 1957; Marlier 1964); the tribe Smicrideini (Hydropsychinae) was established for the genus by Flint (1974a).

Smicridea fasciatella (Oklahoma, Johnston Co., 10 May 1970, ROM 700326)
A, larva, lateral x22; B, fore trochantin, lateral; C, head, ventral; D, head, dorsal; E, segments VIII and IX, ventral

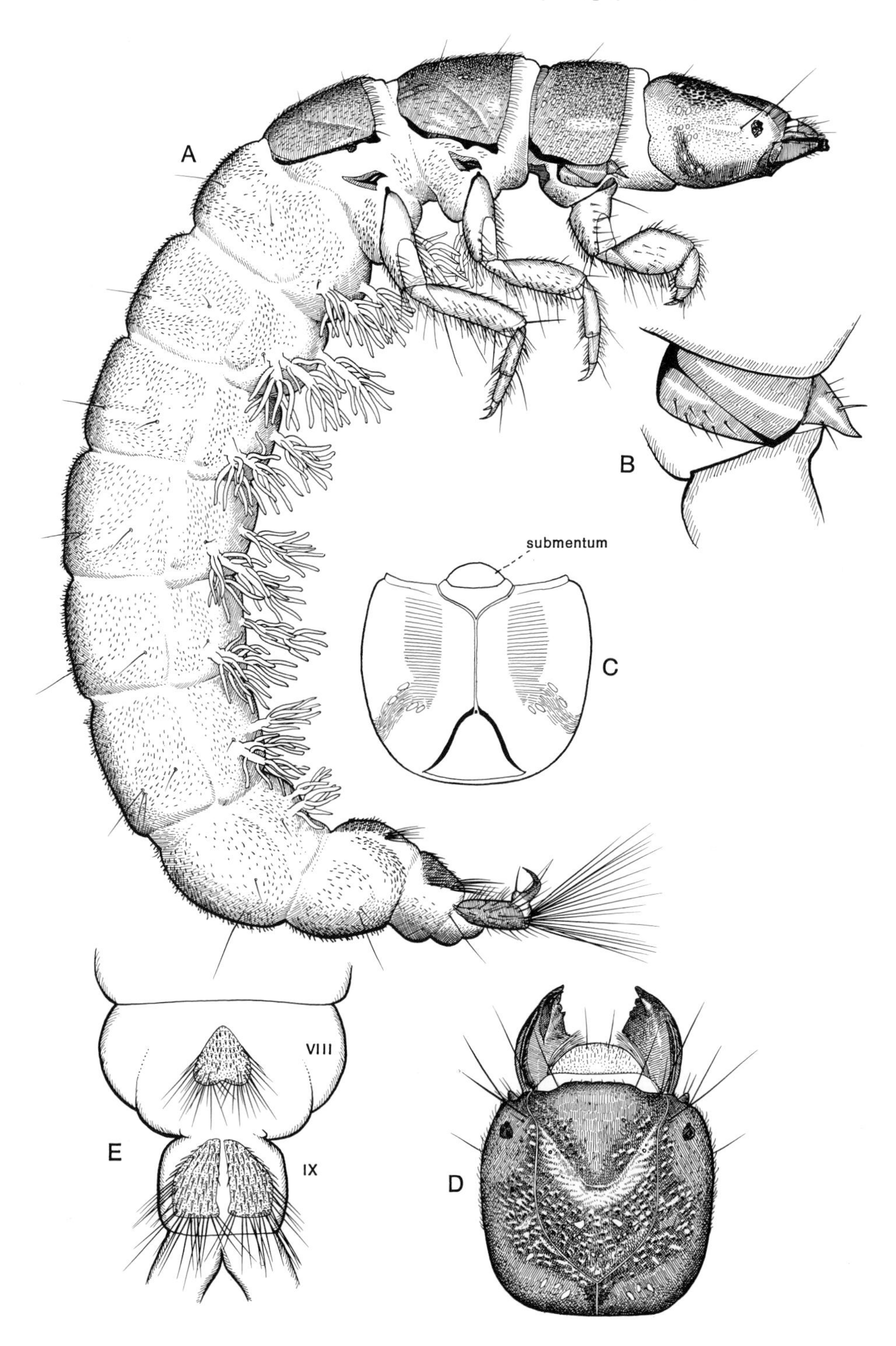
A
B
submentum
C
VIII
E
IX
D

7
Family Hydroptilidae

Often called micro-caddisflies because they include the smallest species of 2 to 3 mm in length, the Hydroptilidae also include genera in which the larvae reach 6 mm and differ little in size from some of the Glossosomatidae and Psychomyiidae. Hydroptilids occur in all faunal regions, and although richly represented in tropical latitudes where they account for a large part of the caddisfly fauna in some areas, they are still an abundant and highly diverse group through much of North America. There are 14 Nearctic genera with approximately 175 species north of Mexico, and there can be little doubt that many more species are still to be discovered.

In the North American fauna there are hydroptilid genera characteristic of all types of permanent waters from cold springs through streams and rivers to lakes, but as in most other parts of the world little study has been devoted to their biology. The classic work by Nielsen (1948) on Danish species in five genera is the source of much that is known about the biology of the family. Information available on food in various genera indicates that the larvae utilize algae to a large extent - consuming the cellular contents from filamentous forms and grazing on periphytic diatoms.

Hydroptilids can be recognized by the presence of sclerotized plates on all three thoracic nota, and by the general absence of gills on abdominal segments. But apart from these features shared in common, they are highly diverse in other ways, such as over-all shape of the body, extent of ecdysial sutures of the head, length of antennae, presence of mid-dorsal ecdysial sutures on the second and third thoracic nota, length of legs, and structure of the anal prolegs. Larvae of several genera bear dorsal chloride epithelia on most abdominal segments, areas specialized for ionic absorption in osmoregulation (Wichard 1976).

The life cycle of hydroptilid caddisflies is unusual because the first four instars are spent as free-living larvae without cases, the onset of case-building behaviour being delayed until the final instar. In most genera the larva up to the beginning of the fifth instar has a slender abdomen; but during the fifth instar membranous parts of the body, especially the abdomen, become enlarged, often enormously. Strategy in most genera seems to be dependence on the final instar, living in its case, to accumulate the bulk of the stored energy reserve. Free-living, early instars occur in the same habitats in which they ultimately live as case-makers, and evidently utilize the same food resources as the older larvae (Nielsen 1948). In temperate latitudes there is a tendency for larvae to overwinter as fifth instars in cases and, in some genera at least (e.g. *Agraylea, Hydroptila*, and *Oxyethira*), to pass rather rapidly through the earlier instars. In his study of Danish species, Nielsen (1948) found that larvae of a species of *Agraylea* passed through the first four instars in 11-21 days, and those of *Oxyethira* in 13-16 days; he also found, however, that larvae of *Orthotrichia* and *Ithytrichia* required longer periods for early larval development as in other Trichoptera.

Diversity among hydroptilid genera is seen also in case-making behaviour. The family is known generally as the purse-case makers, because for the most part final-instar larvae construct portable bivalve cases of silk, often with sand grains, diatoms, or algal filaments incorporated. Most larvae enlarge their cases continually to accommodate the swelling abdomen, and those with bivalve cases cut open the ventral junction of the two edges, add new material, and close the seam again with silk. Other larvae spin portable silken cases shaped like flattened, bottomless bottles; in these cases new silk added to the posterior edges accommodates the growing larva. Still others, in the Leucotrichiinae, spin a flattened case fast to a rock; since all cases in series of this group that we have collected are of approximately the same size, it appears that the larvae accommodate their ultimate space requirements at the beginning of case-making. And finally, larvae of *Mayatrichia* and *Neotrichia* construct cylindrical cases similar in shape to those of most case-making families. According to Nielsen (1948) early fifth instars construct provisional cases of various types as a base for the definitive case, finally cutting it free; in *Hydroptila* the provisional case is a crude tube. Fifth-instar larvae leave a silken thread wherever they go (Nielsen 1948).

Hypermetamorphosis or larval heteromorphosis was first suggested for the Hydroptilidae by Needham (1902); but, as pointed out by Ross (1944), the suggestion arose through failure to recognize that two genera were represented in the material studied rather than the single one supposed. From Needham's illustrations, it is clear that he had pupae and prepupae of *Ithytrichia*, and larvae perhaps of *Oxyethira* based on the row of setae on the dorsum of abdominal segment I. Although this first inference of heteromorphosis in the Hydroptilidae was invalid, Nielsen (1948) reactivated the idea that 'it is almost justifiable to speak of a hypermetamorphosis,' because the first four larval instars differ so greatly from the fifth in structure and behaviour. Although larval heteromorphosis is generally considered an adaptation of parasitic insects, current definition and examples (C.S.I.R.O., *Insects of Australia*, 1970; Snodgrass 1954) include several non-parasitic groups and would accommodate the Hydroptilidae as well.

Subfamilial classification of the Hydroptilidae is still at an elementary stage, but there are useful insights to be gained through its use. Five subfamilies are represented in the Nearctic fauna.

HYDROPTILINAE: *Agraylea, Dibusa, Hydroptila, Ochrotrichia, Oxyethira, Stactobiella*
Larva strongly compressed laterally, frequently with transverse sulcus on venter of middle abdominal segments; chloride epithelia usually present on dorsum of abdominal segments; all legs of approximately same size in running-water forms, middle and hind legs tending

to be longer and more slender in genera inhabiting lentic waters; tarsal claws moderately to very long, tarsi generally same length as claws; tibia and femur of at least fore leg stout, tibia enlarged distoventrally. Larval case usually of two silken valves, often covered with sand grains or algae, carried vertically. Larvae inhabit lotic and lentic waters, often in filamentous algae. The group is widely distributed in all faunal regions.

LEUCOTRICHIINAE: *Alisotrichia, Leucotrichia, Zumatrichia* Larva dorsoventrally depressed; all legs similar, segments rather short and stout; tarsal claws short and stout, tarsus approximately twice as long as claw on all legs; meso- and meta-notal plates usually entire, without mid-dorsal ecdysial line; abdominal segments I–VIII with wide, short, dorsal sclerite (chloride epithelium?), larger than in other genera and not ring-like; fifth instars usually with abdominal segments V–VII enormously enlarged. Larval case usually flattened, ovoid valve of silk resembling egg-case of leech, fixed in one place; in one genus, *Alisotrichia*, fifth instar slender and free-living, making no case until immediately before pupation. Larvae live in running waters. The group occurs only in the New World.

The genus *Anchitrichia* might occur within the southern limits of the Nearctic region; diagnostic characters of the larva were given by Flint (1970).

ORTHOTRICHIINAE: *Ithytrichia, Orthotrichia* Larva somewhat dorsoventrally depressed (*Orthotrichia*) or strongly laterally compressed (*Ithytrichia*); middle and hind legs longer and slenderer than fore legs, tarsal claws moderately long and slender, tarsi of hind legs longer than claws. Larval cases diverse, made entirely of silk, somewhat similar to those in the Hydroptilinae. Larvae live in lentic and lotic waters.

PTILOCOLEPINAE: *Palaeagapetus* Larva dorsoventrally depressed, truncate fleshy tubercle on each side of abdominal segments I–VIII; all legs similar, tibiae and femora stout, tarsi approximately as long as claws. Larval case of two silken valves covered with pieces of liverwort, carried horizontally. Larvae inhabit cold springs. The group is Holarctic, and is regarded as the most primitive subfamily of the Hydroptilidae (Ross 1956).

Mayatrichia and *Neotrichia* On the basis of larval structure and behaviour these two genera comprise a natural unit that probably ranks as a subfamily, but its creation is not proposed here because I have not assessed evidence derived from the adult stage. Larva uniformly cylindrical; head narrowed anteriorly; middle and hind legs longer and their segments slenderer than fore legs; tarsi elongate and slender, those of hind legs at least 1½ times longer than claw; anal prolegs projecting prominently from body. Larval case cylindrical, sometimes of silk alone or with sand grains. Larvae inhabit running water.

The fact that in these two genera the anal prolegs are much longer in relation to the body than in any other North American hydroptilids may be correlated with their having portable cases that lack a slit-like opening through which the posterior end of the abdomen can be extended as an anchor on the substrate. The posterior opening of the cylindrical case in *Mayatrichia* and *Neotrichia* is ovoid and large enough to permit the long anal prolegs to be extended through it, thereby providing contact with the substrate.

The key to genera which follows is based on fifth-instar larvae, for the most part after case-building has begun. The key is not applicable to earlier free-living instars, although certain of the diagnostic characters may still be useful, such as the undivided meso- and meta-notal plates of the Leucotrichiinae, and the caudal gill filaments in *Hydroptila.* Nielsen (1948) has provided a key to early instars of European species in *Agraylea, Hydroptila, Ithytrichia, Orthotrichia*, and *Oxyethira.*

Key to Genera

1 Meso- and meta-notal sclerotized plates usually lacking mid-dorsal ecdysial line, but line present on pronotum (Fig. 7.14A); tarsal claws short and stout, tarsi approximately twice as long as claws (Fig. 7.6C–E); segments V and VI usually abruptly broader than others in dorsal aspect (Fig. 7.6A) (subfamily Leucotrichiinae) **2**

Pro-, meso-, and meta-notal sclerotized plates each with mid-dorsal ecdysial line (Fig. 7.7F); tarsal claws variable, but tarsi approximately twice as long as claws only when claws elongate and slender (Figs. 7.7E, 7.10G, 7.11F) **4**

2 (1) Segments V and VI greatly enlarged in dorsal aspect (Fig. 7.6A); case of silk, flattened and ovoid, fastened immovably to rocks, both ends reduced to small circular opening (Fig. 7.14B,E) **3**

Segments V and VI not enlarged in dorsal aspect more than other segments (Fig. 7.2A); larva free-living without case. Southwestern **7.2 Alisotrichia**

3 (2) Sclerite on dorsum IX with scattered, short stout setae (Fig. 7.14D). Montana **7.14 Zumatrichia**

Sclerite on dorsum IX usually without short, stout setae or, if present, setae in transverse band (Fig. 7.6F). Widespread **7.6 Leucotrichia**

4 (1) Hind tarsi usually short and thick, about same length as tarsal claws (Fig. 7.4H), sometimes shorter (Fig. 7.1F); or if hind tarsi slender and longer than claws (Fig. 7.11F), protibia bears prominent posteroventral lobe (Fig. 7.11D) **5**

Hind tarsi slender, usually about twice as long as tarsal claw (Fig. 7.5E) or longer (Fig. 7.8D), sometimes only a little longer, but protibia without prominent posteroventral lobe as above (Fig. 7.10E) **11**

5 (4) Larva dorsoventrally depressed, segments I–VII with truncate fleshy tubercle on each side (Fig. 7.12A); case of two flattened, elliptical valves covered with liverwort pieces (Fig. 7.12B). Eastern and western mountains (subfamily Ptilocolepinae) **7.12 Palaeagapetus**

Larva more or less laterally compressed, segments I–VIII without lateral tubercles (Fig. 7.9A) (subfamily Hydroptilinae) **6**

6 (5) Tarsal claws stout and abruptly curved, each with thick, blunt spur at base (Figs. 7.3E, 7.13E) **7**

Tarsal claws slender, smoothly curved, each with thin, pointed spur at base (Fig. 7.4F-H) **8**

7 (6) Dorsal abdominal setae stout, each with small sclerotized area around base, dorsal rings of abdominal segments clearly delineated (Fig. 7.3A); larvae occur on red algae, case of two symmetrical valves incorporating this alga (Fig. 7.3B,F). Eastern **7.3 Dibusa**

Abdominal setae slender, their bases without sclerotized area, dorsal rings indistinct (Fig. 7.13A); case two symmetrical valves of silk (Fig. 7.13B,F). Eastern and western **7.13 Stactobiella**

8 (6) Middle and hind legs about 2½ times as long as fore legs (Fig. 7.11D-F); case entirely of silk, shaped like flattened flask open at bottom (Fig. 7.11B,G). Widespread **7.11 Oxyethira**

Middle and hind legs not more than 1½ times as long as fore legs (Figs. 7.1D-F, 7.4F-H) **9**

9 (8) Three filamentous gills arising from posterior end of abdomen, one from dorsomedian position on segment IX, other two at lateral sclerites of anal prolegs (Fig. 7.4A); case laterally compressed, consisting of two silken valves covered usually with sand grains, sometimes with diatoms (Fig. 7.4B,C). Widespread **7.4 Hydroptila**

Posterior end of abdomen without filamentous gills (Fig. 7.1A) **10**

10 (9) Intersegmental grooves on venter of abdomen much deeper than on dorsum, forming prominent ventral lobes on segments II-VII inclusive, each ventral lobe with transverse sulcus (Fig. 7.1A); case laterally compressed, consisting of two silken valves with algal filaments incorporated concentrically (Fig. 7.1B,G). Widespread **7.1 Agraylea**

Intersegmental grooves on venter of abdomen little deeper than those on dorsum, transverse sulcus often present on venter segments II-V inclusive (Fig. 7.9G); case usually laterally compressed, consisting of two silken valves covered with sand grains (Fig. 7.9J,K), but sometimes of only one valve carried like tortoise shell (Fig. 7.9B,C). Widespread **7.9 Ochrotrichia**

11 (4) Anal prolegs elongate and cylindrical, projecting prominently beyond general body outline (Fig. 7.7A) **12**

Anal prolegs short, conforming to general body outline, not projecting prominently (Fig. 7.10A) (subfamily Orthotrichiinae) **13**

12 (11) Abdomen somewhat depressed, intersegmental grooves prominent, lateral fringe of hairs present (Fig. 7.8A); case of fine sand grains, cylindrical (Fig. 7.8F). Widespread **7.8 Neotrichia**

Abdomen more inflated, not depressed, intersegmental grooves not prominent, lateral fringe of hairs absent (Fig. 7.7A); case of silk, cylindrical but usually with transverse or longitudinal ridges (Fig. 7.7B). Widespread **7.7 Mayatrichia**

13 (11) Most abdominal segments with prominent dorsal and ventral projections (Fig. 7.5A); case flattened pouch of silk, posterior end open, anterior end reduced to small circular opening (Fig. 7.5B). Widespread **7.5 Ithytrichia**

Abdominal segments without dorsal and ventral projections (Fig. 7.10A); case of silk, somewhat pod-like in shape with longitudinal ridges (Fig. 7.10B,I). Widespread **7.10 Orthotrichia**

7.1 Genus Agraylea

DISTRIBUTION AND SPECIES *Agraylea* is Holarctic in distribution with three North American species known: *A. costello* Ross, Ontario and Maine, and probably widespread in the northeast; *A. multipunctata* Curtis, Holarctic, transcontinental in Canada and northern states, south to Colorado and Tennessee; and *A. saltesa* Ross, in western states.

Morphological characters for the final-instar larva of *A. multipunctata* were outlined by Betten (1934) and Ross (1944), and for all instars by Nielsen (1948).

MORPHOLOGY Larvae of *A. multipunctata* (A) are strongly compressed and have lightly coloured sclerites. A transverse sulcus, evidently a cuticular inflection serving as a muscle insertion, is well developed on the venter of segments II–VII inclusive. There are no gill filaments on segment IX as there are in *Hydroptila.* The middle and hind legs are substantially longer than the fore legs; there is a single sclerite between the fore coxae (C). Length of larva up to 6 mm.

CASE The larval case in *Agraylea* (B,G) is carried vertically and consists of two valves of silk with algal filaments incorporated concentrically. In lateral aspect the ends of each valve are approximately the same width as the middle, and the valve is generally symmetrical bilaterally about its long axis; this evidence led Nielsen (1948) to conclude that *A. multipunctata* larvae make increments to their cases along both dorsal and ventral margins, a behaviour distinctive among the Hydroptilinae. Length of larval case up to 6.5 mm.

BIOLOGY *Agraylea* larvae occur in lakes, ponds, and areas of reduced current in large rivers. According to Nielsen (1948) larvae of *A. multipunctata* are found in beds of submerged aquatic plants where they feed on the cellular contents of filamentous algae; grasping of algal filaments is believed facilitated by the broadened fore femur and tibia. In other studies, food of *A. multipunctata* in a Kentucky stream was found to include diatoms (Minckley 1963), and Siltala (1907b) found larvae of the same species in the Gulf of Finland to have ingested pieces of the marine alga *Fucus*, along with diatoms and other algae.

Agraylea prob. *multipunctata* (Ontario, Leeds Co., 16 June 1966, ROM)
A, larva, lateral x23, chloride epithelium enlarged; B, case, lateral x12; C, prothorax, ventral; D, fore leg, lateral; E, middle leg, lateral; F, hind leg, lateral; G, case, dorsal

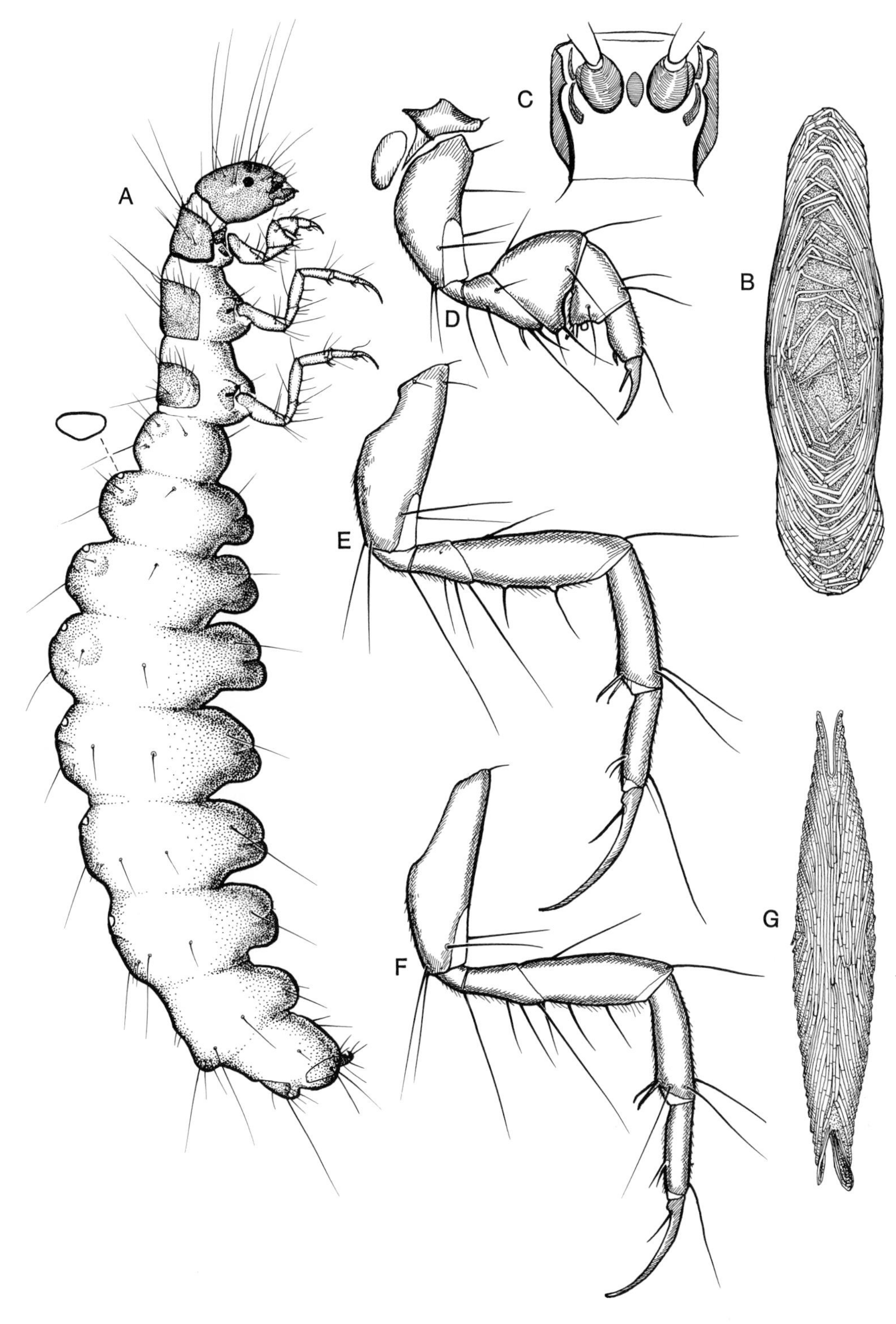
A
B
C
D
E
F
G

7.2 Genus Alisotrichia

DISTRIBUTION AND SPECIES *Alisotrichia* is a New World genus largely confined to Central America and Mexico, but on the basis of a single larva recorded from Utah (Flint 1970, *Alisotrichia* species 3) the genus is included in the present work.

Taxonomic data for larvae and adults in this genus were summarized by Flint (1970).

MORPHOLOGY Although assigned to the Leucotrichiinae (Flint 1970), *Alisotrichia* differs in several respects from other members of the subfamily. Final-instar larvae (A) do not have enlarged abdominal segments as in *Leucotrichia* and *Zumatrichia*, but they are distinctive in bearing long, stout setae on all thoracic and abdominal segments. The dorsum of each abdominal segment is covered by a wide sclerite, that on segment IX longer than the others; along each side of segments I–VIII is a conical projection with two stout setae. A stout seta arises from the convex edge of the anal claw (C). All three thoracic legs are similar, rather short and stout, the tarsi approximately twice as long as the claws (B). Length of larva up to 3 mm.

CASE Here, as well, discordance of this genus from other leucotrichiines is evident. According to Flint (1970), *Alisotrichia* larvae do not construct a larval case at the beginning of the final instar, but continue a free-living existence to the end of that stage; a simple, silken covering with a few lateral openings is then constructed. Evidently there is no feeding activity within the shelter, and shortly after construction of the simple shelter larvae spin a complete pupal cocoon.

BIOLOGY Larvae known for *Alisotrichia* live in running waters. Deferral of case-making until immediately before pupation indicates that the biology of *Alisotrichia* larvae is different from that of other hydroptilids. Perhaps the larvae are predacious and would be hindered by a case, or perhaps they do not require the large energy reserve represented by the enlarged abdomen of final-instar larvae in other hydroptilid genera.

REMARKS The larva illustrated shows evidence beneath the cuticle of stout setae similar to those already on the external surface, and thus probably represents the end of the fourth instar. Even so, the specimen still exhibits the principal diagnostic characters attributed to *Alisotrichia*. The specimen illustrated, provided by O.S. Flint, bears the information: Utah, City Springs, ½ mile north of St George, 2,900 ft (900 m).

Alisotrichia sp. (Utah, Washington Co., USNM)
A, larva, dorsal; B, hind leg, lateral; C, segment IX with anal proleg, lateral

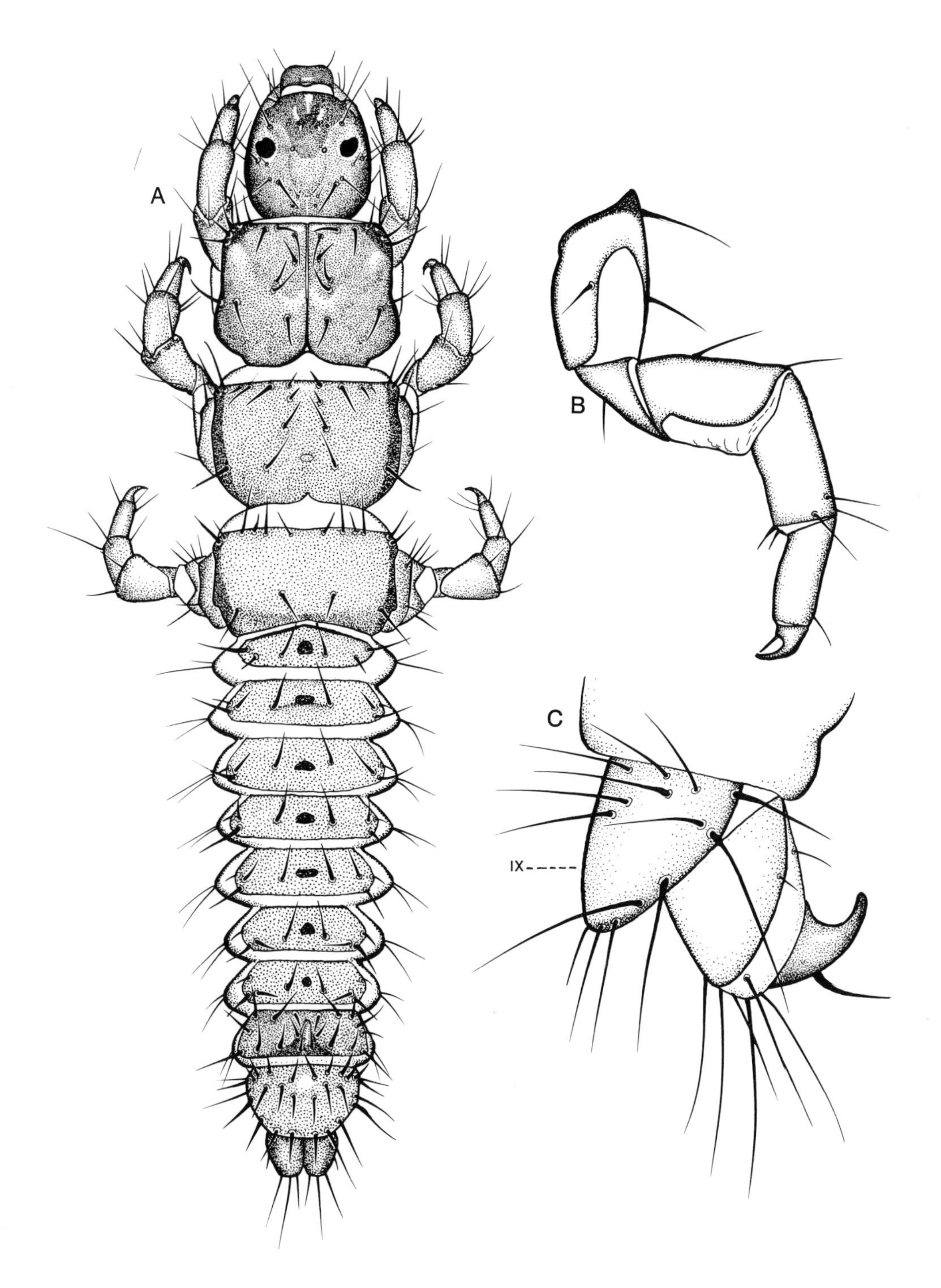
A
B
C
IX

7.3 Genus **Dibusa**

DISTRIBUTION AND SPECIES This monotypic genus comprises only *D. angata* Ross, a species of very local occurrence, but known from North Carolina, Tennessee, Arkansas, Kentucky, and Oklahoma.

The larva has not been identified in the literature up to now; its association here is possible because of collections made available through V.H. Resh and S.E. Neff, University of Louisville.

MORPHOLOGY This is the largest of the North American hydroptilids, and among the Hydroptilinae *D. angata* is distinguished by stout setae on the abdominal segments, each arising from a darkened, sclerotized base (A). The dorsal chloride epithelia of the abdomen are relatively large and darkly pigmented. Although difficult to see unless mounted on a slide and examined under a compound microscope, the tarsal claws are distinctive from those of all other North American Hydroptilidae; the claws are stout and strongly curved (E), suggestive of the claws of a sloth, and at the base of each tarsal claw the spur is thick and blunt, somewhat as in *Stactobiella* but much smaller. Sclerotized parts are medium brown in colour, and the head is spotted with rather large muscle scars. Length of subterminal instar larva illustrated 3.5 mm, but probably final instars reach at least 6 mm judging by length of the pupae.

CASE Final-instar larvae have usual hydroptiline cases of two symmetrical valves, the outer layer of each apparently consisting of elongate, more or less concentrically arranged pieces of the freshwater red alga *Lemanea*. Length of larval case illustrated 5 mm, but pupal cases in series up to 8 mm.

BIOLOGY Observations indicate that larvae of *D. angata* are associated with the freshwater red alga *Lemanea*; pupal cases were found attached to the algal thallus at its base. Pupal cases of this genus collected subsequently in Ontario were also on *Lemanea*.

REMARKS Material studied consists of the larva illustrated, which is not a final instar, a pharate male sufficiently developed to be identified as *Dibusa* prob. *angata* (collected 22 July 1973), a pharate female and many empty pupal cases (collected 6 June 1972) - all from the same stream and on *Lemanea*. Larval sclerites with the pharate adults establish the association.

Dibusa prob. *angata* (Kentucky, Johnson Co., 16 Feb. 1973, V. Resh, ROM)
A, larva, lateral x47, chloride epithelium enlarged; B, case, lateral x16; C, fore leg, lateral; D, middle leg, lateral; E, hind leg, lateral, tarsal claw enlarged; F, case, dorsal

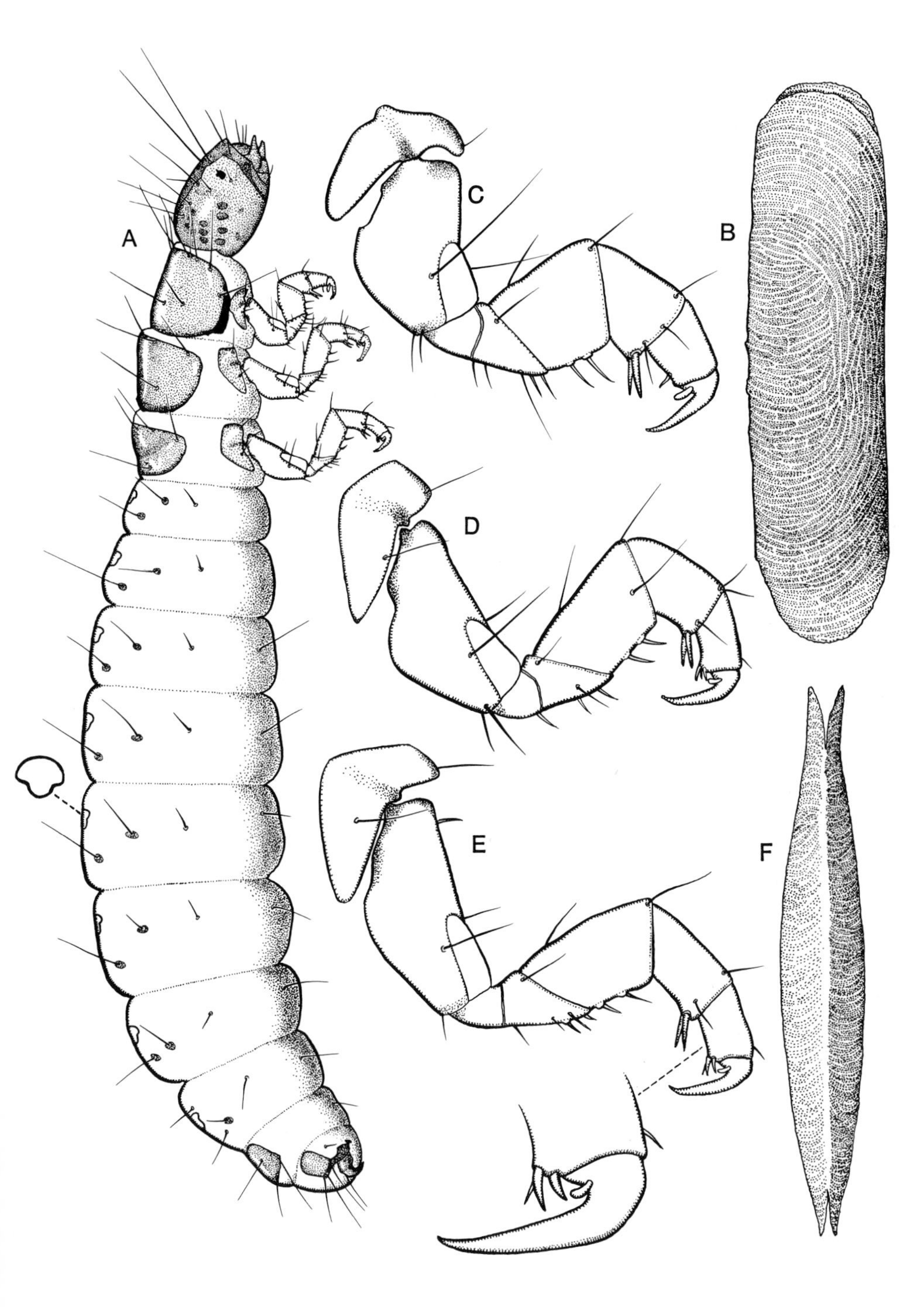
A
B
C
D
E
F

7.4 Genus Hydroptila

DISTRIBUTION AND SPECIES *Hydroptila* is a large genus recorded from all six faunal regions. In North America alone approximately 60 species are now known, and the genus is represented throughout most of the continent.

Larvae were associated for nine species by Ross (1944); colour of the sclerotized areas, the only basis that was found for separating the larvae to species, was variable and not sufficiently precise for unequivocal diagnosis.

MORPHOLOGY The apical abdominal gills (A) characteristic of *Hydroptila* larvae are best seen against a darker background. Larvae of *Hydroptila* studied have three small sclerites on the venter of the prothorax (D); some larvae have a curved, transverse sulcus on the meso- and meta-nota (E), but lack the anterolateral lobes of the metanotum present in *Ochrotrichia*. Length of larva up to 5 mm.

CASE Final-instar larvae have laterally compressed cases of two silken valves covered usually with a single layer of sand grains (B,C), occasionally with diatoms or algae. The case is carried vertically, the ventral edge fairly straight, the dorsal edge more or less curved. The two valves are fastened together except for a slit-like opening at each end as shown. According to Nielsen (1948) the larva enlarges its case by cutting open the ventral connection between the valves and adding materials along the ventral edge and at the ends; the marked asymmetry in the lateral aspect of the case resulting from this behaviour is partly rectified toward the end of the larval period when the larva adds new material only at the ends and no longer cuts open the ventral edge. Length of larval case up to 5.5 mm.

BIOLOGY Larvae of *Hydroptila* live in lakes and also in running waters. According to Nielsen (1948), all instars feed on filamentous algae by piercing the cells and eating the contents; but larvae of this genus in Britain are also reported to feed on diatoms and unicellular algae (Percival and Whitehead 1929). Two North American riverine species were found to have a one-year life cycle (Anderson 1967b; Cloud and Stewart 1974), but a Danish species was reported to have two generations per year (Nielsen 1948); in these studies most larvae overwintered as active fifth instars.

Hydroptila sp. (Ontario, Durham Co., 23 May 1958, ROM)

A, larva, lateral x29, chloride epithelium and segment IX enlarged; B, case, lateral x17; C, case, ventral; D, prothorax, ventral; E, meso- and meta-nota, lateral; F, fore leg, lateral; G, middle leg, lateral; H, hind leg, lateral

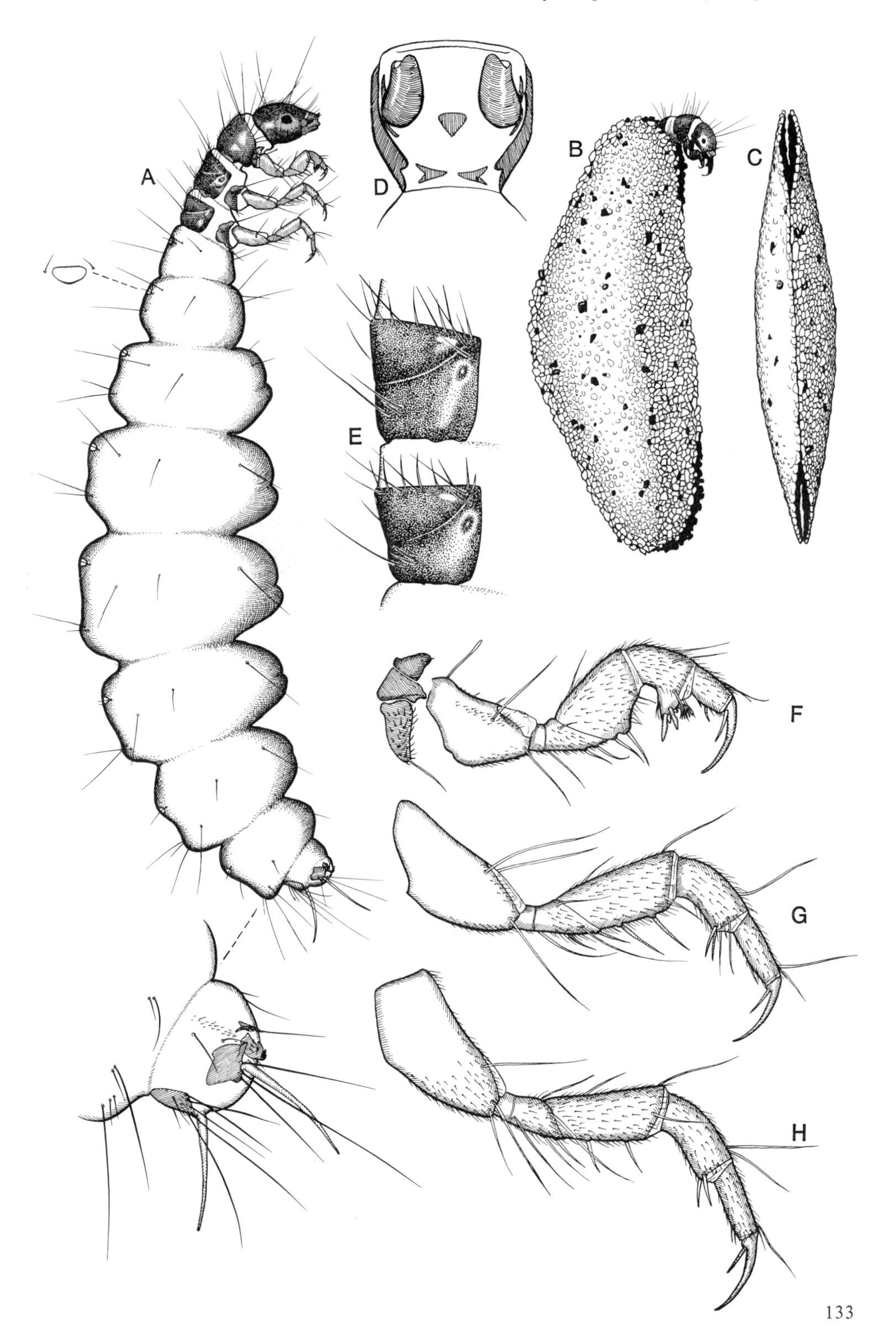
A
B
C
D
E
F
G
H

7.5 Genus **Ithytrichia**

DISTRIBUTION AND SPECIES *Ithytrichia* is a small Holarctic and Oriental genus in which two North American species are known: *I. clavata* Morton, transcontinental and evidently Holarctic (Fischer 1961, 1971); and *I. mazon* Ross in Illinois.

In North America, final-instar larvae of this genus were identified by Ross (1944); several instars of the European species *I. lamellaris* Eaton were described by Nielsen (1948).

MORPHOLOGY Final-instar larvae of *Ithytrichia* are strongly compressed laterally, and are distinctive because of the prominent lobate projection on the dorsum and venter of most abdominal segments (A). The tarsi are slender and elongate (C–E), approximately twice as long as the tarsal claws. A single median gill filament arises on segment IX, and the larva appears to be one of the few hydroptilids with anal papillae (A). Length of larva up to 3.5 mm.

CASE The transparent larval case (B,F), made entirely of silken secretion, resembles a flat pouch open at the posterior end, but with the anterior opening reduced to a small hole through which the head and thorax of the larva can be extended. According to Nielsen (1948) the case of *I. lamellaris* is carried in such a way that the flat sides are dorsoventral, thereby imposing a 90-degree torsion on the thorax and the junction between abdominal segments I and II; an extended account of case-building in this species is given by the same author. The median silken filaments at the posterior end of the case illustrated are the beginning of the pupal case attachment; the anterior end of the developing pupa lies at the posterior end of the larval case. Length of larval case up to 3.5 mm.

BIOLOGY *Ithytrichia* larvae live on rocks and on moss in running-water habitats, for which their ability to keep the flat side of the case toward the substrate is important. The food of *I. lamellaris* is largely diatoms scraped from rocks and other substrates in running water (Nielsen 1948). *I. lamellaris* requires a full year to complete its life cycle in Denmark, overwintering as fifth-instar larvae with cases (Nielsen 1948). In a Texas river, fifth-instar *I. clavata* larvae were found to drift, mostly at night (Cloud and Stewart 1974).

Ithytrichia prob. *clavata* (Ontario, Renfrew Co., 23 July 1969, ROM 690370)
A, larva, lateral x43, segment IX enlarged; B, case, ventral x22; C, fore leg, lateral; D, middle leg, lateral; E, hind leg, lateral; F, case, lateral

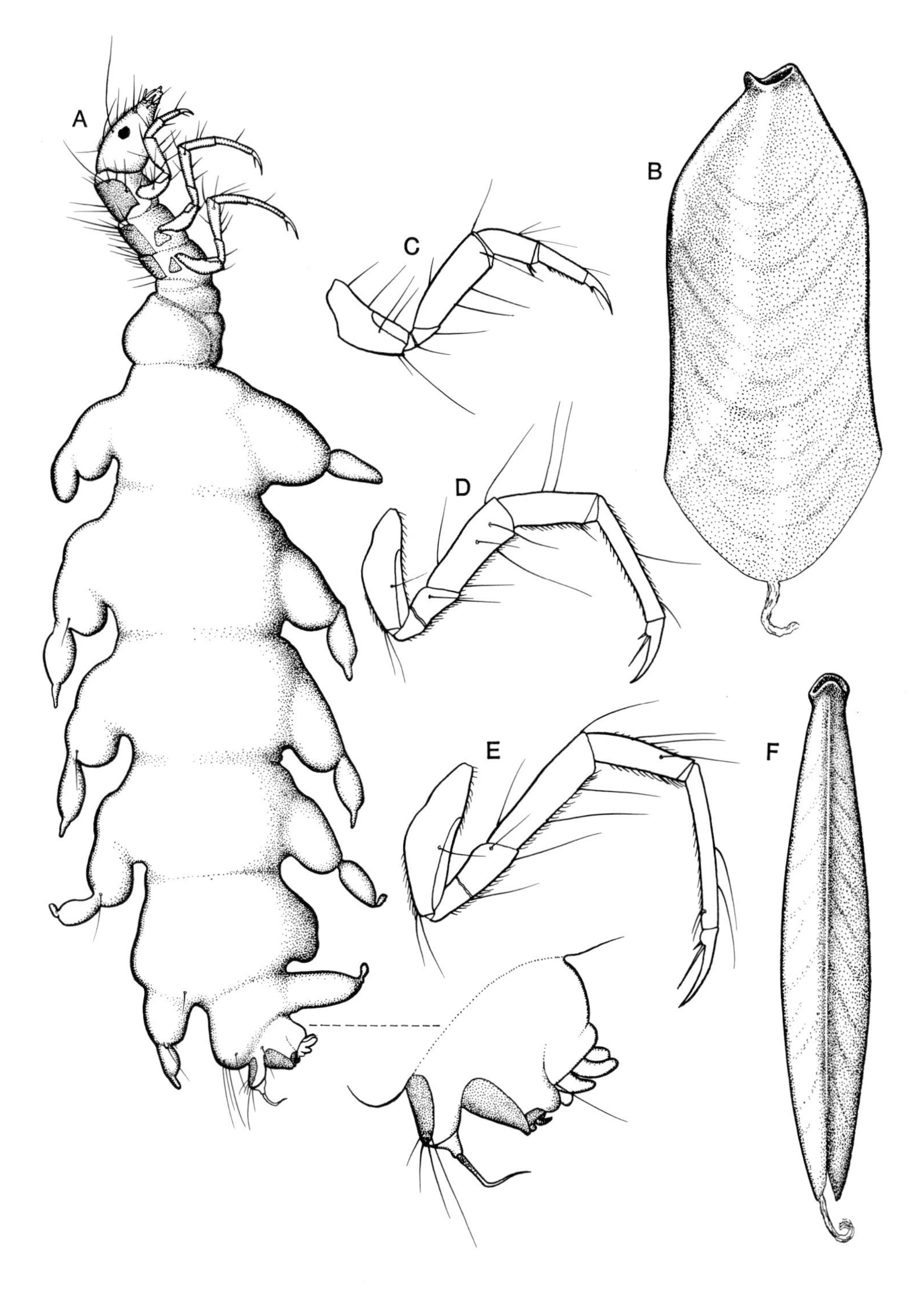
A
B
C
D
E
F

7.6 Genus **Leucotrichia**

DISTRIBUTION AND SPECIES Species of *Leucotrichia* occur only in the New World, for the most part in Mexico and Central America. Three are known north of Mexico: *L. limpia* Ross in Arizona and Texas, and *L. sarita* Ross in Texas, both also in Mexico; and *L. pictipes* (Banks), widespread over much of the United States from Minnesota west to Oregon and California, east to New York and Virginia.

Diagnostic characters for larvae and adults of all three species were given by Flint (1970).

MORPHOLOGY Final, case-dwelling, instars of several of the Leucotrichiinae are very distinctive because of the lateral distention of abdominal segments v, vi, and vii (A). North of Mexico *Leucotrichia* is the common genus, and larvae are distinguished from those of *Zumatrichia* by setal characters of the dorsum of segment ix and by their single tarsal claws. The abdomen is bright green in life. Length of larva up to 5 mm.

CASE The case is flattened, elliptical in outline, made entirely of silk, and fastened immovably around its entire periphery to a rock (B,G). The general aspect is that of the egg-case of a leech. So small and insignificant in appearance are the cases that one is often slow to realize that a single rock under scrutiny may harbour a hundred or more of the larvae, each in its own case. The larva encloses itself within the case at the beginning of the final instar, leaving a circular rimmed opening at each end. Length of larval case up to 5.5 mm.

BIOLOGY Larvae live on rocks in strong currents of running waters, and graze on surrounding periphyton and fine particulate detritus by extending the slender, anterior part of the body through the anterior or posterior openings (Lloyd 1915b, as *I. confusa*). Collection data indicate that most larvae overwinter in the final instar, pupating and emerging as adults from May through August. The capacity of the abdomen for accommodating food reserves is very large; fifth-instar larvae taken in August (Mississippi R., Minn.) before they began to build cases had undistended abdomens approximately equal in over-all size to the head and thorax; in later fifth instars with cases from the same collection, the combined length of the head and thorax was approximately one-quarter the length of the abdomen.

Leucotrichia prob. *pictipes* (Oregon, Umatilla Co., 18 Sept. 1966, ROM)
A, larva, dorsal x24, with dorsal sclerite enlarged; B, case, dorsal x18; C, fore leg, lateral; D, middle leg, lateral; E, hind leg, lateral; F, segment IX, lateral; G, case attached to rock, lateral

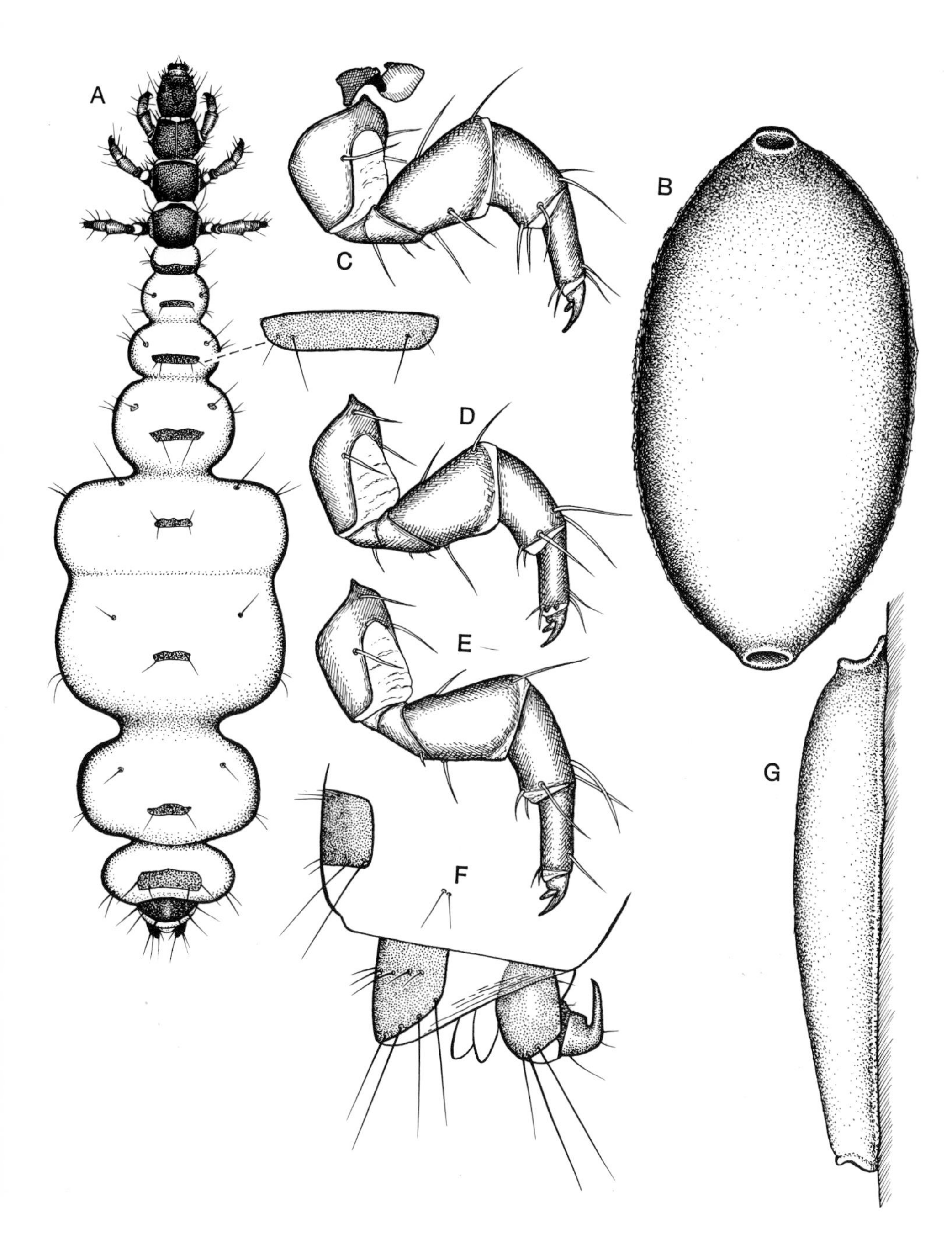
A
B
C
D
E
F
G

7.7 Genus **Mayatrichia**

DISTRIBUTION AND SPECIES *Mayatrichia* is a small North American genus recorded in the literature from Mexico through Wyoming and Montana to Saskatchewan, through Ontario to Maine and south to Florida. Three species are known: *M. acuna* Ross in Texas; *M. ponta* Ross in Oklahoma; and *M. ayama* Mosely with the broad distribution over much of the continent as indicated, extended to Utah where we collected pharate adults. Even as widely distributed as they are, *Mayatrichia* caddisflies must be regarded as local in occurrence.

Diagnostic characters for larvae of *M. ayama* were given by Ross (1944); we associated the larva of *M. ponta*, illustrated here, in Oklahoma.

MORPHOLOGY In the North American fauna, final-instar larvae of *Mayatrichia* are similar only to those of *Neotrichia.* On the basis of the limited material available for study, *Mayatrichia* larvae lack a lateral fringe of abdominal setae, and in general seem to have a more rotund abdomen with the intersegmental grooves less distinct (A). The legs (C–E) are relatively shorter, their segments thicker than in *Neotrichia*, and there are short, stout setae arising over much of each thoracic tergum (F). In examples studied, the head is more attenuate anteriorly than in *Neotrichia*, but this might not be so for the genus as a whole. A free-living, early instar of *M. ayama* was illustrated by Ross (1944, fig. 557). Length of larva up to 2.5 mm.

CASE Cases of final-instar *Mayatrichia* larvae are cylindrical, slightly curved, tapered posteriorly, and made entirely of silken secretion. The walls of the case are reinforced by ridges or ribs of silk. In *M. ponta* there is a pair of external ventrolateral ridges (B); in *M. ayama* there is a series of longitudinal and circular ridges (Ross 1944, fig. 558). Length of larval case up to 2.5 mm.

BIOLOGY Larvae occur on rocks in rapid sections of rivers and streams, generally in running waters of rather large size. The attenuate head suggests some specialized feeding behaviour, although gut contents of larvae (3) examined were almost entirely fine organic particles; the narrowed anterior part of the head might also be related to construction of the unusual cases characteristic of *Mayatrichia* species. Emergence of adults was recorded in Illinois from June through early September.

Mayatrichia ponta (Oklahoma, Murray Co., 7-8 May 1970, ROM 700320)
A, larva, lateral x76, segments VIII, IX, and anal prolegs enlarged; B, case, ventrolateral x39; C, fore leg, lateral; D, middle leg, lateral; E, hind leg, lateral; F, head and thorax, dorsal

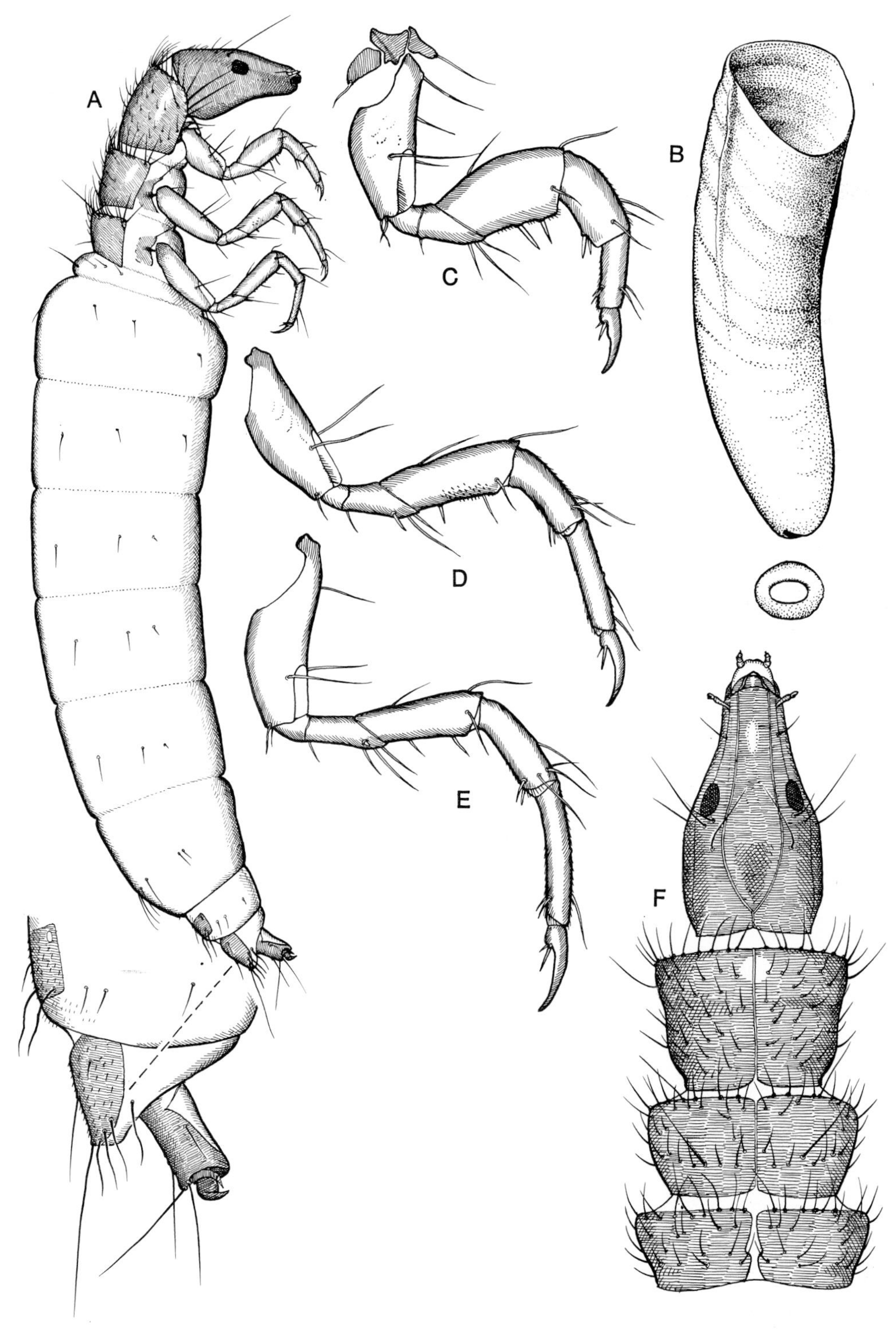
A
B
C
D
E
F

7.8 Genus **Neotrichia**

DISTRIBUTION AND SPECIES Species of *Neotrichia* occur only in the Neotropical and Nearctic regions. Sixteen species are known north of Mexico, and are recorded over much of the continent as far west as Oregon and north to Saskatchewan and Maine.

Larvae were associated for several species of *Neotrichia* by Ross (1944), but reliable characters for their separation were not found.

MORPHOLOGY *Neotrichia* includes the smallest North American caddisflies. Final-instar larvae are similar to those of *Mayatrichia*, both genera distinctive in having long and slender tarsi on the middle and hind legs, the tibia of the first legs also slender and without a ventral lobe; and they are further distinctive in having the anal prolegs projecting free from the body. *Neotrichia* larvae can be distinguished by the lateral fringe of abdominal setae, and by the longer, sparser setae on the thoracic terga (E). The legs in at least some *Neotrichia* larvae (B-D) are relatively longer and slenderer than in *Mayatrichia*, and the intersegmental grooves of the abdomen more clearly defined (A). Length of larva up to 2 mm.

CASE It is likely that any difficulty in distinguishing *Neotrichia* from *Mayatrichia* larvae can be resolved by the larval cases. Cases known for *Neotrichia* (F) are covered with tiny sand grains, although in general shape they are cylindrical and slightly tapered as in *Mayatrichia*. Length of larval case up to 2.5 mm.

BIOLOGY As in *Mayatrichia*, larvae occur on rocks in rapid sections of rivers and streams. In Ontario we have taken larvae of both *Neotrichia* and *Mayatrichia* at the same site.

Neotrichia sp. (Ontario, Hastings Co., 24 June 1973, ROM 730110)
A, larva, lateral x62; B, fore leg, lateral; C, middle leg, lateral; D, hind leg, lateral; E, head and thorax, dorsal; F, case, ventrolateral x29

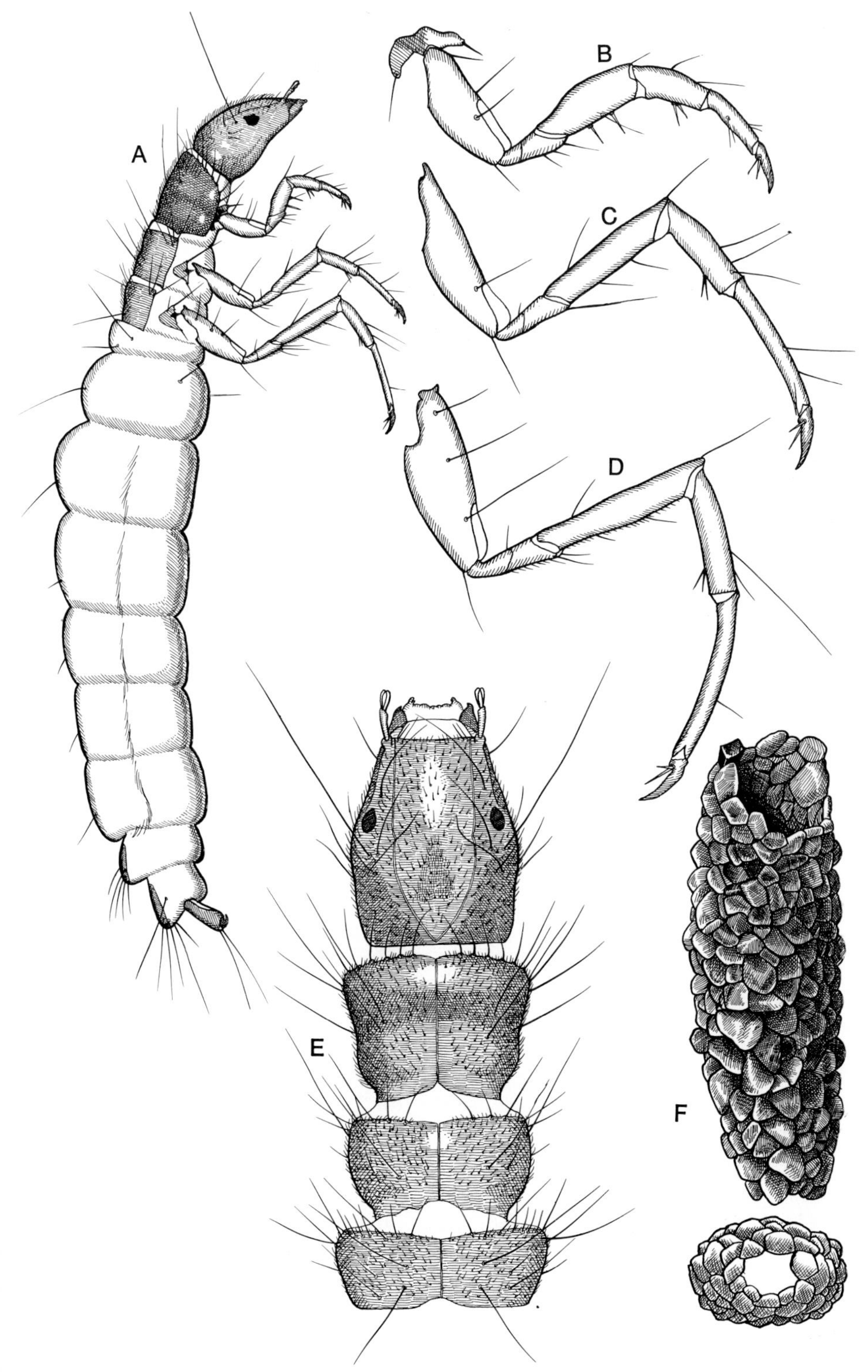
A
B
C
D
E
F

7.9 Genus **Ochrotrichia**

DISTRIBUTION AND SPECIES Species of this genus occur only in the New World. In North America north of Mexico approximately 42 species are known, largely in western and southeastern sections, with fewer species in the northeast.

Larvae were identified for some species by Ross (1944), and characterized generally by Flint (1972).

MORPHOLOGY Final-instar larvae of *Ochrotrichia* are generally similar to those of *Hydroptila*, but lack the apical filamentous gills of that genus. The meso- and meta-nota in *Ochrotrichia* are extended as anterolateral lobes on each side, the lobes extending beyond the point where the transverse sulcus terminates at the end of the anterior row of notal setae (H). Prosternal sclerites are present, and there may be three as in *Hydroptila*, or only two (F). Abdominal segments frequently bear a dorsomedian sclerotized ring (A) presumably circumscribing the chloride epithelium, and there is often a transverse sulcus on the venter of segments II–V inclusive (G). Intersegmental grooves between abdominal segments ventrally are not as prominent as in *Agraylea.* There is little difference in size of the three pairs of legs (G). Length of larva up to 5.5. mm.

CASE Most *Ochrotrichia* larvae construct cases that are similar to those of *Hydroptila*, laterally compressed and of two silken valves covered with sand grains (I,J,K), or occasionally filamentous algae. Rarely, cases consist of a single, more convex valve carried tortoise-like, the ventral valve evidently represented by a flat sheet of silk (B,C,E). Length of case up to 6 mm.

BIOLOGY *Ochrotrichia* larvae live in running waters of wide diversity from rivers and warm streams to cold spring runs and evidently in temporary streams as well (Ross 1944). Our series with single-valve cases was collected in California from a population where larvae were living in a thin film of water flowing over rocks of a small spring stream; similar larvae found by Vaillant (1965) in both east and west were said to feed by scraping diatoms from the rock surface.

REMARKS *Metrichia*, recognized by some authors as a genus, was placed as a subgenus of *Ochrotrichia* by Flint (1972), to a large extent because larvae of the two groups were indistinguishable. Taxonomy of adults was reviewed by Denning and Blickle (1972).

Ochrotrichia sp.
A–F (larva with univalve case; California, El Dorado Co., 18 July 1966, ROM)
A, larva, dorsal x40, dorsal abdominal sclerite enlarged; B, case, dorsal x23; C, case with larva, ventrolateral; D, fore leg, lateral; E, case, diagrammatic cross-section, dorsal surface uppermost; F, prothorax, ventral
G–K (larva with bivalve case; California, Marin Co., 7 June 1961, ROM)
G, larva, lateral x47; H, meso- and meta-nota, lateral; I, case, diagrammatic cross-section near end; J, case with larva, lateral x21; K, case, ventral

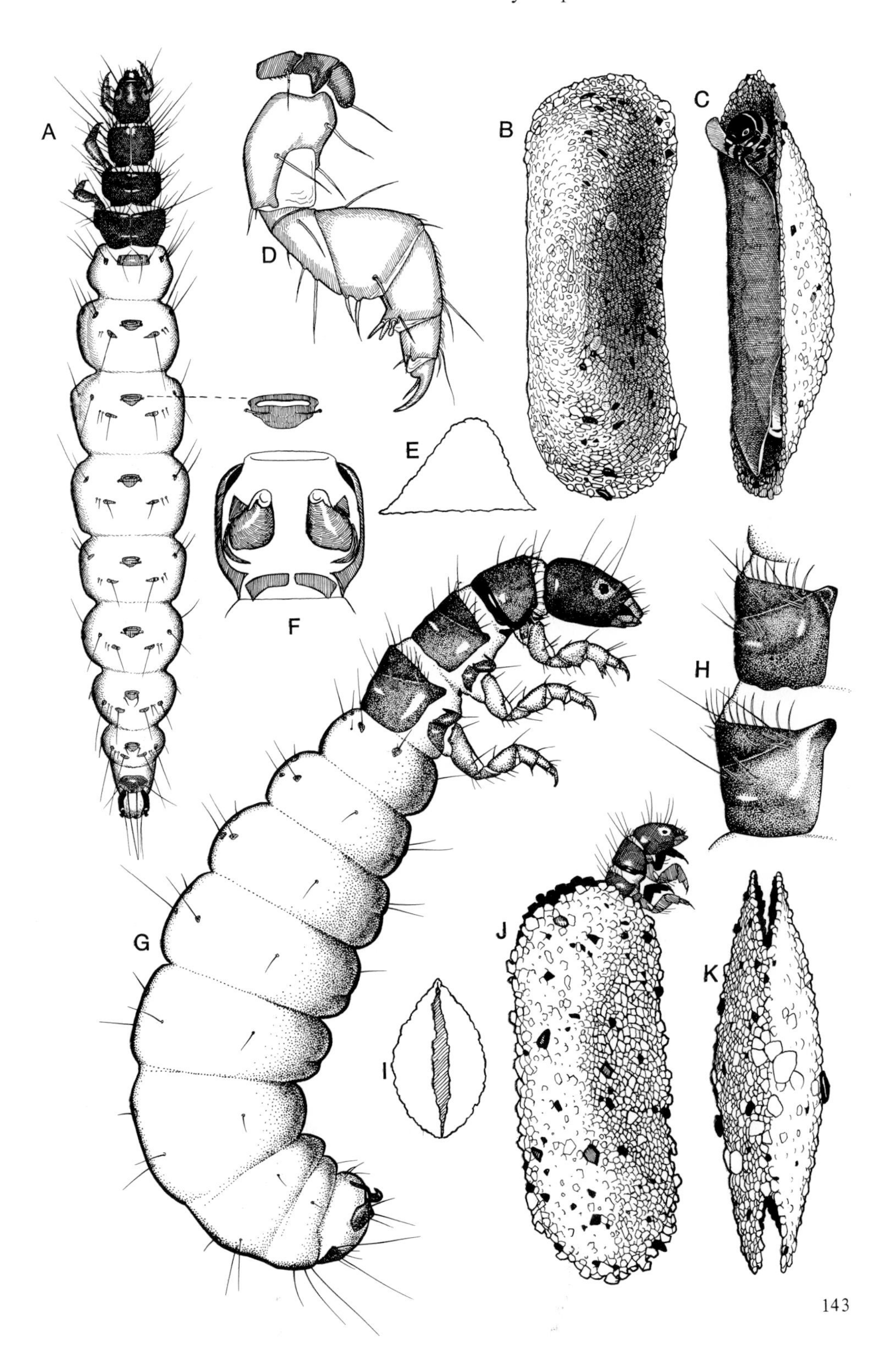
A
B
C
D
E
F
G
H
I
J
K

7.10 Genus **Orthotrichia**

DISTRIBUTION AND SPECIES *Orthotrichia* is a relatively small genus of wide distribution, species having been recorded from the Nearctic, Palaearctic, Oriental, and Ethiopian regions. Six species are now known in North America, largely in the eastern half of the continent; one species, *O. cristata* Morton, shows a westerly extension of this range to Montana and British Columbia.

Diagnostic characters for final-instar larvae of *Orthotrichia* were provided by Ross (1944, 1959), and morphological data for all stages of the European species *O. tetensii* (Kolbe) by Nielsen (1948).

MORPHOLOGY In final-instar larvae of *Orthotrichia* tarsi of the middle and hind legs are slender and elongate (E,F,G), only slightly longer than the tarsal claw, and the fore tibia (E) lacks the ventral enlargement and stout apical setae of such genera as *Hydroptila*; near the distal end of each tarsus is a flattened plate (F) apparently derived from a modified spur. The fore coxa bears rows of spines (C). The labrum is strangely asymmetrical with a median sclerotized point (D). According to Nielsen (1948), larvae of *O. tetensii* have three caudal gill filaments, much as in *Hydroptila*, but these become atrophied and disappear early in the fifth instar. Length of larva up to 3.5 mm.

CASE Cases of final-instar *Orthotrichia* larvae are distinctive from any others in the North American fauna. The case (B,I) is made of silk alone, and although basically purse-like in design with slit-like openings at each end between two valves, in cross-section (H) it is actually depressed. Characteristic of *Orthotrichia* cases alone are the dorsolateral, longitudinal ridges or keels on each side of the mid-dorsal line. Length of larval case up to 3.5 mm.

BIOLOGY Larvae of *Orthotrichia* live in submerged beds of aquatic plants along lake margins or in slowly flowing sections of rivers. Nielsen (1948) observed that the pointed labrum enabled larvae of *O. tetensii* to puncture and feed on the contents of more robust algal filaments than larvae of other genera could.

Orthotrichia sp. (Ontario, Carleton Co., 28 June 1971, ROM 710492)
A, larva, lateral x43, segment IX and anal proleg enlarged; B, case, dorsal x24; C, fore coxa, ventral; D, labrum, dorsal; E, fore leg, lateral; F, middle leg, lateral, end of tarsus enlarged; G, hind leg, lateral; H, case, diagrammatic cross-section; I, case, lateral

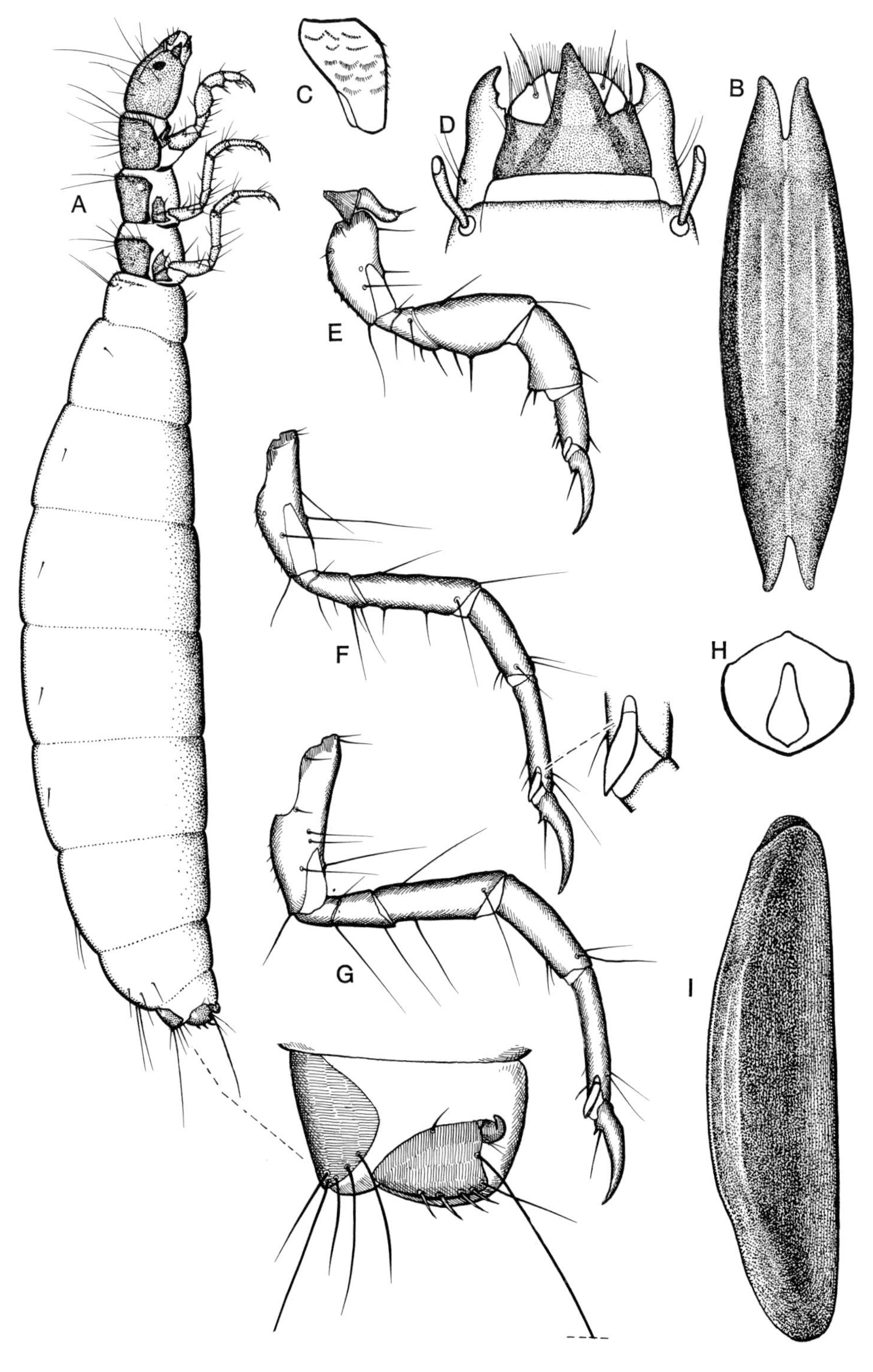
A
B
C
D
E
F
G
H
I

7.11 Genus **Oxyethira**

DISTRIBUTION AND SPECIES Species of *Oxyethira* have been recorded from all six major faunal regions of the world. About 30 species are known in North America, and they range widely over much of the continent from east to west.

Larvae were associated with species in the genus by Ross (1944), and a pupal case illustrated in fig. 498.

MORPHOLOGY Larvae are distinguished primarily by the unusually long and slender legs of the last two thoracic segments - each about 2½ times longer than the fore leg and very different from it in structure (A). Also distinctive is the long distoventral lobe of the fore tibia (D). Clusters of small, sharp spines are borne at various points on the legs, as on the tarsus in D. The antennae are long (C). Length of larva up to 4 mm.

CASE In North America final-instar *Oxyethira* larvae can be immediately recognized by their flattened, bottle-shaped cases made entirely of silk with no other materials added (B,G). The head and thorax are extended through the thickened anterior neck of the case, as illustrated; the truncate posterior end is not closed, although the free edges usually lie close together. It is to these free edges that the larva adds silk to accommodate its growing abdomen. A detailed account of case construction by the European *O. costalis* Curtis was given by Nielsen (1948); early fifth-instar larvae fashion a mass of detritus into a ring that serves as the form for spinning the anterior end of the case, with increments added to the posterior end as required. Length of larval case up to 5 mm.

BIOLOGY Larvae of *Oxyethira* live in lakes and other standing waters or in areas of slow current in rivers. They frequent submerged beds of aquatic plants where, according to Nielsen (1948), they feed on filamentous algae by puncturing the cells and eating the contents in the same manner as larvae of *Agraylea* and *Hydroptila* do; diatoms and entire algal filaments were also found in larval guts (Siltala 1907b). Two generations per year were recorded in Denmark for *O. costalis* (Nielsen 1948).

Oxyethira sp. (Wyoming, Teton Co., 14 Sept. 1966, ROM)
A, larva, lateral x49; B, case with larva, lateral x15; C, head, dorsal; D, fore leg, lateral, distal portion enlarged; E, middle leg, lateral; F, hind leg, lateral; G, case, ventral

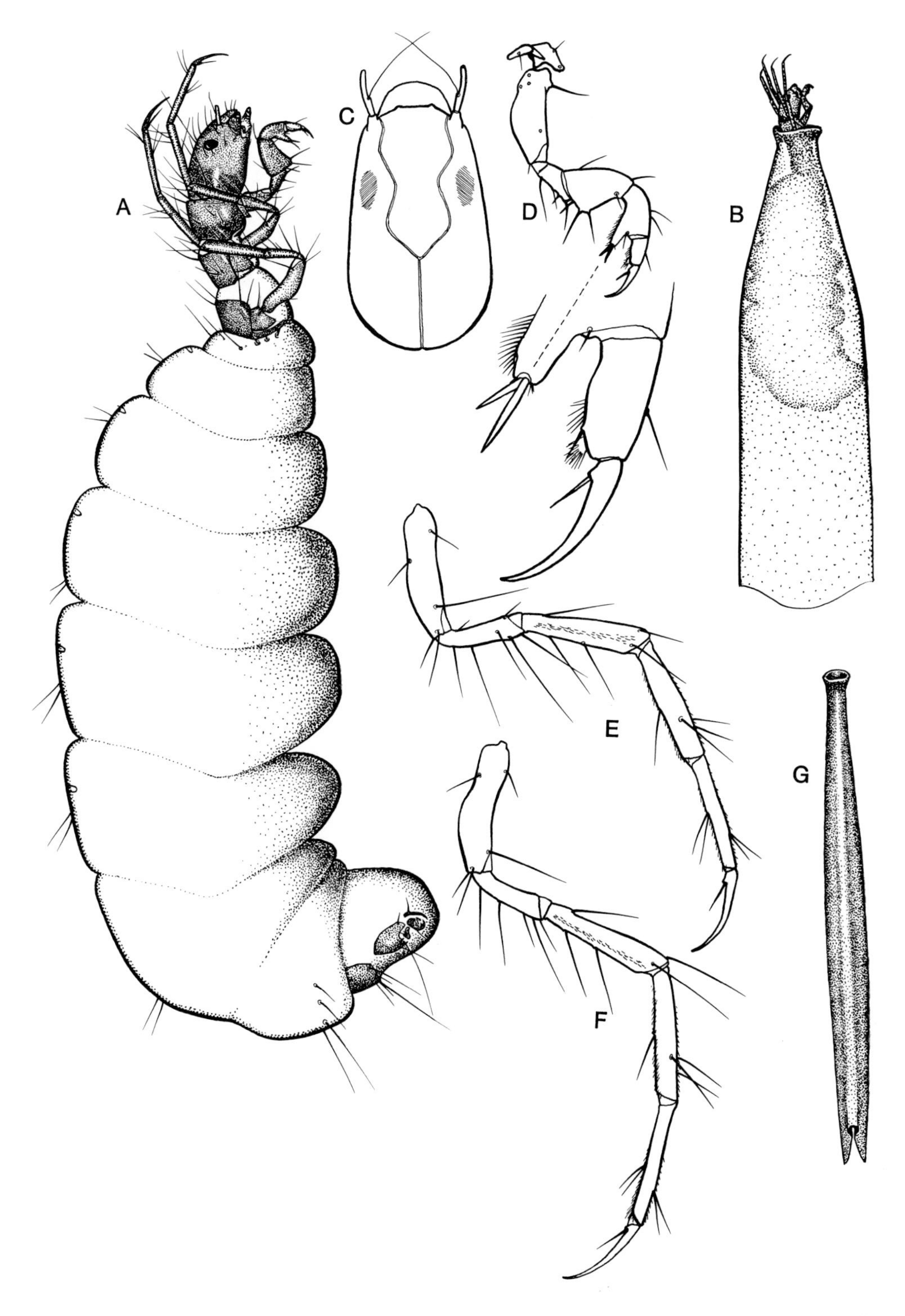
A
B
C
D
E
F
G

7.12 Genus **Palaeagapetus**

DISTRIBUTION AND SPECIES *Palaeagapetus* is confined to mountainous sections of North America where it is represented by three species. *P. nearcticus* Banks and *P. guppyi* Schmid occur in the west from California to Vancouver Island. *P. celsus* (Ross) is recorded from the Appalachian Mountains of North Carolina and Tennessee, and the Laurentians of Quebec (Roy and Harper 1975); since colonies are evidently local in occurrence, it is likely that *P. celsus* will be found in montane areas between these two extremes.

The larva of *P. celsus* was described by Flint (1962a). Similar larvae from Oregon were provided for study here by N.H. Anderson.

MORPHOLOGY Larvae of *P. celsus* are depressed dorsoventrally, and are distinguished from all other North American hydroptilids by the truncate fleshy tubercles on each side of abdominal segments I–VIII (A). All legs are similar in structure (C–E). Length of larva up to 5 mm.

CASE The case of final-instar *P. celsus* is carried with the flattened surfaces held dorsoventrally (B); but, as in many other hydroptilids, the essential design is two silken valves with slit-like openings at front and rear. Cases of *P. celsus* are distinctive, however, in being covered with small pieces of liverwort. Length of case up to 6 mm.

BIOLOGY Larvae of *P. celsus* occur in small, cold seepage springs in growths of liverwort on rocks and wood, often above the water surface. Other observations were provided by Flint (1962a).

REMARKS This is the only North American genus in the Ptilocolepinae, considered the most primitive subfamily in the Hydroptilidae.

Palaeagapetus celsus (Tennessee, Sevier Co., 20 May 1970, ROM)
A, larva, dorsal x30; B, case with larva, dorsal x12; C, fore leg, lateral; D, middle leg, lateral; E, hind leg, lateral; F, segment IX and anal proleg, lateral; G, case, lateral

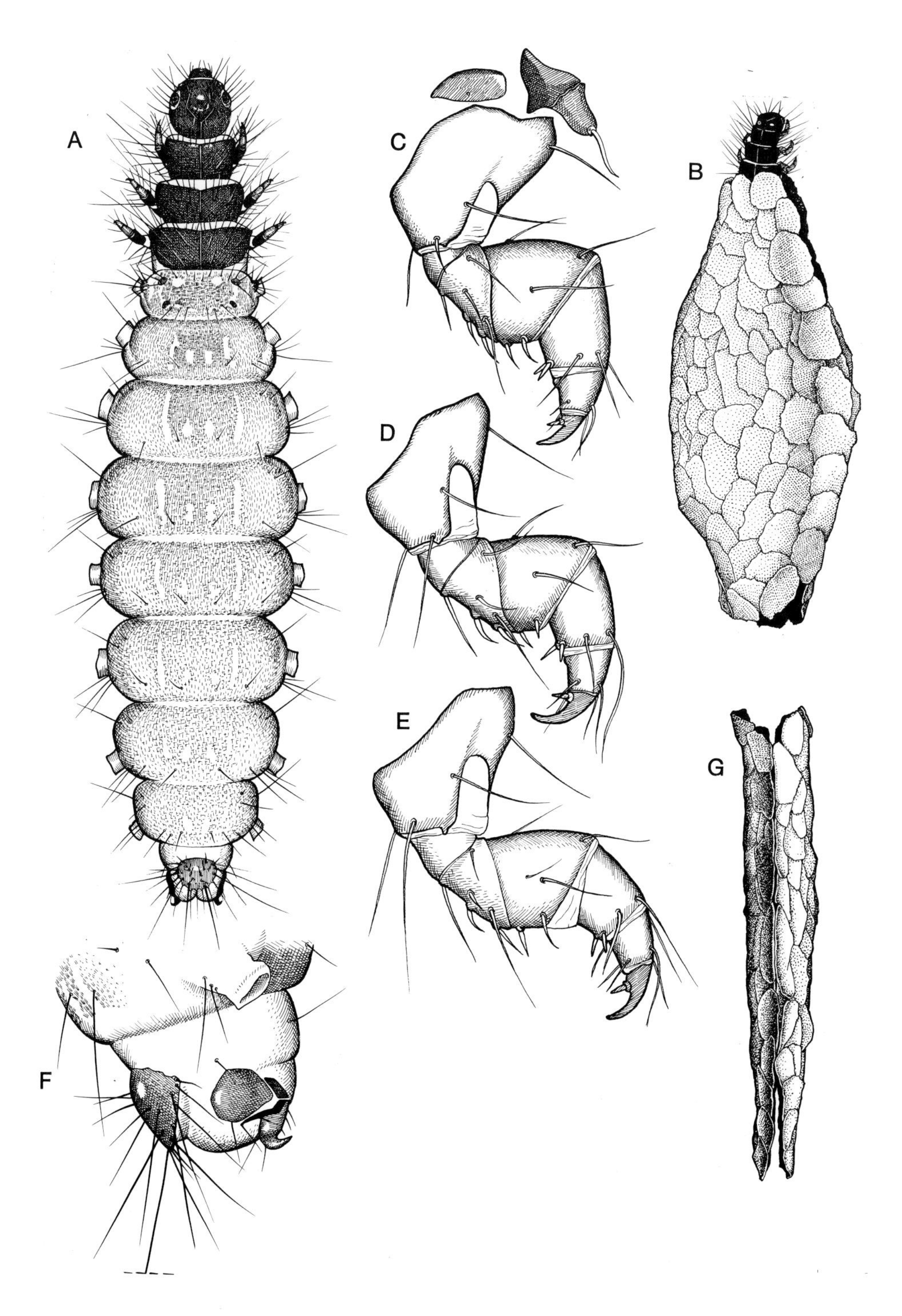
A
B
C
D
E
F
G

7.13 Genus **Stactobiella**

DISTRIBUTION AND SPECIES *Stactobiella* is a small Holarctic genus with three North American species recorded over much of the continent west to the Rocky Mountain states, and north to Minnesota, Ontario, and Maine. Although the genus was not previously recorded from Pacific coast states, we have collected larvae, illustrated here, in the Coast Range of Oregon. Populations of all of the *Stactobiella* species are evidently local in occurrence.

Larvae were associated for *S. palmata* (Ross) by Ross (1944), under the generic name *Tascobia.*

MORPHOLOGY Final-instar larvae are generally similar to those of other genera in the Hydroptilinae, especially to *Dibusa* because of the stout, sharply curved tarsal claws, each with a thickened basal spur (C–E). In larvae examined, *Stactobiella* have smaller abdominal setae than do *Dibusa*, and the setae lack the sclerotized basal areas of that genus (A). Although basal spurs of the tarsi in both genera are unusually stout and blunt, the spurs in *Stactobiella* are nearly half as long as the claw (E), but relatively much smaller in *Dibusa*. Length of larva up to 3 mm.

CASE Also typical for the Hydroptilinae, the larval case of *Stactobiella* consists of two silken valves with little or no additional materials, and with a slit-like opening at each end (B,F). Although no detailed observations have been made on case-building, it is likely that the larva enlarges the case by slitting the seams, adding silk to the free edges, and binding them together again. The case is carried vertically. Length of larval case at least up to 3 mm.

BIOLOGY Larvae of *Stactobiella* live on rocks in rapid, often small, streams. According to Ross (1944), pupation and emergence of adults of *S. palmata* occur in early spring in Illinois, and therefore it seems likely that this species overwinters as final-instar larvae in cases.

REMARKS Diagnostic characters for males and females of two species of *Stactobiella* were given by Ross (1944) under the name *Tascobia*, and for all males by Ross (1948a).

Stactobiella sp. (Oregon, Benton Co., 10 April 1964, ROM)
A, larva, lateral x63; B, case, lateral x31; C, fore leg, lateral; D, middle leg, lateral; E, hind leg, lateral, tarsal claw enlarged; F, case, dorsal

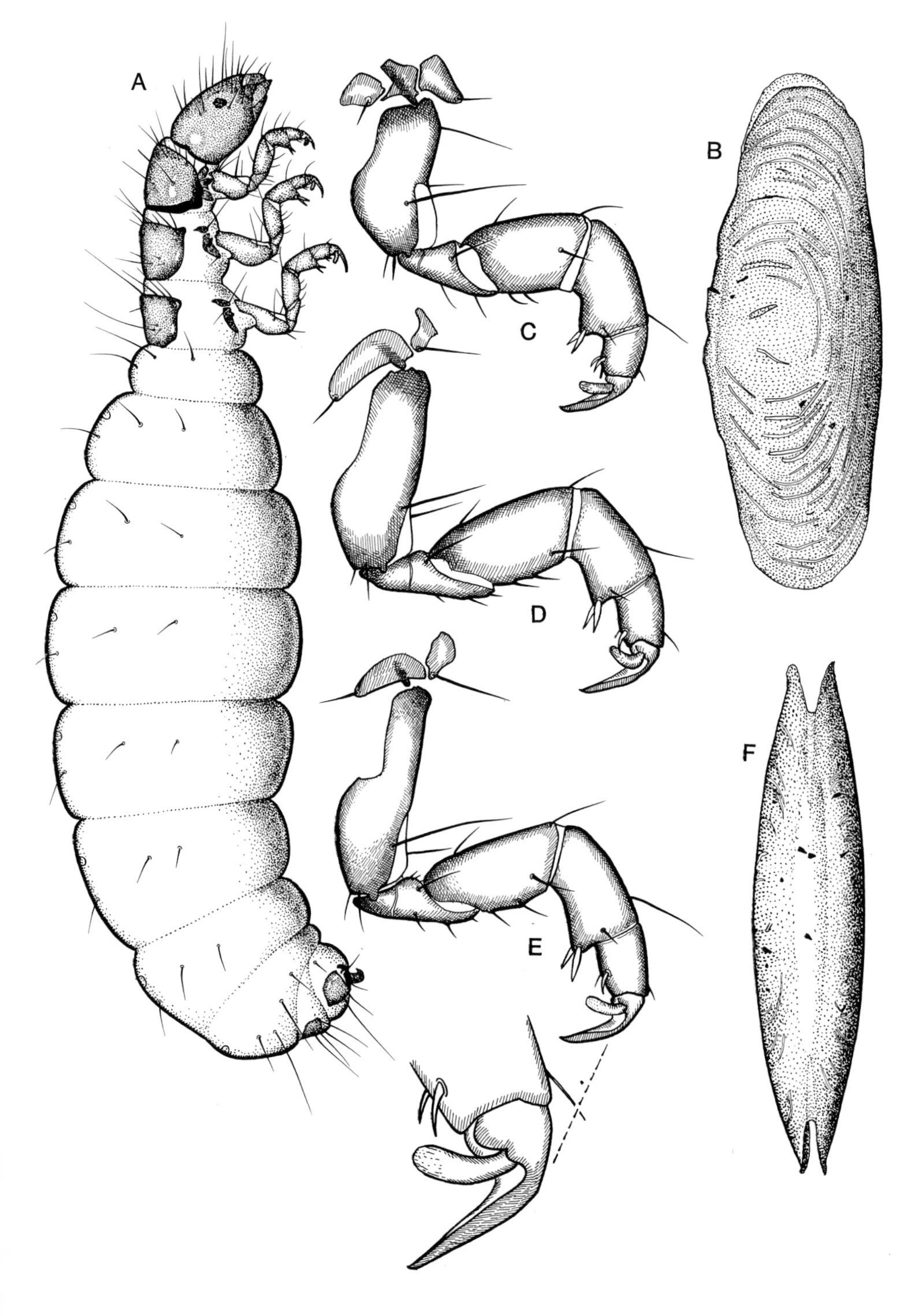
A
B
C
D
E
F

7.14 Genus Zumatrichia

DISTRIBUTION AND SPECIES The genus is represented only in the New World, and mainly in Central America and Mexico. *Zumatrichia notosa* (Ross), the only species known to occur north of Mexico, has been collected in Montana (Flint 1970).

The larva of *Z. notosa* is unknown but that of a Dominican species provided by O.S. Flint, Smithsonian Institution, is illustrated here. Taxonomic data available for larvae and adults in the genus were summarized by Flint (1970).

MORPHOLOGY From the larva of *Z. antilliensis* Flint, illustrated here, it is apparent that larvae of *Z. notosa*, when found, will be generally similar to those of *Leucotrichia*, but probably distinguishable by the stout setae on the dorsal sclerite of segment IX (D), as indicated in the key. The larva illustrated also differs from *Leucotrichia* in having bifid tarsal claws (C), and this may prove to be diagnostic for *Z. notosa* as well. As in *Leucotrichia*, abdominal segments V and VI become enormously enlarged in the fifth instar, and anterior portions of the frontoclypeal sutures are lacking. Length of larva up to 3 mm.

CASE The larval cases of *Zumatrichia* species (B,E) are generally similar to those of *Leucotrichia*, and there appears to be no consistent basis for distinguishing between them. The cases are constructed entirely of silk in a fixed position on rocks. Length of larval case up to 4 mm.

BIOLOGY Larvae of *Zumatrichia* species live in fast-flowing parts of running waters, evidently large rivers for the most part (Flint 1968a). Adults of *Z. notosa* were collected along the Missouri River in Montana. Larvae graze upon periphyton reached by extending head and thorax through the small case openings.

Zumatrichia antilliensis (Dominica, 5 Dec. 1964, USNM)
A, larva, dorsal x55, dorsal abdominal sclerite enlarged; B, case, dorsal x21; C, hind leg, lateral, tarsal claw enlarged; D, segments VIII, IX, and anal proleg, lateral; E, case on rock, lateral

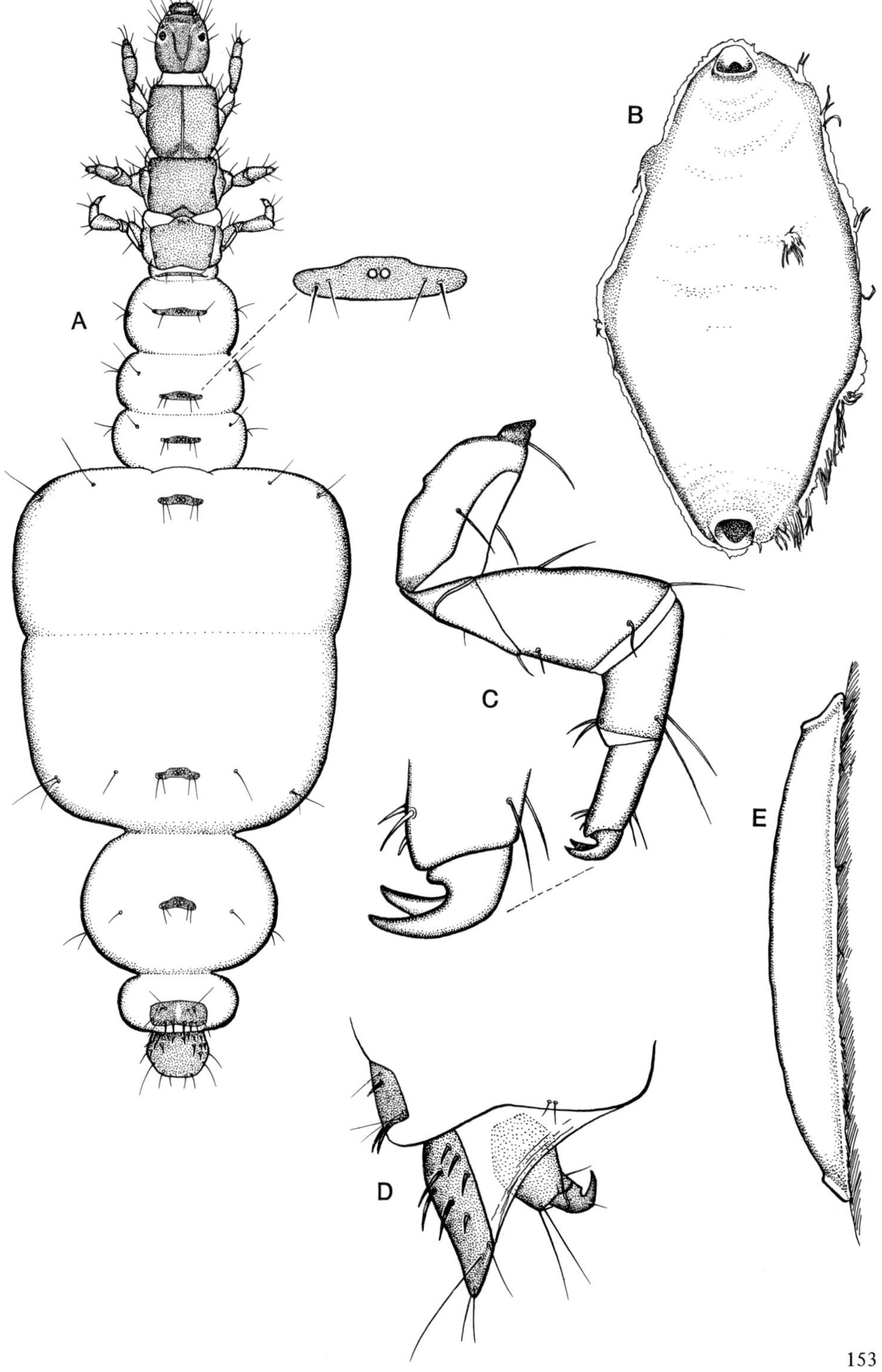
A
B
C
D
E

8
Family Lepidostomatidae

The family is represented in all faunal regions except the Neotropical and Australian. Some 70 species are now recognized on this continent north of Mexico. These are assigned to the two Nearctic genera currently recognized, *Lepidostoma* and *Theliopsyche*, although up to 15 genera have been proposed for the North American species, largely on the basis of what appear to be secondary sexual characters of the males (Ross 1946).

The majority of lepidostomatid larvae in North America are generally concordant with the one illustrated here as *Lepidostoma. Theliopsyche*, for which larvae are described here for the first time, has several distinctive larval characters. But in some of our collections from the west, the density of setation on meso- and meta-nota is much greater than in typical *Lepidostoma*, or the shape of the head is flattened much as in *Theliopsyche*, or bulbous as in the limnephilid genus *Ecclisocosmoecus*; different larval cases (see below) are associated with some of these distinctive morphological characters. For the present, all of these larval types, apart from *Theliopsyche*, are assigned provisionally to *Lepidostoma* pending further data. It can be noted that adults of the eastern *Lepidostoma togatum* (Hagen) were assigned to the genus *Goerodes* by Corbet, Schmid, and Augustin (1966) and by Nimmo (1966); associated larval material of that species in the ROM collection indicates little discordance from typical *Lepidostoma*. There is little doubt that data from larvae will contribute substantially to resolving the taxonomy of the Lepidostomatidae when their identity has been established for more species.

Larvae of the Lepidostomatidae are generally similar to those of the Limnephilidae but are distinguished by the position of the antennae close to the eyes, by the absence of a median dorsal hump on abdominal segment I, and also by the absence of chloride epithelia enclosed by sclerotized oval rings on the abdominal segments. North American lepidostomatid larvae that we have studied possess a dense patch of pectinate spines on the distal end of the hind coxa (Fig. 8.1A). The mandibles have separate teeth, and the prosternal horn is well developed. Abdominal gills are single and arranged in dorsal and ventral rows only, or are absent; the lateral abdominal fringe is sparse and often absent, and lateral tubercles occur on several segments. Segment VIII bears a broad lobe at each side. Anal papillae are usually present.

8 Family **Lepidostomatidae**

The typical larval case associated with this family is four-sided, and constructed of quadrate pieces of bark and leaves (Fig. 8.1C). In four-sided cases of the Lepidostomatidae the component pieces are squarish, distinguishing them from cases of the Brachycentridae in which the pieces are elongate. But the North American lepidostomatid fauna also includes larvae that build several other types of cases - of bark and leaf pieces arranged irregularly, of leaf pieces arranged spirally (Fig. 8.1G), of pieces of plant stems arranged transversely (Fig. 8.1H), and of sand grains, too (Fig. 8.1F).

Most larvae of this family occur in small, cold streams. In larger rivers they tend to frequent sections of slower current, and they are also found along lake shores. Usually the larvae are associated with accumulations of leaves and other plant materials, and from the studies that have been made it appears that lepidostomatid larvae are part of the detritivorous fauna.

Taxonomic data for adults of the North American Lepidostomatidae were reviewed by Ross (1946), but many species have been described since then; data for males of the California species assigned to *Lepidostoma* were summarized by Denning (1956), and a review of the *vernalis* group of that genus given by Flint and Wiggins (1961).

Key to Genera

1 Head with ventral apotome as long as, or longer than, median ventral ecdysial line (Fig. 8.1E); cases of plant pieces usually 4-sided (Fig. 8.1C), but pieces also arranged irregularly, transversely (Fig. 8.1H), or spirally (Fig. 8.1G); cases also of sand grains (Fig. 8.1F). Widespread **8.1 Lepidostoma**

Head with ventral apotome shorter than median ventral ecdysial line (Fig. 8.2D); cases of sand grains. Eastern **8.2 Theliopsyche**

8.1 Genus **Lepidostoma**

DISTRIBUTION AND SPECIES *Lepidostoma* is a Holarctic genus to which are assigned 65 North American species known north of Mexico. The group is widespread over much of the continent, although most of the species are western.

Larvae have been described for very few species: *L. griseum* (Banks) by Sibley (1926); *L. liba* Ross by Ross (1944); *L. bryanti* (Banks) (as *L. wisconsinensis* Vorhies) by Vorhies (1909). We have associated larval material for about two dozen other species.

MORPHOLOGY The relatively long ventral apotome of the head (E) used in the key as a diagnostic character for *Lepidostoma* subsumes a considerable range in larval morphology, as indicated in the introduction to this family. Most North American larvae are, however, concordant with the one illustrated here (A–E). Length of larva up to 12.5 mm.

CASE Cases of late-instar larvae in most species of *Lepidostoma* are usually four-sided and constructed of quadrate pieces of bark or leaf (C); observations indicate that in at least some of these species early-instar cases are of sand grains and cylindrical, with the four-sided case of bark and leaves built on the anterior end during later instars (Vorhies 1909; Hansell 1972). Final instars in some other species assigned to *Lepidostoma* have cases of plant materials placed spirally or transversely (G,H), or sand grains (F). Length of larval case up to 15 mm.

BIOLOGY *Lepidostoma* larvae most often occur in cool springs and streams, usually in areas of little current, but they are also found in lakes and in temporary streams (Denning 1958a; Mackay 1969). Food studies of *Lepidostoma* larvae show that detritus comprises the major part of what is ingested (Chapman and Demory 1963; Winterbourn 1971a; Anderson and Grafius 1975), and larvae are usually found in association with these materials; larvae are also attracted to dead fish (Brusven and Scoggan 1969). Life history data published by Anderson (1967b) and Mackay (1969) show a single generation per year for five species studied, and indicate as well that there is temporal and spatial separation among larvae of different species living in the same habitat.

Lepidostoma spp.
A–E (Ontario, Hastings Co., 15 Oct. 1968, ROM)
A, larva, lateral x17, apex of hind coxa enlarged; B, head and thorax, dorsal; C, case x10; D, eye and antenna, lateral; E, head, ventral
F (Arizona, Cochise Co., 23 June 1966, ROM), case x5
G (Oregon, Clatsop Co., 14 July 1963, ROM), case x4
H (Oregon, Douglas Co., 4 July 1961, ROM), case x5

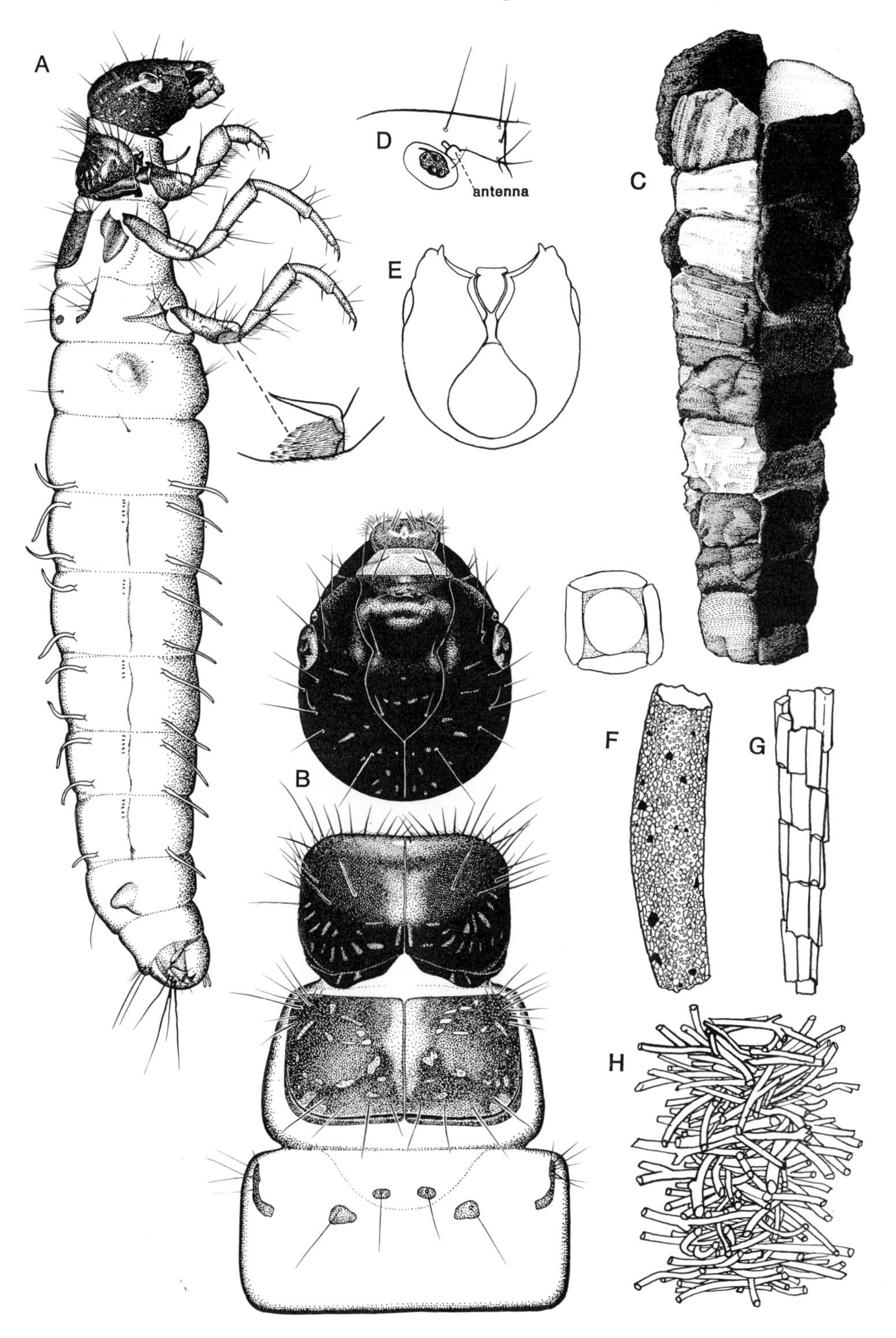
A
B
C
D
antenna
E
F
G
H

8.2 Genus **Theliopsyche**

DISTRIBUTION AND SPECIES The genus *Theliopsyche* is represented only in the Nearctic region, where five species are recorded from the east: Tennessee, North Carolina, New Jersey, New York, New Hampshire, and Quebec. Since the species are exceedingly local in occurrence, it is likely that colonies of these and perhaps additional ones as well will be discovered over much of the eastern part of the continent.

Larvae were not previously known for any species of *Theliopsyche.* We collected an associated series of *T. melas* Edwards at the type locality in Tennessee, providing the basis for the data presented here.

MORPHOLOGY The genus is consistent with others of the family in having larvae with antennae situated close to the eye, with abdominal gills in single filaments, and without a median dorsal hump on abdominal segment I. In the larva of *T. melas* the dorsum of the head is flattened and bordered by a carinate ridge that is particularly pronounced laterally; although this character separates it from most *Lepidostoma* larvae, we have collected in western North America several series of larvae, here assigned to *Lepidostoma*, in which the head is dorsally flattened. Study of our larval material leads to the prediction that *Theliopsyche* can best be distinguished by the length of the ventral apotome of the head which is shorter than the median ventral ecdysial line (D). Length of larva up to 6.5 mm.

CASE The larval case of *T. melas* (C) is of smooth outline, constructed of fine rock fragments, curved and slightly tapered. Length of larval case up to 7 mm.

BIOLOGY Larvae and pupae of *T. melas* were collected in the clean gravel bed of a small spring run. Localities for other species of *Theliopsyche* indicate that they probably live in similar habitats.

Theliopsyche melas (Tennessee, Franklin Co., 14-15 May 1970, ROM 700337)
A, larva, lateral x21; B, head and thorax, dorsal; C, case x 15; D, head, ventral; E, sclerite of segment IX, dorsal

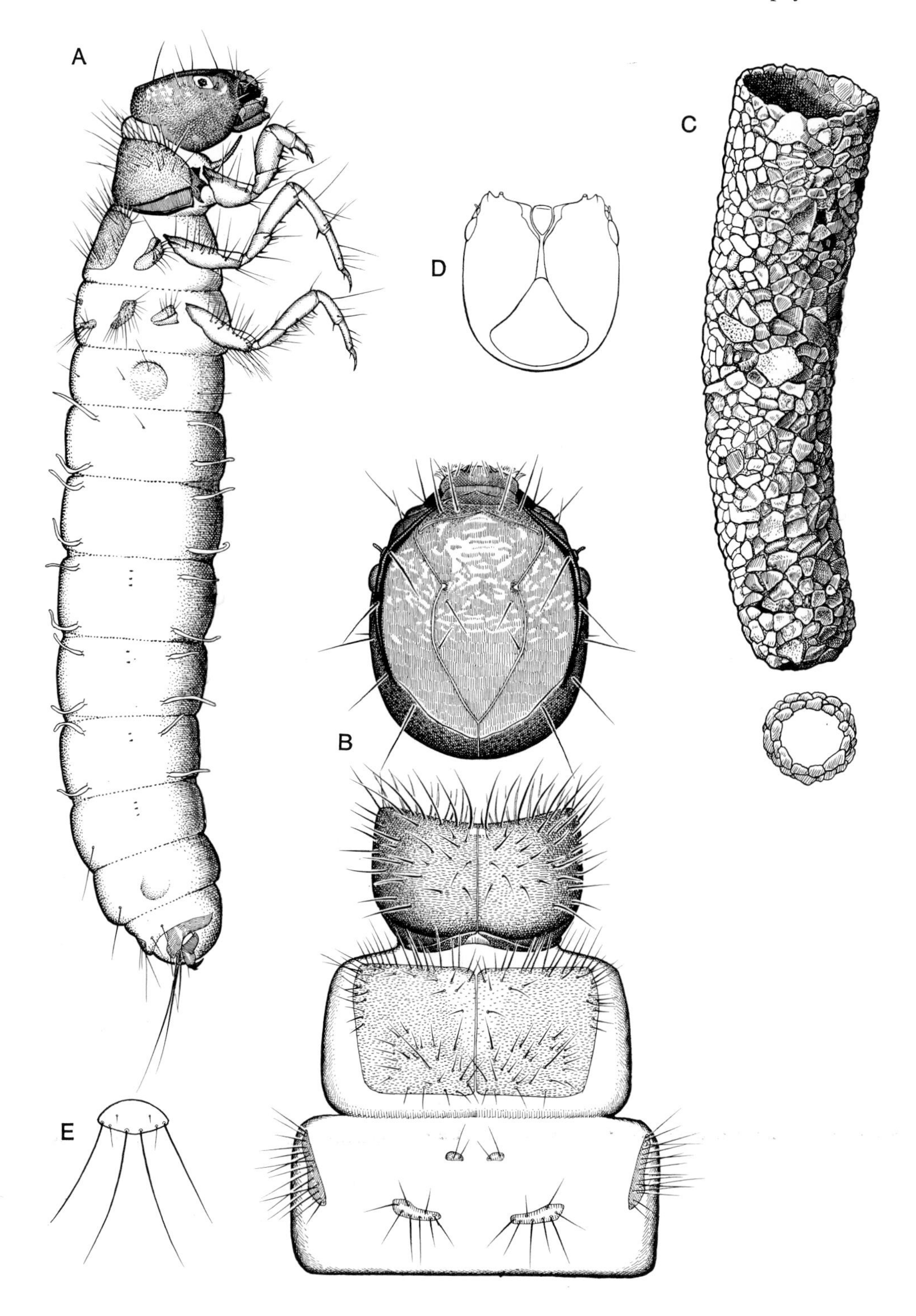
A
B
C
D
E

9
Family Leptoceridae

The Leptoceridae are a large and flourishing family represented in all of the world's faunal regions. Approximately 100 species are known north of Mexico in the seven Nearctic genera; they are common and widespread as a group and often abundant locally.

Almost all Nearctic leptocerid larvae can be distinguished by their long antennae; the only exceptions are some sponge-feeding species of *Ceraclea* in which the antennae have probably become secondarily shortened. Leptocerid larvae are further distinguished by additional unpigmented lines of weakness at which primary sclerites usually subdivide at ecdysis. All of the Nearctic larvae of the family have a subocular line along each side of the head, and some species of *Ceraclea* have an additional supraocular line (Fig. 9.1D); there are pronotal lines of weakness in *Leptocerus, Mystacides,* and *Nectopsyche.* The advantage of secondary separation of these sclerites could be that they are more easily ejected from the posterior opening of the pupal case after larval-pupal ecdysis, a rather unusual aspect in the behaviour of all of the Leptoceridae. I suspect that the fact that the cast larval sclerites are not retained in the posterior end of the pupal case as they are in most other families is related in some way to preventing damage to the long antennae of the pharate adult which are coiled around the anal processes of the pupa; that interpretation suffers, however, from the fact that the same behaviour occurs in the Molannidae, where adults do not have particularly long antennae. The hind legs of leptocerid larvae are much longer than the others, and the segments somewhat modified: the trochanter is lengthened, the femur is subdivided into a short proximal and a longer distal section, the tibia is lengthened and often subdivided into two parts by a constriction near the middle, and in some groups the tarsus is similarly subdivided. The mesonotal plates are lightly sclerotized and frequently have little colour. The metanotum is largely membranous, *sa*3 is usually present, *sa*1 and *sa*2 variously reduced or modified. Humps of the first abdominal segment are present, but often not prominent; characteristic spinose sclerites occur on the lateral humps in some genera. Abdominal gills are usually single, sometimes in groups or lacking. The lateral fringe is usually present, though reduced, and a line of lateral tubercles is usually present on segment VIII. Segment IX bears a small dorsal sclerite with setae. Spines occur on the base of the anal prolegs in some genera.

9 Family **Leptoceridae**

Leptocerids have penetrated most types of warmer, permanent waters in North America. Larvae of most genera seem to be omnivorous feeders, but specialization as a predator is evident in the mandibles of *Oecetis*. In several genera larvae swim through the water with their cases, hair fringes on the long hind legs improving their effectiveness as paddles. Since those seen to swim most often - *Leptocerus* and *Triaenodes* - are normally residents of plant beds near the surface, it seems likely that swimming is a means of dispersing to food resources through the upper levels independent of the bottom substrates to which most other trichopteran larvae are restricted. Larval cases are of several types and materials, but over all the architecture is characteristic for each genus.

Taxonomic problems in the genera of the Leptoceridae have made the biological literature on these insects difficult to interpret. Much of the uncertainty arises from broad application of the name *Leptocerus* to groups such as *Athripsodes* and *Setodes* which in recent years have been more narrowly defined. Taxonomic refinement has continued with the recent segregation of *Ceraclea* from *Athripsodes* (Morse 1975); a second major nomenclatural change in the North American leptocerids is synonymy of *Leptocella* with *Nectopsyche* (Flint 1974c).

Two subfamilies are recognized, the Triplectidinae which are not represented in North America, and the Leptocerinae to which all of our genera are assigned. Grouping of the Nearctic genera by tribes provides some useful insights.

ATHRIPSODINI: *Ceraclea* Ventral apotome of head trapezoidal or triangular, separating genae completely or nearly so; mandibles short and wide, teeth grouped around central concavity; maxillary palpi extending little if at all beyond anterior edge of labrum; each mesonotal plate with longitudinal, dark, curved bar; gills in clusters.

LEPTOCERINI: *Leptocerus, Nectopsyche* Ventral apotome of head triangular, not separating genae posteriorly, ventral ecdysial line indefinite posteriorly; mandibles and maxillary palpi as in Athripsodini; gills single or absent.

MYSTACIDINI: *Mystacides, Setodes, Triaenodes* Ventral apotome of head rectangular and completely separating genae in late instars, but not clearly defined in early instars of *Mystacides* and *Setodes*, and triangular in early instars of *Triaenodes*; mandibles and maxillary palpi as in Athripsodini; hind tibia in late instars usually with constriction near centre, apparently dividing it into two parts; basal segment of each anal proleg with two ventral patches of spines; gills single or absent.

OECETINI: *Oecetis* Ventral apotome of head trapezoidal, not separating genae posteriorly, ventral ecdysial line indefinite posteriorly; mandibles long and blade-like, a single cutting edge without central concavity, sharp apical tooth separated by gap from remainder of teeth; maxillary palpi extending far beyond anterior edge of labrum; labrum with many dorsal setae in addition to primary setae; gills single.

Key to Genera

1 Middle legs with tarsal claw hook-shaped and stout, tarsus curved (Fig. 9.2A); case of transparent silk, slender (Fig. 9.2C). Northern and eastern **9.2 Leptocerus**

Middle legs with tarsal claw normal, slightly curved and slender, tarsus straight (Fig. 9.1A) **2**

2 (1) Anal prolegs with sclerotized, concave plate on each side of anal opening, each plate with marginal spines, and extended into ventral lobe (Fig. 9.6G); case cylindrical, of small stones (Fig. 9.6C). Eastern **9.6 Setodes**

Anal prolegs without sclerotized spiny plates as above, although patches of spines or setae may be present (Fig. 9.7A,E) 3

3 (2) Maxillary palpi extending far beyond anterior edge of labrum, mandibles elongate and blade-like, sharp apical tooth separated by gap from remainder of teeth (Fig. 9.5E); cases of several types and materials. Widespread **9.5 Oecetis**

Maxillary palpi extending little, if any, beyond anterior edge of labrum, mandibles short and wide, teeth grouped closely together around central concavity (Fig. 9.3D) 4

4 (3) Mesonotum with pair of dark, curved bars on weakly sclerotized plates (Fig. 9.1B); abdomen disproportionately thick, gills usually in clusters (Fig. 9.1A); cases of several types and materials. Widespread **9.1 Ceraclea**

Mesonotum without pair of dark bars on plates, abdomen slenderer, gills single or entirely absent (Fig. 9.4A,B) 5

5 (4) Ventral apotome of head rectangular (Fig. 9.7D); tibia of hind leg with constriction near centre, apparently dividing it into two parts (Fig. 9.7A) 7

Ventral apotome of head triangular (Fig. 9.4E) or not apparent; tibia of hind leg usually without middle constriction, no apparent subdivision (Fig. 9.4A) 6

6 (5) Base of each anal proleg with only ventral band of small spines on each side of anal opening (Fig. 9.4F), or without spines in this position; case long and slender, of various materials. Widespread **9.4 Nectopsyche**

Base of each anal proleg with ventral patch of longer spines in addition to band of small spines (Fig. 9.3F) 7

7 (5,6) Hind legs with close-set fringe of long hairs (Fig. 9.7A); case long, slender spiral of plant pieces (Fig. 9.7C). Widespread **9.7 Triaenodes**

Hind legs with only few scattered long hairs (Fig. 9.3A); case irregular, of plant and rock materials (Fig. 9.3C). Widespread **9.3 Mystacides**

9.1 Genus Ceraclea

DISTRIBUTION AND SPECIES All North American species formerly assigned to *Athripsodes* have been transferred to *Ceraclea*, a genus of Holarctic distribution (Morse 1975). Approximately 34 species are known on this continent, and the group as a whole is widespread and common.

The larvae of *C. ancylus* (Vorhies) and *C. diluta* (Hagen) were described by Vorhies (1909), and that of *C. cancellata* (Betten) by Elkins (1936). Diagnostic characters for larvae of 11 species were given by Ross (1944); descriptions and keys to larvae of 19 species were given by Resh (1976).

MORPHOLOGY Larvae of *Ceraclea* (A) are stout-bodied, the first abdominal segment widest and the abdominal gills usually in clusters. The ventral apotome of the head is crescent-shaped and wider than long (E). In some species a parafrontal area is delimited between each frontoclypeal arm of the dorsal ecdysial line and a lightly pigmented supraocular line (B,D). Most species associated with freshwater sponges lack supraocular lines but apparently all of them have short antennae (F), the latter modification probably facilitating their burrowing in sponge colonies. Length of larva up to 12 mm.

CASE To accommodate their stout-bodied architects, cases of *Ceraclea* are unusually wide anteriorly, tapered sharply and curved toward the rear. They are made of sand grains and usually have an overhanging dorsal lip, which in some species is extended along both sides as a flange (C) similar to molannid cases. In species associated with sponges, cases are made almost entirely of silk alone (G), sometimes with pieces of sponge incorporated (Resh 1976). Length of larval case up to 13 mm.

BIOLOGY *Ceraclea* larvae inhabit both lentic and lotic waters, and most species are restricted to rather narrow limits within this range (Resh and Unzicker 1975). Larvae are usually found on bottom substrates and many species are detritus feeders (Resh 1976); but larvae of a number of *Ceraclea* species feed on freshwater sponges (Lehmkuhl 1970; Resh 1976; Resh, Morse, and Wallace 1976), and can be found burrowing in sponge colonies.

Ceraclea spp.

A-E (Saskatchewan, Lac La Ronge Prov. Park, June 1959, ROM)

A, larva, lateral x15, lateral hump sclerite enlarged; B, head and thorax, dorsal, long antenna enlarged; C, case, ventral x9, posterior opening in lateral view to show dorsal position; D, head, lateral; E, head, ventral

F, G (Minnesota, Clearwater Co., 17 Aug. 1972, ROM)

F, head, dorsal, short antenna enlarged; G, case, ventral x5

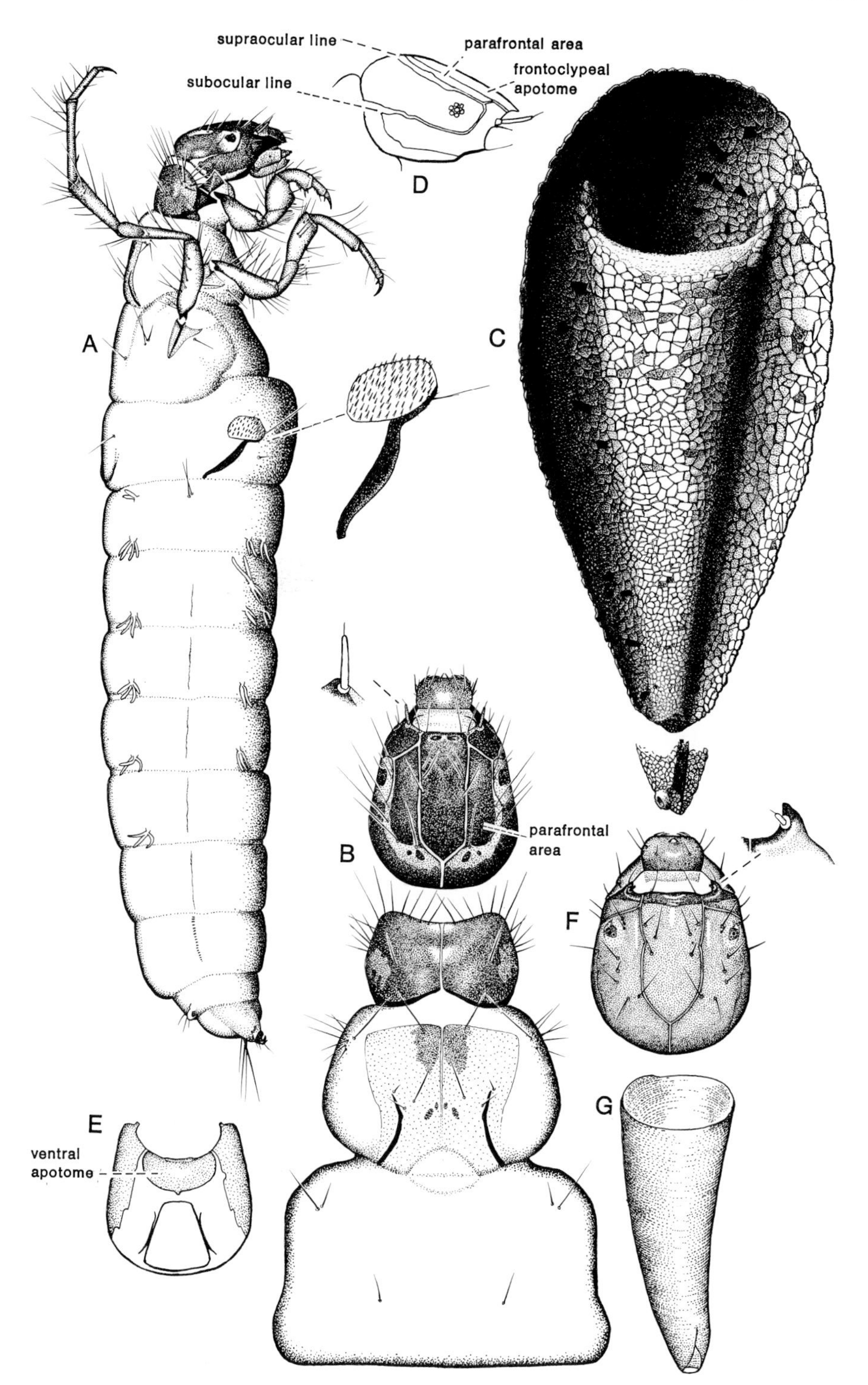
supraocular line
parafrontal area
frontoclypeal
apotome
subocular line
D
A
C
B
parafrontal
area
F
G
E
ventral
apotome

9.2 Genus **Leptocerus**

DISTRIBUTION AND SPECIES *Leptocerus* comprises a group of Holarctic and Oriental distribution; a single species, *L. americanus* (Banks), is known in North America, recorded from Minnesota to Maine and south to Texas and Tennessee.

Larval characters have been described by several workers: Vorhies (1909) and Lloyd (1921) as *Setodes grandis*; and Ross (1944).

MORPHOLOGY *Leptocerus* larvae are distinguished primarily by the structure of the apical segments of the middle legs (A): the tibia and tarsus are thickened and bear ventrally a row of teeth with stout setae; the tarsal claw is thickened and curved with two apical points. The hind legs are densely setate, furthering their effectiveness as swimming paddles. The bases of the anal prolegs bear stout setae (E). In living specimens, membranous parts of the body are bright green. Length of larva up to 7 mm.

CASE Larval cases are made entirely of silk, strongly tapered and slightly curved (C). In contrast with most other case-making genera, *Leptocerus* larvae are highly dependent on their cases; in observations I have made on early instars, they do not crawl or swim when removed from their cases and seem incapable of constructing new ones. Length of larval case up to 9.5 mm.

BIOLOGY By rapid movements of their hind legs, larvae swim rapidly among aquatic plants that grow in lentic waters. A net sample from beds of a plant such as *Ceratophyllum* placed in a large container will often yield astonishing numbers of the larvae that would not otherwise be seen. The strangely modified middle leg, especially the hooked tarsal claw, probably enables the larva to hold a firm resting position on plants. Gut contents of larvae (3) were almost entirely fine particulate matter.

Leptocerus americanus (New York, Tompkins Co., 15 July 1958, USNM)
A, larva, lateral x20, portion of middle leg enlarged; B, head and thorax, dorsal; C, case, lateral x13; D, head, ventral; E, anal prolegs, dorsal

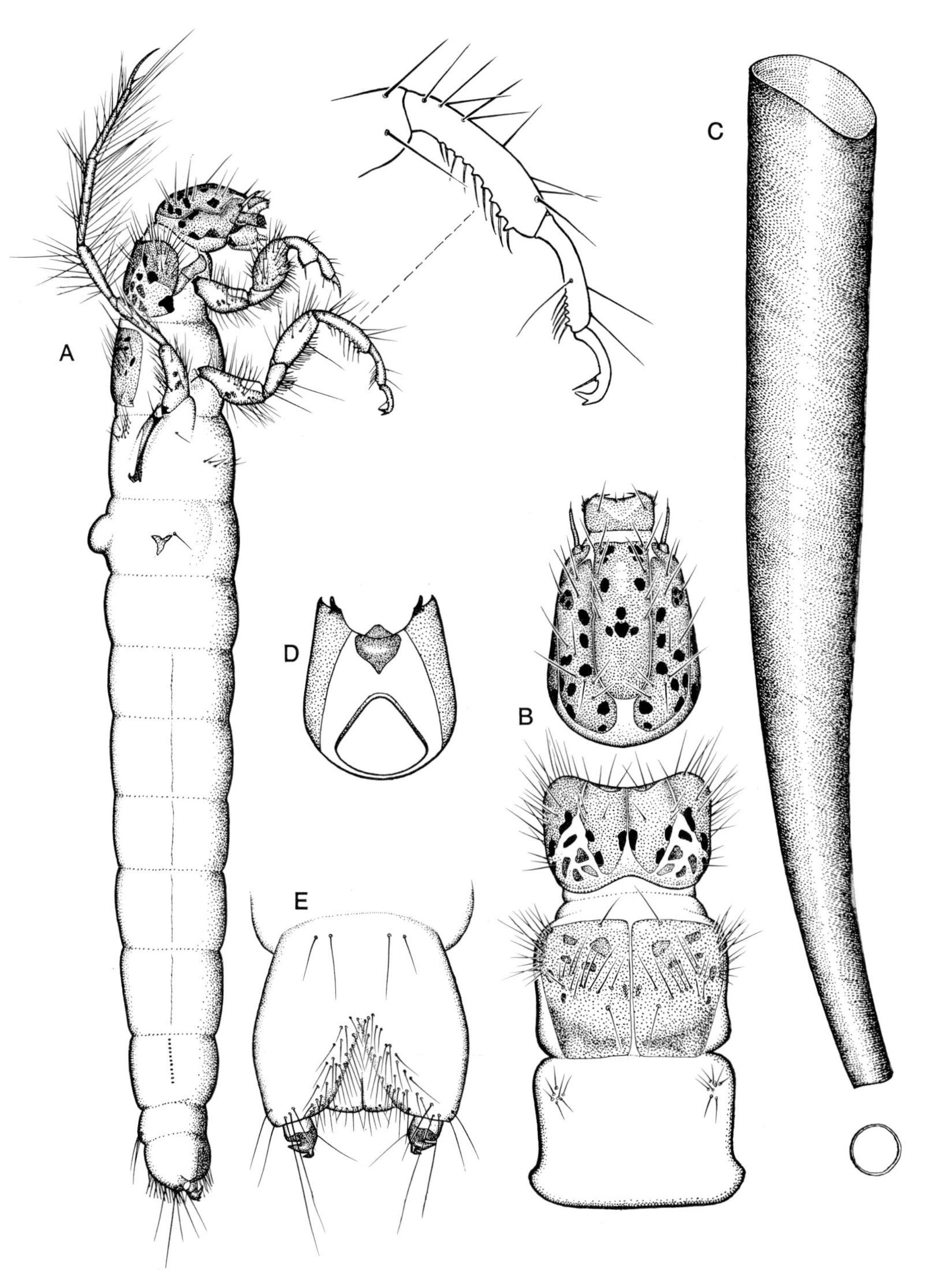
A
B
C
D
E

9.3 Genus **Mystacides**

DISTRIBUTION AND SPECIES *Mystacides* is largely a Holarctic genus, with some representatives in the Oriental region. Three species are currently recognized in North America: *M. interjecta* (Banks) (treated in most references as *M. longicornis* L.) and *M. sepulchralis* (Walker), both northern and transcontinental; and *M. alafimbriata* Hill-Griffin, western. Larval diagnoses for all three were given by Yamamoto and Wiggins (1964).

MORPHOLOGY *Mystacides* larvae are typical of the Mystacidini, and have in addition a secondary division in the hind tarsus (A). Larvae of the genus are readily recognized in the field by their head and thoracic markings of strongly contrasting black spots or blotches on a light background. Single abdominal gill filaments are present in *M. interjecta*, but totally absent in the other two North American species. Two areas of spines occur on the venter of the anal proleg, the distal set rather long (F). Length of larva up to 10 mm.

CASE Larval cases in *Mystacides* are distinctive because of the twigs or conifer needles that usually extend well beyond the front end of a coarse-textured, straight tube (C) of fragments of rocks, mollusc shell, or plant material. Length of larval case up to 30 mm.

BIOLOGY Larvae are usually found in the shallows along shores of lakes and ponds, or in areas of slow current of rivers. It is not uncommon for two of the species to occur close together in the same habitat (Ross 1944; Yamamoto and Wiggins 1964); records suggest a single generation per year for all three, with emergence and flight period extending over much of the summer. Gut contents for *M. sepulchralis* were found by Lloyd (1921) to be fine particles of plant origin; we found high proportions of arthropod remains in larvae (3 of each) of both *M. sepulchralis* and *M. interjecta.*

REMARKS Systematic data for the three North American species were summarized by Yamamoto and Wiggins (1964) and Yamamoto and Ross (1966).

Mystacides sepulchralis (Manitoba, Duck Mountain Prov. Park, 12 June 1962, ROM)
A, larva, lateral x16, hind leg and lateral hump sclerite enlarged; B, head and thorax, dorsal; C, case, ventral x6; D, mandible and maxilla, ventral; E, head, ventral; F, anal proleg, ventral

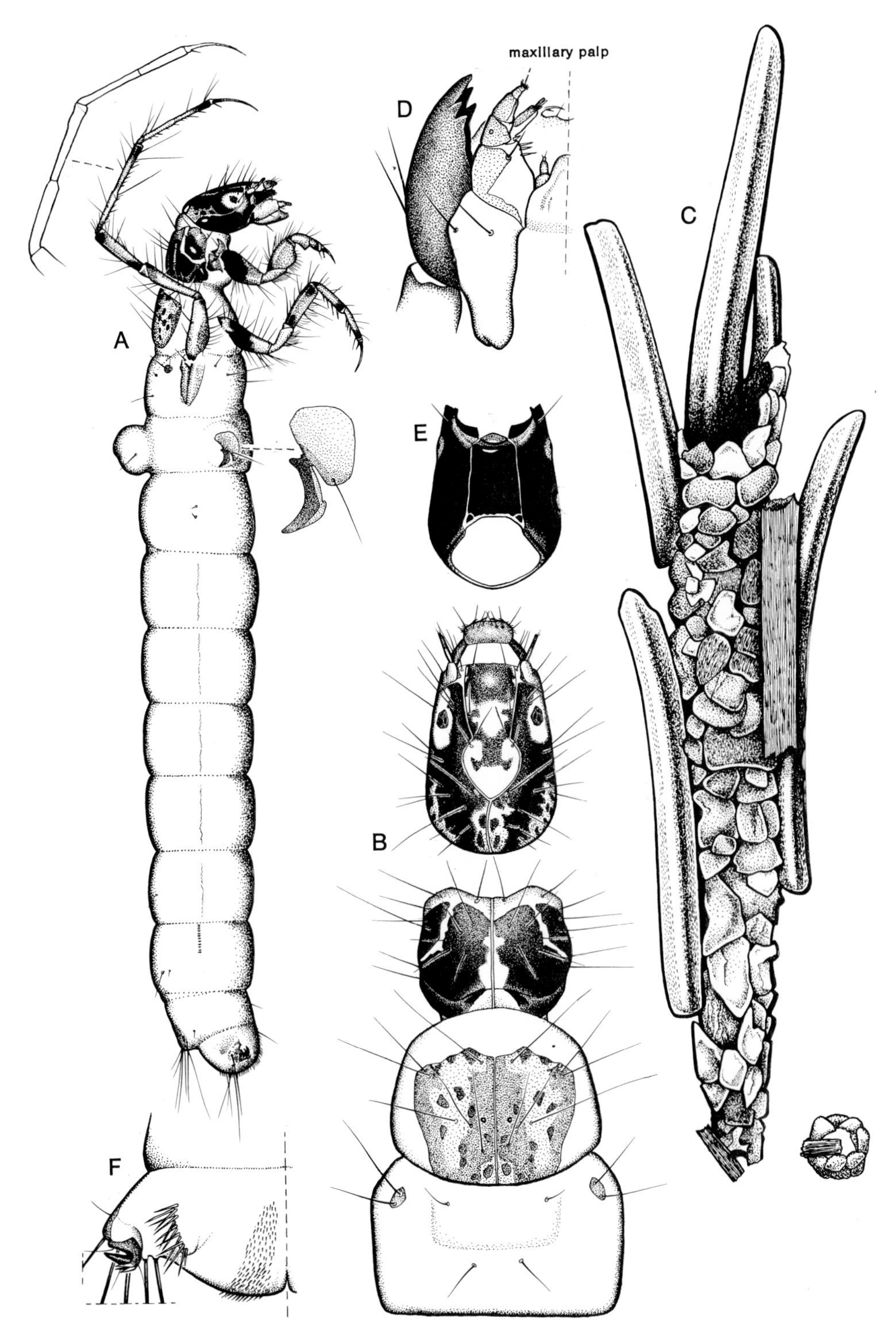
maxillary palp
D
C
A
E
B
F

9.4 Genus Nectopsyche

DISTRIBUTION AND SPECIES All Nearctic species formerly treated under *Leptocella* have been assigned to *Nectopsyche* (Flint 1974c), a genus restricted to the New World. Approximately a dozen species have been described from this continent north of Mexico, but diagnostic characters in the adult stage are indefinite for several of them (Ross 1944).

Under the name *Leptocella* larvae were characterized for six species by Ross (1944), and described in detail for *N. albida* (Walk.) (as *L. uwarowii* Kolenati) by Vorhies (1909) and Elkins (1936).

MORPHOLOGY The ventral apotome is triangular (E) and the hind tibia is not secondarily subdivided (A). Other characters for *Nectopsyche* include the sclerotized bar and circular roughened area on each lateral hump of abdominal segment I (A) and the unpigmented lines delimiting the anterolateral corners of the pronotum (B, G). Head markings range from marbled blotches to discrete bands (Ross 1944, figs. 744-8), and the ends of the leg segments are dark in some species; abdominal gills may be present (A) or absent. A ventral band of tiny spines lies on the basal segment of each anal proleg adjacent to the anal opening in at least some species (F). In a series of larvae from Ontario, the anterolateral margins of the pronotum are extended into toothed lobes (G). Length of larva up to 15 mm.

CASE Larval cases (C) are usually long and slender, and made of plant and mineral fragments with twigs or conifer needles extending beyond one end; we have one pupal case from Ontario similar to the flattened cases of the limnephilid genus *Chyranda.* In some species, cases are almost entirely of mineral materials (Ross 1944, fig. 762), or of diatoms (Wallace et al. 1976). Length of larval case up to 31 mm.

BIOLOGY *Nectopsyche* larvae inhabit lakes and slower currents of rivers on the bottom substrate or on plants. Larvae of at least some species can swim. Gut contents of larvae (3) we examined were largely fine organic particles and vascular plant tissue; Elkins (1936) found a larva gorged with ostracods. Observations on case recognition (Merrill 1969) indicate that larvae of this genus are among the few that back into their cases, although not to the complete exclusion of entry head first.

Nectopsyche intervena (Banks) (California, Inyo Co., 16 May 1969, ROM)
A, larva, lateral x11, lateral hump sclerite enlarged; B, head and thorax, dorsal; C, case, lateral x7; D, head, lateral; E, head, ventral; F, anal prolegs, ventral
G, *Nectopsyche* sp. (Ontario, Algoma Dist. 25 Sept. 1959, ROM), pronotum, dorsal

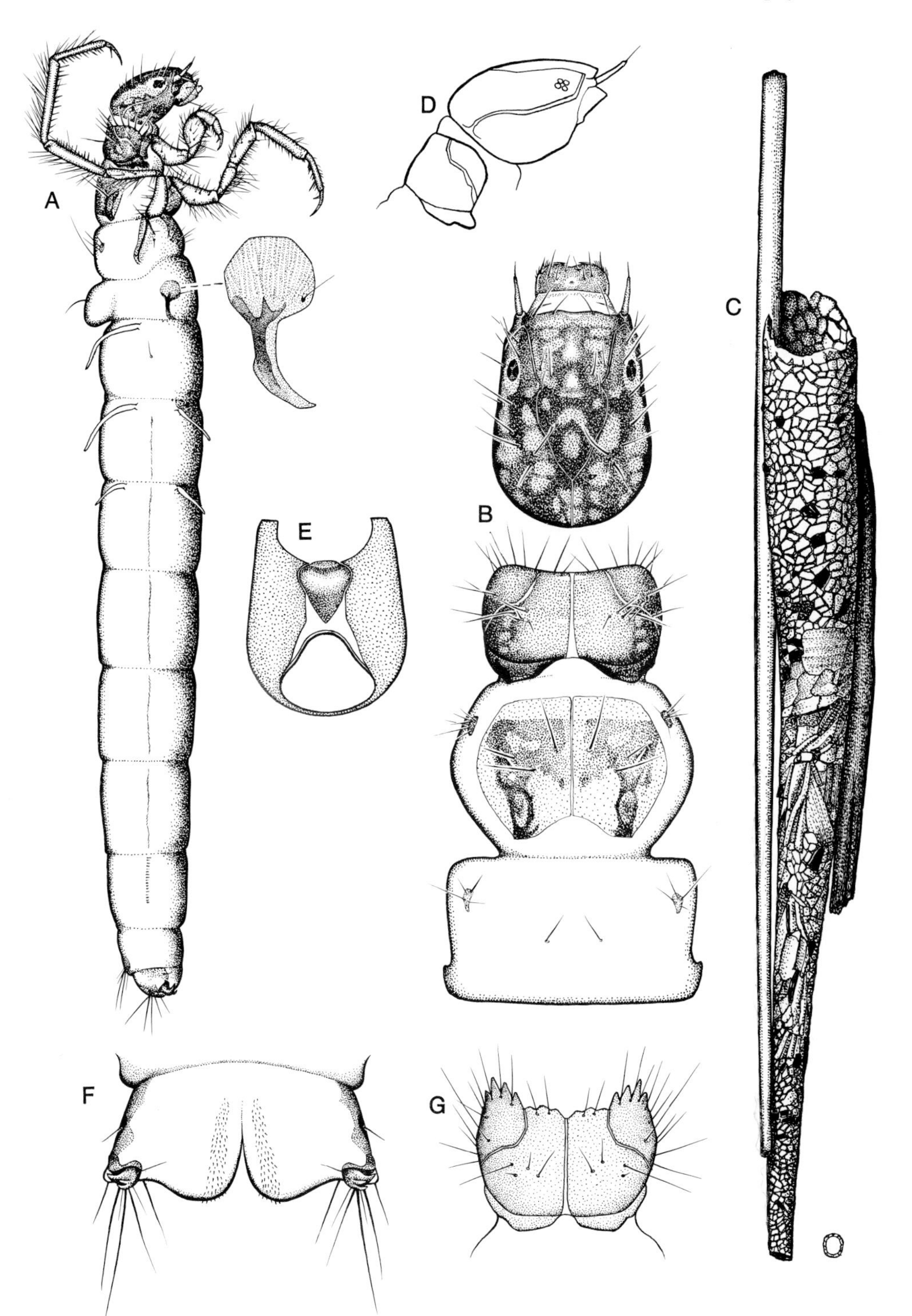
A
B
C
D
E
F
G

9.5 Genus **Oecetis**

DISTRIBUTION AND SPECIES *Oecetis* is a large genus represented in all faunal regions of the world. In North America approximately 20 species are known north of Mexico; as a group they are widely distributed over much of the continent, and are often abundant.

Diagnostic characters for larvae of five species were given by Ross (1944).

MORPHOLOGY *Oecetis* larvae are distinguished from all other North American leptocerids by the long maxillary palpi and single-blade mandibles (B, E) and by the supernumerary setae on the labrum (D). The genus is the only Nearctic member of the tribe Oecetini. Length of larva up to 12.5 mm.

CASE Larval cases of *Oecetis* species are varied in both form and materials: small fragments of rock (C), often combined with bark or leaves; and short lengths of stems and twigs placed transversely (Ross 1944, fig. 833). Length of larval case up to 15 mm.

BIOLOGY *Oecetis* larvae are bottom dwellers, living in both lentic and lotic waters; some Palaearctic species are reported from brackish water (Lepneva 1966). Early-instar larvae of at least some species can swim, but we have not observed later instars doing so. The long mandibles identify *Oecetis* larvae as predators; gut contents of larvae from a lake in British Columbia showed that animals were the dominant food (Winterbourn 1971a). An Asian species of *Oecetis* is reported to feed on rice plants in Japan (see Balduf 1939).

Oecetis sp. (Manitoba, Duck Mountain Prov. Park, 12 June 1962, ROM)
A, larva, lateral x15; B, head and thorax, dorsal; C, case, ventrolateral x10; D, labrum, dorsal; E, mandible and maxilla, ventral; F, head, ventral

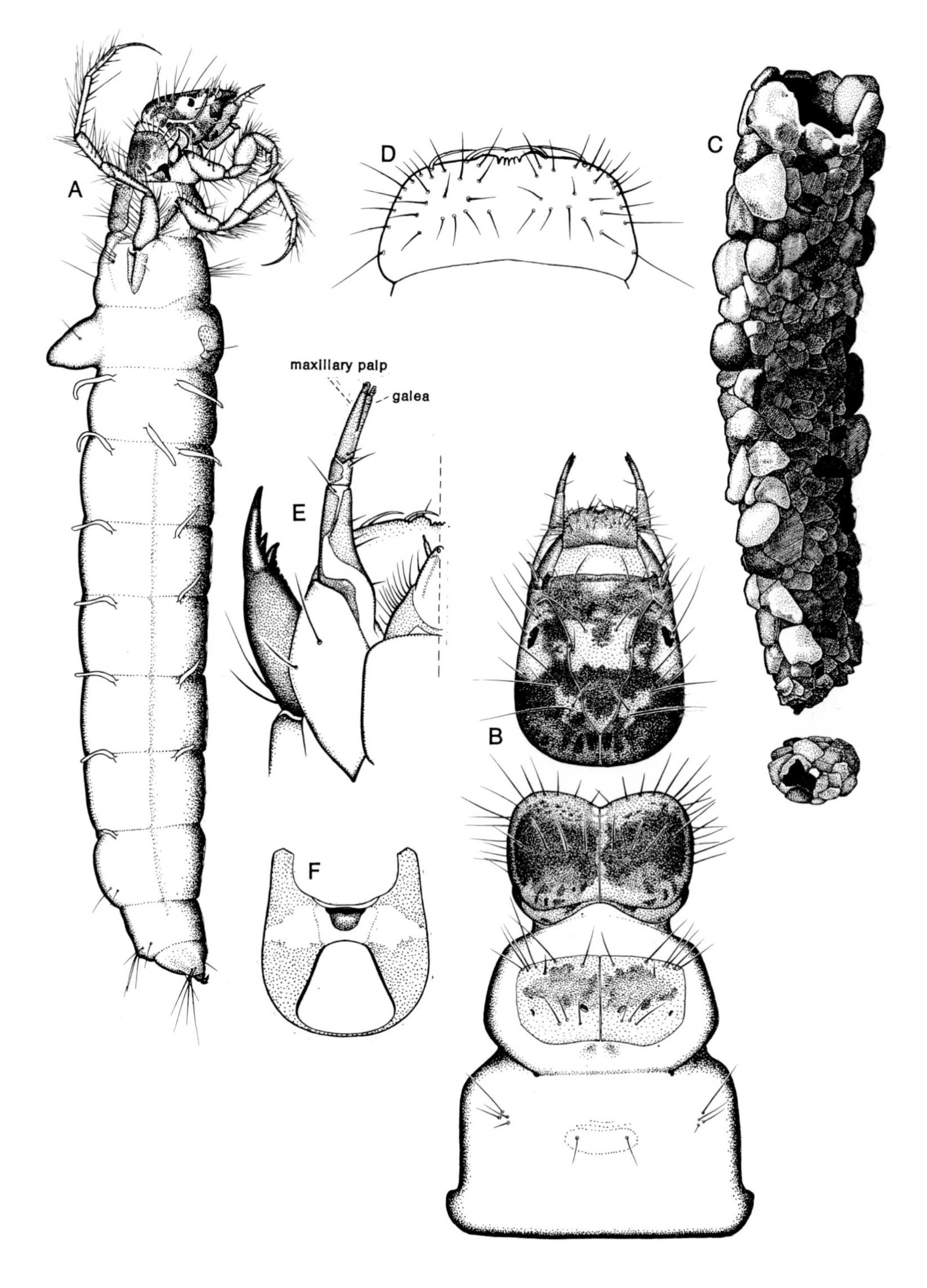
A
D
C
maxillary palp
galea
E
B
F

9.6 Genus **Setodes**

DISTRIBUTION AND SPECIES The North American species assigned to *Setodes* seem to represent a concordant group; beyond that, the genus is at least Oriental and Palaearctic in distribution. Six species are known north of Mexico; records are confined to the eastern part of the continent, from Minnesota and Maine to Oklahoma and North Carolina.

Larvae have been associated with certainty only for *S. incerta* (Walk.) (Merrill and Wiggins 1971).

MORPHOLOGY Larvae are generally concordant with other members of the Mystacidini but distinguished from all leptocerids by the sclerotized, spiny anal plates (G); these plates are clearly specializations of the lateral sclerite and the two patches of spines evident in other mystacidine genera such as *Triaenodes.* Colours of head and thorax are pale (B) or dark (Merrill and Wiggins 1971, fig. 12). Length of larva up to 8.5 mm.

CASE Larval cases (C) are constructed of rock fragments, slightly curved but with little taper. The posterior opening is of approximately the same diameter as the anterior opening, and not restricted with silk or rock. Length of larval case up to 8.5 mm.

BIOLOGY *Setodes* larvae occur in running waters and also along lake shores. Observations on living larvae in dishes of sand (Merrill and Wiggins 1971) indicate that the larvae burrow into loose substrates to conceal all but the case opening at which the larva stations itself; larvae can reverse their position within the case, which perhaps explains why there is essentially no difference between anterior and posterior ends. The armoured posterior end of the larva may then be a protective adaptation for repelling intruders from what would otherwise be an exposed rear flank, thus compensating for the specialized case-building behaviour. Larval guts contained fine particles, vascular plant fragments, and arthropod sclerites; in captivity larvae fed readily on enchytraeid worms. While feeding, larvae were observed to reverse their respiratory current. These and other aspects of the biology of *Setodes* were considered by Merrill and Wiggins (1971).

Setodes incerta (Michigan, Presque Isle Co., 1 July 1969, ROM)
A, larva, lateral x21, lateral hump sclerite enlarged; B, head and thorax, dorsal; C, case, lateral x14; D, hind leg, lateral; E, mandibles, dorsal; F, head, ventral; G, anal prolegs, caudolateral, anal claw enlarged

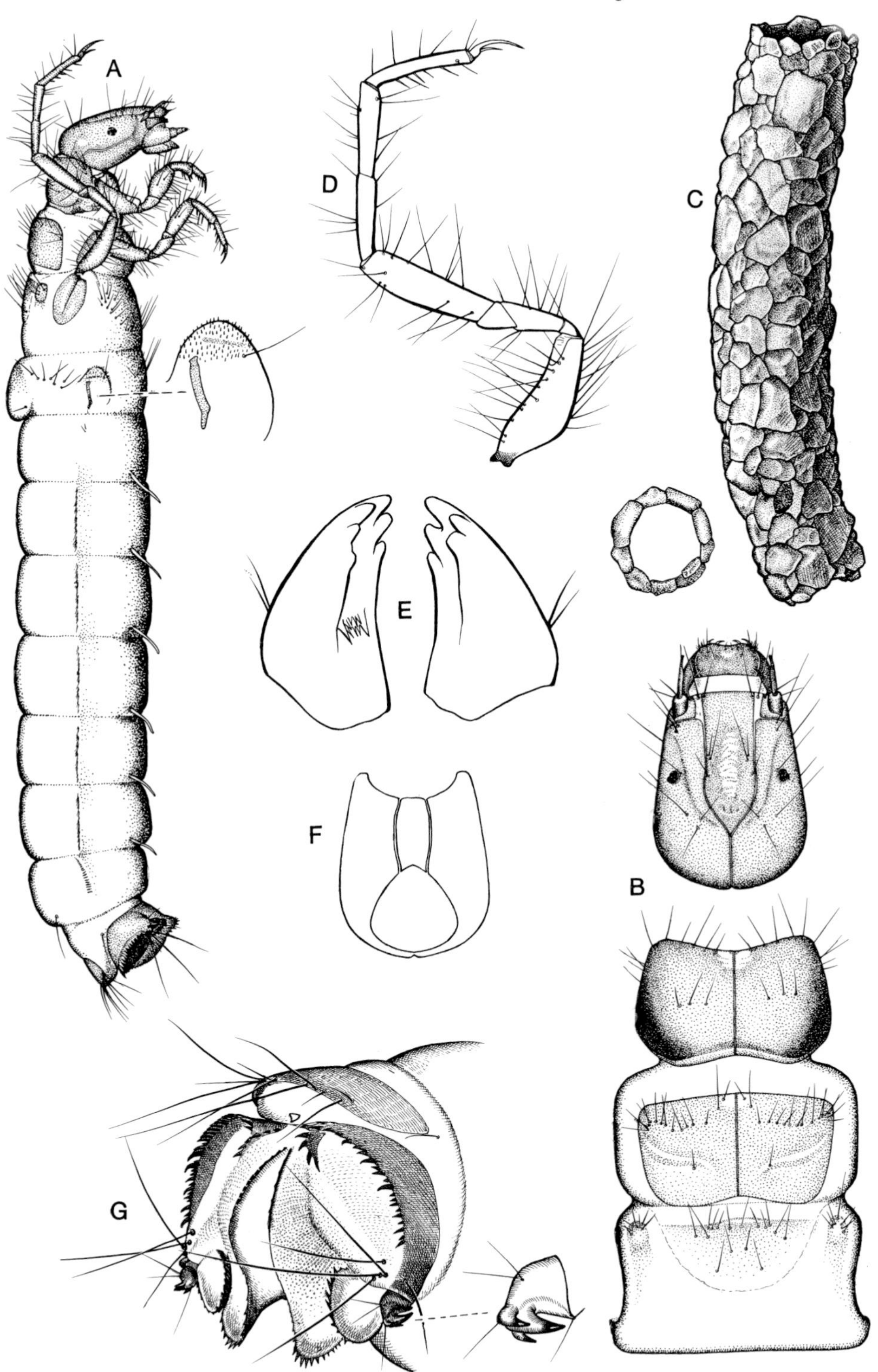
A
D
C
E
F
B
G

9.7 Genus **Triaenodes**

DISTRIBUTION AND SPECIES Species of *Triaenodes* are predominantly Holarctic, but extend also into the Ethiopian and Oriental faunal regions and into the northernmost part of the Neotropical region (Flint 1967a). The group is widely distributed over North America, and approximately 25 species are now known north of Mexico; most occur in the eastern half of the continent but several are transcontinental.

A key to larvae of five species was given by Ross (1944); larval characters for *T. flavescens* Banks were given by Vorhies (1909) and for *T. marginata* Sibley by Sibley (1926).

MORPHOLOGY In young larvae of at least some species of *Triaenodes* the ventral apotome is triangular, the rectangular condition (D) developing in later instars. *Triaenodes* larvae are distinguished from others in the Mystacidini by the long, dense hairs on the hind tibia and tarsus (A), an asset in their free-swimming behaviour. Head markings are usually dark bands or blotches contrasting with a light ground colour. Length of larva up to 14.5 mm.

CASE Slender, tapered cases of pieces of green plants arranged spirally (C) are diagnostic for *Triaenodes* alone in the North American fauna. Larvae construct cases of dextral or sinistral spirals (Merrill 1969), but the behaviour pattern is fixed for one or the other in an individual larva (Tindall 1960). Length of larval case up to 33 mm.

BIOLOGY *Triaenodes* larvae occur in plant beds in both lotic and lentic waters where they swim with their cases, propelled by their hind legs. This behaviour allows independence from the bottom substrate, enabling the larvae to move freely in the upper levels among dense beds of submerged aquatic plants. Green plant tissues are ingested (Berg 1949; McGaha 1952; Tindall 1960), and are used also in building cases; gut contents of larvae (3) were largely vascular plant fragments and fine organic particles. *Triaenodes* larvae exploit green plants so successfully that at least one species, *T. bicolor* (Curtis), has become a pest of cultivated rice in Italy (Moretti 1942); this is also testimony to the high water temperatures that can be tolerated - 28 to 31°C (Moretti 1942).

Triaenodes sp. (Ontario, Kent Co., 3 June 1965, ROM)
A, larva, lateral x14, hind tibia and lateral hump sclerite enlarged; B, head and thorax, dorsal; C, case, lateral x7; D, head, ventral; E, anal prolegs, ventral

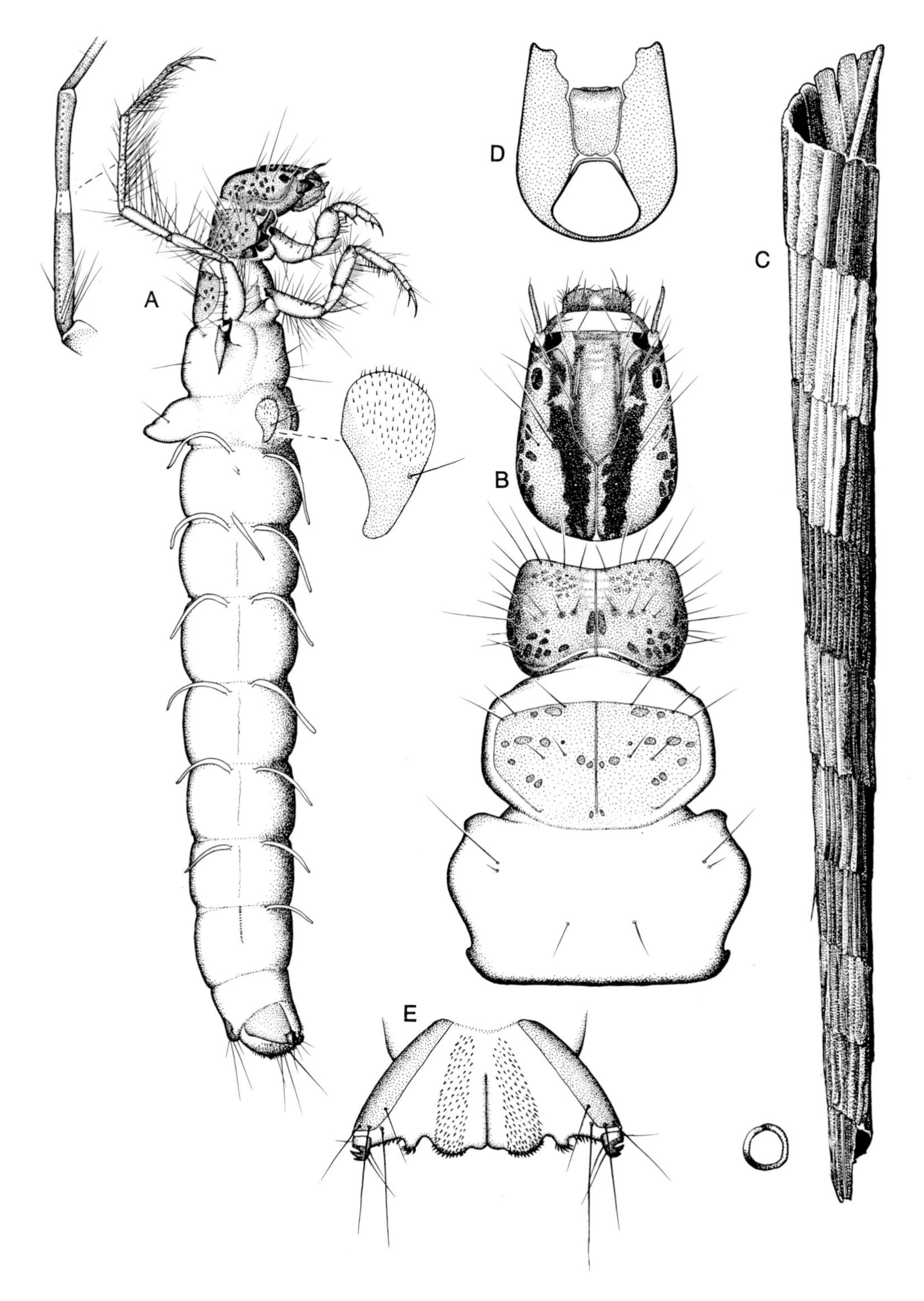
D
C
A
B
E

10
Family Limnephilidae

The Limnephilidae are the largest family of the Trichoptera in North America, with more than 300 species now assigned to 52 genera. The group is dominant at higher latitudes and elevations through much of the northern hemisphere.

Larvae of the Limnephilidae occupy a wider range of habitats than any other family in the Trichoptera, and it is remarkable that more than one-third of all the Nearctic caddisfly genera are members of this single family. There are genera characteristic of spring streams, rivers, lakes, and marshes; there are some whose larvae live in temporary pools and streams or in the organic muck of spring seepage areas, and a few limnephilid larvae live in moist terrestrial sites. Plant materials are the principal food. The ecological role of larvae in the subfamilies Pseudostenophylacinae and Limnephilinae is reduction of large pieces of plant debris to small particles, although recent work has demonstrated that fungi growing on the dead plant materials are the primary attraction (Mackay and Kalff 1973). Limnephilid larvae occasionally found clustered on the body of dead fish (Brusven and Scoggan 1969) or mammals may be attracted to the microflora associated with decomposition of these tissues. Most larvae of the other subfamilies – Apataniinae, Neophylacinae, and Goerinae – have scraper-like mandibles in which the separate teeth have become fused into an entire edge; these larvae feed chiefly by scraping periphyton and fine organic particles from rock surfaces. Although lacking these modified mandibles, many larvae of the Dicosmoecinae appear to feed largely by scraping.

Diapause occurs in the life cycle of a number of species in the Limnephilidae; development is suspended for a time and reactivated, largely by changing photoperiod, with the result that the life cycle coincides more effectively with seasonal changes and food resources in a habitat (see under Life Cycles). A life cycle of one year is the normal condition, but collection data indicate that larval development extends beyond one year in certain species.

Larvae in some limnephilid genera construct cases of rock fragments, and in others plant materials are used. In the Nearctic fauna, at least, these behavioural patterns bear a broad relationship to the subfamilial groupings, as indicated in the outline of characters that follows. In a general way, larvae living in currents of cool waters use rock materials for case-building, and those in more lentic habitats, especially in warmer waters, use plant materials.

Because the Limnephilidae are such a large group and their larvae diverse structurally, an accurate description for the family would require that almost every character be qualified; it is more useful to consider character states for each subfamily and these are outlined at the end of this introduction. Larval diagnosis for the family as a whole is based on a combination of several characters: consistent location of the antenna approximately mid-way between the eye and the anterior edge of the head capsule, with prosternal horn and chloride epithelia almost always present. It can be added that the extensive development of setae on abdominal segment I in many Limnephilidae is unusual in the Trichoptera, and almost entirely unique among Nearctic caddisfly families. Arrangement of these setae has proved to be of particular taxonomic value in working out diagnostic characters for the genera; and it is likely that the chaetotaxy of segment I will be useful for recognizing larvae to species. A descriptive basis for these setae is outlined in the general chapter on morphology. As in other families of the limnephilid branch of the Limnephiloidea, lateral tubercles are usually present on most segments.

All of the diverse elements of the Limnephilidae were resolved into a single comprehensive classification of the world fauna in an important study by Schmid (1955). Although based almost entirely on characters of the adults, that classification was shown to have a high level of concordance with larval morphology (Flint 1960). Further study, especially of the western Nearctic fauna, has demonstrated that there are, however, some phyletic lineages still discordant with the subfamilial groupings of that classification (Wiggins 1973c). The classification proposed by Schmid is followed here for the most part, but several genera are listed as *incertae sedis* because their affinities based on larval morphology are not consistent with those that have been interpreted from adults. Until these discordant aspects are resolved, it is difficult to construct satisfactory larval diagnoses or keys to the Limnephilidae at the subfamily level although, in a group so large even on this continent alone, precise subfamilial diagnoses would be of considerable practical value; certainly the limnephilid subfamilial groupings have biological relevance. Thus, in what follows, the general comparative features of each subfamily are given, but these are not necessarily diagnostic. Larval characters in existing limnephilid subfamilies have been analysed phylogenetically elsewhere (Wiggins 1973c), and my purpose here is to outline only the group affinities of the Nearctic genera in so far as they are supported by larval morphology.

Only one subfamily, the Drusinae (Ecclisopteryginae of Nielsen 1943b), does not occur in North America; species of this group, largely European, occur in cold, running waters.

APATANIINAE: *Apatania* Head with or without secondary setae; bases of setae nos. 14 and 15 close together, distance between them approximately equal to one-half diameter of eye; mandibles with entire scraping edge; labrum with anterolateral margins extended as membranous lobes; labium with sclerite of palpiger reduced and hook-shaped; maxilla with galea a rounded membranous lobe; ventral apotome vase-shaped. Thorax with primary setal areas of mesonotum discrete or confluent; metanotal *sa*1 without sclerite, setae arising in transverse line; metanotal *sa*2 sclerites frequently reduced or absent; trochanteral brush lacking on middle and hind legs; major ventral setae on femora of middle and hind legs variable; basal seta of tarsal claws long, reaching almost to tip of claw. Abdominal gills single; lateral fringe present; chloride epithelia present ventrally; dorsal sclerite of segment IX with many setae; anal claw with accessory hook reduced or absent. Cases of rock pieces only.

The Apataniinae are inhabitants of northern or montane parts of the Holarctic and Oriental regions. Larvae live in cool streams in montane areas, and in lakes at more northerly latitudes.

DICOSMOECINAE: *Allocosmoecus, Amphicosmoecus, Cryptochia, Dicosmoecus, Ecclisocosmoecus, Ecclisomyia, Ironoquia, Onocosmoecus* Head with primary setae only; bases of setae nos. 14 and 15 not close together, distance between them approximately equal to diameter of eye; mandibles with separate tooth-like points; labrum with anterolateral margins entirely sclerotized; labium with sclerite of palpiger ring-like, extending more than half-way around base of labial palp; maxilla with galea sclerotized, elongate, and finger-like (shorter in *Cryptochia*); ventral apotome vase-shaped. Thorax not modified; primary setal areas of mesonotum tend to be confluent; metanotal *sa*1 and *sa*2 each represented by pair of sclerites with setae (*sa*1 sclerites fused in *Amphicosmoecus*); trochanteral brush present on all legs; femora of middle and hind legs with three or more major setae on ventral edge; basal seta of tarsal claws much shorter than claw. Abdominal gill filaments single or multiple; lateral fringe present; chloride epithelia ventral only; dorsal sclerite of segment IX with many setae; anal claw with accessory hook. Cases constructed of rock materials in some genera, of plant materials in others.

Represented over most of the Holarctic region, the Dicosmoecinae are the only group of the Limnephilidae in South America (Schmid 1955, fig. 5); the only limnephilid known in the Australian region, *Archaeophylax ochreus* Mosely, was assigned to the Dicosmoecinae (Flint 1960) on the basis of larval morphology (Neboiss 1958). Except for some species of *Ironoquia* which live in temporary pools and streams, North American larvae of this subfamily are generally characteristic of cool, running waters. This subfamily is believed to contain the most primitive living members of the Limnephilidae.

GOERINAE Head with or without secondary setae; labium with sclerite of palpiger reduced and hook-shaped; maxilla with galea a rounded membranous lobe. Thorax with pronotum variously thickened laterally, but normal in *Goereilla*; mesonotum subdivided into two or three pairs of sclerites, in most genera with transverse ridge; mesepisternum modified as toothed lobe or prominent process; trochanteral brush usually lacking on legs; major ventral setae on middle and hind femora variable. Abdominal gills usually single, sometimes branched or absent; lateral fringe present or absent; chloride epithelia usually ventral only, also dorsal in *Goera*; dorsal sclerite of segment IX with many setae; dorsal plate present at posterior margin of lateral sclerite of anal proleg in most genera; anal claw with accessory hook lacking in most genera, present in *Goereilla*. Cases are constructed of rock materials.

Although the range of character states in the Goerinae is very broad, all larvae of the subfamily can be recognized by the subdivided mesonotal sclerites and by the modified mesepisternum. The modifications to mesonotum and to pronotum in most goerine genera serve one function at least - closing off the anterior opening of the case; when the larva withdraws into its case, the pronotum, the anterior part of the mesonotum, and the mesepisterna are pressed together to form an operculum-like structure, with the head curled ventrally. Although the selective advantage for this highly specialized function is not clear, the need for closure seems further confirmed by the unusually small opening in the posterior end of the case in such genera as *Goera, Goeracea,* and *Goerita.* Recognized as a distinct family by many workers (e.g. Malicky 1973), this group was placed as a subfamily of the Limnephilidae by Nielsen (1942, 1943b) and Flint (1960); additional evidence was introduced by Wiggins (1973b, 1976), and this classification is adopted provisionally here.

GOERINI: *Goera, Goeracea, Goerita* Mandibles with entire scraping edge; labrum with anterior margin membranous; head setae nos. 14 and 15 close together; basal seta of tarsal claws long, extending nearly to tip of claw. Larvae live in running waters, from cold springs to large streams, where they scrape algae and organic particles from exposed rock surfaces. The group is widely distributed through the Holarctic, Oriental, and Ethiopian faunal regions.

LEPANIINI: *Goereilla, Lepania* Mandibles with separate tooth-like points; labrum with anterior margin sclerotized; head setae nos. 14 and 15 normally distant; basal seta of tarsal claws short. Larvae live in organic ooze of spring seepage areas, and evidently feed largely on pieces of decaying plant materials. The two genera are known only in the western Nearctic region, and are believed to be the most primitive members known in the subfamily.

LIMNEPHILINAE Characters generally as in the Dicosmoecinae, except that: mesonotal setae reduced in number and primary setal areas tend to be separate; middle and hind femora usually have only two major setae on ventral edge; and setae on dorsal sclerite of segment IX usually reduced in number. Cases in most genera are constructed of plant materials, although rock materials are used in some.

This is the dominant group of the Limnephilidae, comprising approximately one-half of all species in the family. Four tribes are recognized, and all but the Chaetopterygini, a small Palaearctic group, are represented in North America.

CHILOSTIGMINI: *Chilostigma* (larva unknown), *Chilostigmodes* (larva unknown), *Frenesia, Glyphopsyche, Grensia, Homophylax, Phanocelia* (larva unknown), *Psychoglypha* Abdominal gills branched in some genera, single in others; chloride epithelia occur ventrally only. Genera are characteristic of both cool lotic and warm lentic waters.

LIMNEPHILINI: *Anabolia, Arctopora, Asynarchus, Clistoronia, Grammotaulius, Halesochila, Hesperophylax, Lenarchus, Leptophylax* (larva unknown), *Limnephilus, Nemotaulius, Philarctus, Platycentropus, Psychoronia* Most abdominal gills of dorsal and ventral series arise in clusters of three or more filaments; chloride epithelia frequently occur on side and dorsum of abdominal segments as well as venter. Larvae of most genera use plant materials for cases, but rock materials are used by some. This tribe includes most of the lentic water genera of the Limnephilidae, a fact correlated generally with the occurrence of multiple gill filaments throughout the group (Wichard 1974).

STENOPHYLACINI: *Chyranda, Clostoeca, Desmona, Hydatophylax, Philocasca, Pycnopsyche* All abdominal gills single; chloride epithelia present only on venter of abdominal segments. Most species in this tribe live in cool, lotic waters.

NEOPHYLACINAE: *Farula, Neophylax, Neothremma, Oligophlebodes* Head with or without secondary setae; bases of setae nos. 14 and 15 very close together, distance between them less than one-half diameter of eye; mandibles with entire scraping edge; labrum with membranous anterolateral margins; labium with sclerite of palpiger reduced and hook-shaped; maxilla with galea a rounded membranous lobe; ventral apotome T-shaped in *Neophylax* and *Oligophlebodes.* Thorax with anterior margin of pronotum rounded in dorsal aspect; prosternal horn reduced in *Neophylax*; mesonotum emarginate anteromesially; metanotal *sa*1 sclerite lacking; trochanteral brush lacking on middle and hind legs; generally two major setae on ventral edge of middle and hind femora; basal seta of tarsal claws long, reaching almost to tip of claw. Abdominal gills single or absent; lateral fringe reduced or absent; chloride epithelia present ventrally, the number reduced in some genera; dorsal sclerite of segment IX with setae reduced in number; anal claw usually with small accessory hook. Cases of rock pieces only.

The Neophylacinae are a small group of the Nearctic, Asian Palaearctic, and Oriental regions where larvae are confined to running waters. Relationships of the genera *Farula* and *Neothremma*, both considerably different from *Neophylax* and *Oligophlebodes*, have been uncertain (Schmid 1955, 1968b); their larval morphology shows close similarity to the Asian genus *Uenoa* (syn. *Eothremma*).

PSEUDOSTENOPHYLACINAE: *Pseudostenophylax* Characters generally as in Dicosmoecinae, except abdominal gills single and setae present on membrane between metanotal *sa*2 sclerites. Case of sand grains.

This group is predominantly one of the Oriental and Asian Palaearctic regions, with only a small Nearctic element (Schmid 1955, fig. 7). Larvae of the Nearctic genus live in small streams.

Incertae Sedis: *Imania, Manophylax, Moselyana, Pedomoecus, Rossiana* Conflicting evidence for subfamilial classification of *Imania, Manophylax,* and *Moselyana* was reviewed elsewhere (Wiggins 1973c). *Pedomoecus* was referred to the Dicosmoecinae (Schmid 1955) as was *Rossiana* (Schmid 1968b); perhaps these assignments will prevail. But until the larval morphology in all of these genera can be interpreted consistently with that of the adults, I prefer to regard their position as uncertain.

A satisfactory generic key to limnephilid larvae is difficult to produce and often tedious to use. The key which follows is based on close comparative study of a great deal of material. Consistent diagnosis for many genera was found possible only with characters requiring careful microscopic study under good illumination. Final instars should always be used where possible because details of setae and small sclerites may not be fully developed in younger larvae. Since a large number of species, and several genera, are not yet known in the larval stage, certain specimens may not be identifiable with the key even when the characters are carefully interpreted; further evolution of this key is virtually certain. It is especially important in the Limnephilidae to confirm each generic identification with the corresponding summary of structural, biological, and distributional information.

10 Family **Limnephilidae**

Key to Genera*

1 Mesonotum with setal areas of each half separated into two or three separate sclerites (Figs. 10.18B, 10.31B) — 2

Mesonotum with the three setal areas of each half on large single plate, the pair of plates closely appressed along mid-dorsal line (Fig. 10.2B) — 8

2 (1) Mesepisternum enlarged anteriorly, forming either sharp, elongate process (Fig. 10.31B), or short, rounded, spiny prominence (Fig. 10.20G) — 3

Mesepisternum not enlarged anteriorly (Fig. 10.48B) — 7

3 (2) Gills mostly three-branched (Fig. 10.18A); case of small rock fragments, usually with two larger pebbles along each side (Fig. 10.18C). Eastern and western — **10.18 Goera**

Gills single (Fig. 10.20A), sometimes restricted to segments III and IV or III alone (Fig. 10.19A), or absent entirely (Fig. 10.21A) — 4

4 (3) Gills single (Fig. 10.20A) — 5

Gills absent entirely (Fig. 10.21A); case of small rock fragments, tapered and slightly curved, outline smooth (Fig. 10.21C). Southeastern — **10.21 Goerita**

5 (4) Most abdominal segments with dorsal and ventral gills (Fig. 10.20A); mesepisternum extended anteriorly only as short, rounded, spiny prominence (Fig. 10.20A,G); case of fine rock fragments, tapered and slightly curved, outline smooth (Fig. 10.20C). Western — **10.20 Goereilla**

Gills restricted to segments III and IV or III alone; mesepisternum extended anteriorly as prominent pointed process (Fig. 10.31B) — 6

6 (5) Mesepisternum laterally compressed, dorsum of pronotum flat, each metanotal *sa*1 consisting of one or two setae without sclerite or with very small sclerite (Fig. 10.19B); case of small rock fragments, several larger pieces along each side (Fig. 10.19C). Western — **10.19 Goeracea**

Mesepisternum dorsoventrally depressed, dorsum of pronotum convex, each metanotal *sa*1 consisting of several setae on distinct sclerite (Fig. 10.31B); case of small rock fragments without lateral pieces, tapered and slightly curved (Fig. 10.31C). Western — **10.31 Lepania**

*Larvae are unknown in four genera: *Chilostigma*, known from Minnesota (see Wiggins 1975); *Chilostigmodes*, known from Alaska to Labrador (see Krivda 1961); *Leptophylax*, known from north-central states (see Ross 1944); and *Phanocelia*, known from Manitoba and the Northwest Territories (see Schmid 1968b).

7 (2) Dorsum of head with many primary setae unusually stout and prominent (Fig. 10.40B); case of rock fragments, tapered and slightly curved, outline smooth (Fig. 10.40C). Western **10.40 Pedomoecus**

Dorsum of head with primary setae largely unmodified, but posterolateral portions of head extended as flanges, pronotum with pair of concavities (Fig. 10.48B); case of rock fragments, slightly curved, texture rough (Fig. 10.48C). Western **10.48 Rossiana**

8 (1) Basal seta of tarsal claws extending to, or almost to, tip of claw (Fig. 10.4A); mandibles with apical edge entire and not subdivided into tooth-like points (Fig. 10.4D) **9**

Basal seta of tarsal claws extending far short of tip of claw (Fig. 10.34A); mandibles with apical edge subdivided into tooth-like points (Fig. 10.34D) **15**

9 (8) Metanotal *sa*1 sclerites absent, the setae numerous and arranged in transverse band across mid-dorsal line (Fig. 10.4B); case of rock fragments, strongly tapered, anterodorsal lip tending to overhang anteroventral lip of case in final instar (Fig. 10.4C). Widespread **10.4 Apatania**

Metanotal *sa*1 sclerites present (Fig. 10.33B) or, if absent, the setae on each side not arranged in continuous band across mid-dorsal line as above (Fig. 10.38B) **10**

10 (9) Metanotal *sa*1 sclerites absent (Fig. 10.38B), or very small and restricted to base of one seta (Fig. 10.36B) **11**

Metanotal *sa*1 sclerites well developed, each bearing several setae (Fig. 10.33B) **14**

11 (10) Larva and case extremely slender, abdominal gills lacking (Fig. 10.37) **12**

Larva and case not unusually slender as above, abdominal gills present (Fig. 10.38) **13**

12 (11) Segment II with lateral setal fringe extending full length of segment (Fig. 10.37A); anterior margin of pronotum weakly dentate (Fig. 10.37B); case mainly of sand grains, long and slender, exterior surface with thin silken covering (Fig. 10.37C). Western **10.37 Neothremma**

Segment II with lateral fringe reduced to patch of hairs at anterior margin of segment (Fig. 10.15A); anterior margin of pronotum smooth (Fig. 10.15B); case of sand grains, exterior covered with silk, similar to the preceding but even more slender (Fig. 10.15C). Western **10.15 Farula**

13 (11) Pronotum with prominent longitudinal ridges (Fig. 10.38B); case of rock fragments, strongly tapered and curved, outline smooth (Fig. 10.38C). Western **10.38 Oligophlebodes**

10 Family **Limnephilidae**

Pronotum without prominent longitudinal ridges, mesonotum with antero-median emargination (Fig. 10.36B); case of coarse rock fragments, larger pebbles along each side (Fig. 10.36C). Eastern and western **10.36 Neophylax**

14 (10) Dorsum of head evenly rounded, lacking carina (Fig. 10.33B); venter I bearing sclerite with circular hole (Fig. 10.33E); case of rock fragments, tapered and curved, with plant material attached (Fig. 10.33C). Western **10.33 Manophylax**

Dorsum of head flattened, frequently with prominent carina (Fig. 10.28A,B); venter I lacking sclerite; case of rock fragments, tapered and curved (Fig. 10.28C), occasionally with row of pebbles along each side. Western **10.28 Imania**

15 (8) Abdominal gills lacking (Fig. 10.34A); case of fine rock fragments with shiny silken covering, tapered and curved (Fig. 10.34C). Western **10.34 Moselyana**

Abdominal gills present **16**

16 (15) Anterior margin of pronotum densely fringed with long hairs (Fig. 10.10A); dorsum of head flattened and bearing two bands of closely packed scale-hairs (Fig. 10.10B); case of wood and bark, flattened dorsoventrally and tapered (Fig. 10.10C). Western **10.10 Cryptochia**

Anterior margin of pronotum and dorsum of head without dense hairs as above (Fig. 10.43B) **17**

17 (16) Most abdominal gills single (Fig. 10.9A) **18**

Most abdominal gills of dorsal and ventral rows multiple (Figs. 10.32A, 10.25A), although lateral gills sometimes single (Fig. 10.17A) **29**

18 (17) Metanotal *sa*1 and *sa*2 sclerites large in relation to metanotum, distance between *sa*2 sclerites no more than twice the maximum dimension of one *sa*2 sclerite (Fig. 10.14B); case a slender tube of rock fragments frequently incorporating long pieces of plant material (Fig. 10.14C). Western **10.14 Ecclisomyia**

Metanotal *sa*1 and *sa*2 sclerites smaller than above in relation to metanotum, distance between *sa*2 sclerites more than twice the maximum dimension of one *sa*2 sclerite (Fig. 10.7B); *sa*1 sclerites sometimes fused into single median sclerite (Fig. 10.27B) **19**

19 (18) Lateral humps of segment I with one or two sclerites adjacent to base of hump, the sclerites often only lightly pigmented but distinguishable by the smooth and relatively shinier surface (Figs. 10.7A, 10.26A, 10.45A, 10.47A) **20**

Lateral humps of segment I without sclerites (Fig. 10.13A) **25**

20 (19) Large single sclerite at base of lateral hump enclosing posterior half of hump and extending posterodorsad as irregular lobe (Fig. 10.7A); case of leaves or bark formed into flattened tube with lateral seam and narrow flange along each side (Fig. 10.7C). Transcontinental **10.7 Chyranda**

One or two smaller sclerites at base of lateral hump (Figs. 10.11A, 10.26A, 10.45A, 10.47A) **21**

21 (20) Two sclerites, variously shaped and positioned, at base of each lateral hump (Figs. 10.11A, 10.45A) **22**

One elongate sclerite at posterior edge of base of each lateral hump (Fig. 10.27A) **23**

22 (21) Two small, ring-shaped sclerites located posterodorsally at base of lateral hump (Fig. 10.11A); case largely of rock fragments with some small pieces of wood (Fig. 10.11C). Western **10.11 Desmona**

One rounded posterior sclerite and one elongate dorsal sclerite at base of lateral hump (Fig. 10.45A); case a straight tube of rock and wood fragments (Fig. 10.45C). Transcontinental **10.45 Psychoglypha**

23 (21) Sclerite at base of lateral hump elongate, its longest dimension almost equivalent to basal width of hump (Figs. 10.27A, 10.47A) **24**

Sclerite at base of lateral hump shorter, its longest dimension equal to approximately one-half basal width of hump (Fig. 10.26A); case usually cylinder of smooth outline and thin walls, occasionally three-sided, consisting largely of bark pieces irregularly arranged (Fig. 10.26C). Western **10.26 Homophylax**

24 (23) Metanotal *sa*1 sclerites fused on mid-dorsal line into single sclerite (Fig. 10.27B,D); segment II with ventral chloride epithelium, in which case segment IX bears only a single seta laterad of dorsal sclerite (*Hydatophylax argus*); or venter II without chloride epithelium, in which case segment IX bears tuft of 3–6 setae laterad of dorsal sclerite (*H. hesperus*, Fig. 10.27A); case of wood pieces or leaves, irregular in outline (Fig. 10.27C). Eastern and western **10.27 Hydatophylax**

Metanotal *sa*1 sclerites not fused on mid-dorsal line (Fig. 10.47B), although often close together; venter II without chloride epithelium, segment IX usually with single seta laterad of dorsal sclerite (Fig. 10.47A); case of twigs or gravel, or of leaves and occasionally three-sided (Fig. 10.47C). Widespread in east, extending to Rocky Mountains **10.47 Pycnopsyche**

25 (19) Anterior margin of pronotum with flattened scale-hairs, dorsum of head flattened (Fig. 10.42B,D); case of coarse rock fragments, slightly curved (Fig. 10.42C). Western **10.42 Philocasca**

Anterior margin of pronotum without flattened scale-hairs, setae normal, dorsum of head usually convex, rarely flattened (only in the western species *Pseudostenophylax edwardsi*, Fig. 10.44D) **26**

26 (25) Mesonotal *sa*1 and *sa*2 separated by gap free of setae (Fig. 10.9B) **27**

Mesonotal *sa*1 and *sa*2 connected by continuous band of setae (Fig. 10.13B) **28**

27 (26) Pronotum covered with fine spines (Fig. 10.23B); case of plant and rock fragments (Fig. 10.23C). Northern **10.23 Grensia**

Pronotum smooth and shiny, lacking fine spines (Fig. 10.9B); case of leaf pieces fastened together to form wing-like flanges at each side of flattened tube (Fig. 10.9C). Western **10.9 Clostoeca**

28 (26) Head and pronotum strongly inflated and with pebbled texture (Fig. 10.13A,B); case of small rock fragments, tapered and curved, outline smooth (Fig. 10.13C). Western **10.13 Ecclisocosmoecus**

Head and pronotum not unusually inflated (Fig. 10.44A,B), although dorsum of head flat in the one western species (Fig. 10.44D); sclerotized areas not pebbled; case of rock fragments, tapered and curved, outline smooth (Fig. 10.44C). Eastern and western **10.44 Pseudostenophylax**

29 (17) Most gills with three branches (Fig. 10.32A), none with more than four (Fig. 10.12A) **30**

At least some gills with more than four branches (Figs. 10.25A, 10.30A, 10.46A) **46**

30 (29) Dorsum of head with two bands of contrasting colour extending from coronal suture to base of each mandible (Fig. 10.24B), and/or narrowed posterior portion of frontoclypeal apotome with three light areas - one along each side and one at posterior extremity (Figs. 10.8B, 10.32B) **31**

Dorsum of head lacking bands or other well-defined contrasting areas, usually largely uniform in colour (Fig. 10.1B), with prominent spots at points of muscle attachment (Fig. 10.3B) **35**

31 (30) Venter I usually with more than 100 setae over all (Fig. 10.8D); small spines on head and pronotum (Fig. 10.8B); short, stout setae on lateral sclerite of anal proleg (Fig. 10.8A); case cylindrical, usually of pieces of twigs and bark, sometimes of small rock fragments (Fig. 10.8C). Western **10.8 Clistoronia**

Venter I with fewer than 100 setae over all, usually without spines on head and pronotum; short, stout setae on lateral sclerite of anal proleg usually lacking; cases highly variable (Figs. 10.32, 10.41) **32**

32 (31) Mesonotal *sa*1 consisting of single seta (Fig. 10.24B) **33**

Mesonotal *sa*1 consisting of more than one seta (Fig. 10.6B) **34**

33 (32) Chloride epithelia on dorsum as well as venter of several abdominal segments (Fig. 10.24A); case of wood or leaf fragments, changed to fine gravel before pupation. Western **10.24 Halesochila**

Chloride epithelia absent dorsally, present only on venter of most abdominal segments (Fig. 10.35A); case of leaf pieces. Northern and transcontinental **10.35 Nemotaulius**

34 (32) Chloride epithelia present both dorsally and ventrally on most abdominal segments (Fig. 10.6A); case of plant and rock materials (Fig. 10.6C). Northern, extending south in western mountains **10.6 Asynarchus** (in part)

Chloride epithelia usually entirely absent dorsally but present ventrally on most abdominal segments (Figs. 10.32A, 10.41A); case of wide range of materials (Figs. 10.32C-F, 10.41C). Widely distributed throughout continent **10.32 Limnephilus** (in part)
10.41 Philarctus

35 (30) Femur of hind leg with two major setae arising from ventral edge (Fig. 10.17A) **39**

Femur of hind leg with more than two major setae arising from ventral edge (Fig. 10.12A) **36**

36 (35) Metanotal *sa*1 sclerites fused along mid-dorsal line into single sclerite (Fig. 10.2B); case a hollow twig with ring of bark pieces at anterior end (Fig. 10.2C), or entire case of wood fragments. Western, but extending into Saskatchewan **10.2 Amphicosmoecus**

Metanotal *sa*1 sclerites clearly separate (Fig. 10.3B) **37**

37 (36) Dorsum I with transverse row of setae posterior to median dorsal hump (Fig. 10.1D); scale-hairs on dorsum of head (Fig. 10.1B); venter II with two chloride epithelia (Fig. 10.1E); case of small pebbles arranged into slightly curved and flattened cylinder of irregular outline (Fig. 10.1C). Western **10.1 Allocosmoecus**

Dorsum I usually lacking setae posterior to median dorsal hump (Fig. 10.12D); scale-hairs absent from dorsum of head; venter II with single chloride epithelium (Fig. 10.12E), sometimes with smaller epithelium at each side, or epithelia lacking on II **38**

38 (37) Tibiae with several pairs of stout spurs (Fig. 10.12A); case of rock fragments (Fig. 10.12C), with plant materials in cases of younger larvae. Western montane areas **10.12 Dicosmoecus**

Tibiae never with more than one pair of stout spurs (Fig. 10.39A); case usually of pieces of wood and bark (Fig. 10.39C), of rock fragments in 1 western species. Eastern and western **10.39 Onocosmoecus**

39 (35) Dorsum VIII with continuous transverse band of setae (Fig. 10.31G) **40**

Dorsum VIII with median gap in transverse band of setae (Fig. 10.22F); case of sedge or similar leaves arranged longitudinally to form cylinder (Fig. 10.22C). Northern and transcontinental **10.22 Grammotaulius**

40 (39) Pronotum, especially anterior margin, and lateral sclerite of anal proleg both with short, stout setae (Fig. 10.17A,B) **41**

Pronotum usually without short, stout setae (Fig. 10.3B), lateral sclerite of anal proleg never with such setae **42**

41 (40) Tibiae and tarsi of all legs with dark, contrasting band (Fig. 10.17A); case of twigs and bark arranged to form smooth cylinder (Fig. 10.17C). Transcontinental **10.17 Glyphopsyche**

Tibiae and tarsi lacking dark band (Fig. 10.16A); case mostly of small stones with wood fragments forming a smooth cylinder (Fig. 10.16C). Eastern **10.16 Frenesia**

42 (40) Metanotal *sa*2 represented by a few setae, usually two, without sclerite (Fig. 10.5B); case of elongate leaf pieces arranged in smooth cylinder (Fig. 10.5C). Northeastern and western **10.5 Arctopora**

Metanotal *sa*2 with more than two setae, and with sclerite (Fig. 10.6B) **43**

43 (42) Dorsum of head with numerous large spots, coalescing in places especially on frontoclypeal apotome into diffuse blotches (Fig. 10.3B); anterolateral corner of pronotum with small patch of spines (Fig. 10.3D); case of plant materials usually cylindrical but sometimes three-sided (Fig. 10.3C). Eastern and northern **10.3 Anabolia**

Dorsum of head with varied markings, including spots (Fig. 10.32G) but not coalescing as above; anterolateral corner of pronotum usually lacking patch of spines **44**

44 (43) Prosternal horn unusually long, extending ventrally well beyond distal edge of head capsule, approximately to mentum of labium (Fig. 10.43D); case cylindrical, of short, narrow pieces of plant material arranged transversely (Fig. 10.43C). Eastern and northern **10.43 Platycentropus**

Prosternal horn not unusually long, extending ventrally approximately to distal edge of head capsule (Fig. 10.32A); cases diverse, of plant or rock materials, (Fig. 10.32C–F). Widely distributed throughout the continent **45**

45 (44) Chloride epithelia present dorsally, laterally, and ventrally on most abdominal segments (Fig. 10.6A); case of plant and rock materials (Fig. 10.6C). Northern, extending south in western mountains **10.6 Asynarchus** (in part)

Chloride epithelia usually entirely absent dorsally but present ventrally and laterally on most abdominal segments (Figs. 10.32A, 10.41A); case of wide range of materials (Fig. 10.32C-F). Widely distributed throughout continent **10.32 Limnephilus** (in part)

46 (29) Femora of second and third legs with approximately five major setae along ventral edge (Fig. 10.29A); case of bark and leaves (Fig. 10.29C) or of sand. Eastern **10.29 Ironoquia**

Femora of second and usually third legs with two major setae along ventral edge (Fig. 10.25A) **47**

47 (46) Metanotum with all setae confined to primary sclerites (Fig. 10.30B); case of lengths of sedge leaves arranged longitudinally to form cylinder (Fig. 10.30C), or of fragments of bark and leaves. Northern and transcontinental, higher elevations in western mountains to California **10.30 Lenarchus**

Metanotum with at least a few setae arising between primary sclerites (Figs. 10.25B, 10.46B) **48**

48 (47) Lateral series of abdominal gills on segment II and occasionally III, and of single filaments only (Fig. 10.46A); case of small rock fragments (Fig. 10.46C). High elevations in New Mexico, Colorado, and probably adjacent states **10.46 Psychoronia**

Lateral series of abdominal gills usually extending to segment V, and at least some with more than one branch (Fig. 10.25A); case mostly of small rock fragments, sometimes with plant pieces as well (Fig. 10.25C). Widespread **10.25 Hesperophylax**

10.1 Genus **Allocosmoecus**

DISTRIBUTION AND SPECIES *Allocosmoecus partitus* Banks, the single species known in this genus, was known only from Idaho (Flint 1966); we have collected it in British Columbia, California, Idaho, Oregon, and Washington.

Larvae in *Allocosmoecus* have not been identified previously in the literature; we reared specimens from California and Oregon.

MORPHOLOGY Larvae are similar to those of *Dicosmoecus* but are distinguished by short scale-hairs scattered over the head and pronotum (B), and by a transverse row of setae posterior to the median dorsal hump on abdominal segment I (D). *Allocosmoecus* is also distinctive in having three ventral chloride epithelia on each of segments III–V (A,E), and in lacking sclerotized tubercles on the side of abdominal segments. As in *Dicosmoecus* the middle and hind tibiae bear several pairs of stout, spur-like setae (A), and the femur of at least the hind legs bears more than two major setae (A). Sclerotized parts of the head and thorax are dark brown to black. For the most part, abdominal gills have three or four branches, occasionally one or two. Length of larva up to 25 mm.

CASE The case of the final instar is constructed entirely of small stones but, as in *Dicosmoecus*, cases of the earlier instars show a transition from plant to rock components. Superficially the final-instar case is similar to that of *Dicosmoecus* spp., but usually is constructed of coarser pieces, giving it a rougher, cruder aspect. Length of larval case up to 31 mm.

BIOLOGY We collected larvae of *A. partitus* in small, cool streams, but did not find them in the larger rivers often frequented by some species of *Dicosmoecus*. Gut contents of larvae (3) examined were largely vascular plant fragments, fine organic particles, and filamentous algae. Our collections indicate that a generation is completed in one year; larvae seal up their cases in June and July, and adults appear in September and October, the length of the intervening period suggesting that the larvae undergo diapause similar to *Dicosmoecus* (q.v.).

Allocosmoecus partitus (Idaho, Kootenai Co., 8 June 1967, ROM)
A, larva, lateral x6, middle tibia enlarged; B, head and thorax, dorsal, portion of head capsule enlarged; C, case, ventral x4; D, segment I, dorsal; E, segments I–III, ventral

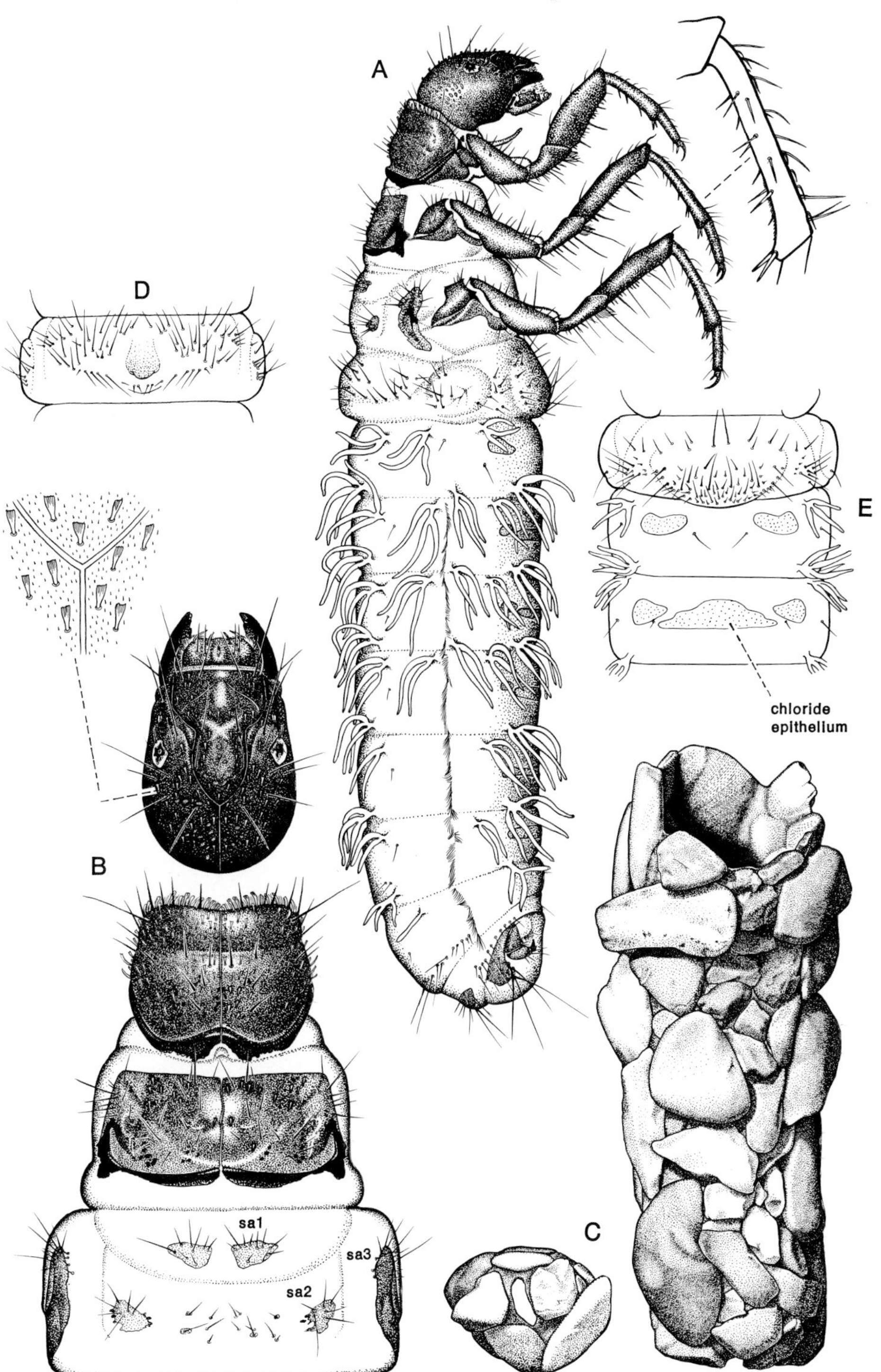
A
B
C
D
E
sa1
sa2
sa3
chloride
epithelium

10.2 Genus **Amphicosmoecus**

DISTRIBUTION AND SPECIES A single species, *Amphicosmoecus canax* (Ross), is known in this genus. Originally discovered in Utah (Ross 1947), this species was subsequently recorded from Alberta and British Columbia (Nimmo 1965).

The larva of *A. canax* has not previously been recorded; we have collected series of a distinctive larva, here illustrated, in Alberta, British Columbia, California, Oregon, and Saskatchewan; and we reared from one of the British Columbia larvae a pharate female that seems generally concordant with females of *A. canax*.

MORPHOLOGY Structurally, these larvae resemble those in *Onocosmoecus*, but can be distinguished by the fused *sa*1 sclerites on the metanotum (B). Sclerotized parts of the head and thorax are golden brown in colour and have the light median band characteristic of several genera in the Dicosmoecinae; the hind femur bears more than two major setae, and the middle and hind tibae bear only apical spurs (A). Length of larva up to 23 mm.

CASE Larvae believed to be of this genus have been found in cases that are of two types. One (C) is a hollow twig or stem, frequently with a turret of bark pieces around the anterior end; in some cases the twig shows evidence of having been tapered at the ends presumably by the mandibles, as illustrated. The other case type is constructed entirely of small chunks of wood suggestive of the case in *Onocosmoecus* except that in *Amphicosmoecus* the pieces of bark are thicker. Length of larval case up to 33 mm.

BIOLOGY These larvae live in small, cool streams although we have collected some in lakes (Saskatchewan). Our records are consistent with one generation per year; final-instar larvae occur as early as June and July, adults in September and October. Evidence suggests that the species is very local in distribution, and even where it does occur, larvae are not abundant.

Amphicosmoecus canax (Oregon, Klamath Co., 8 June 1968, ROM)
A, larva, lateral x7; B, head and thorax, dorsal; C, case x5

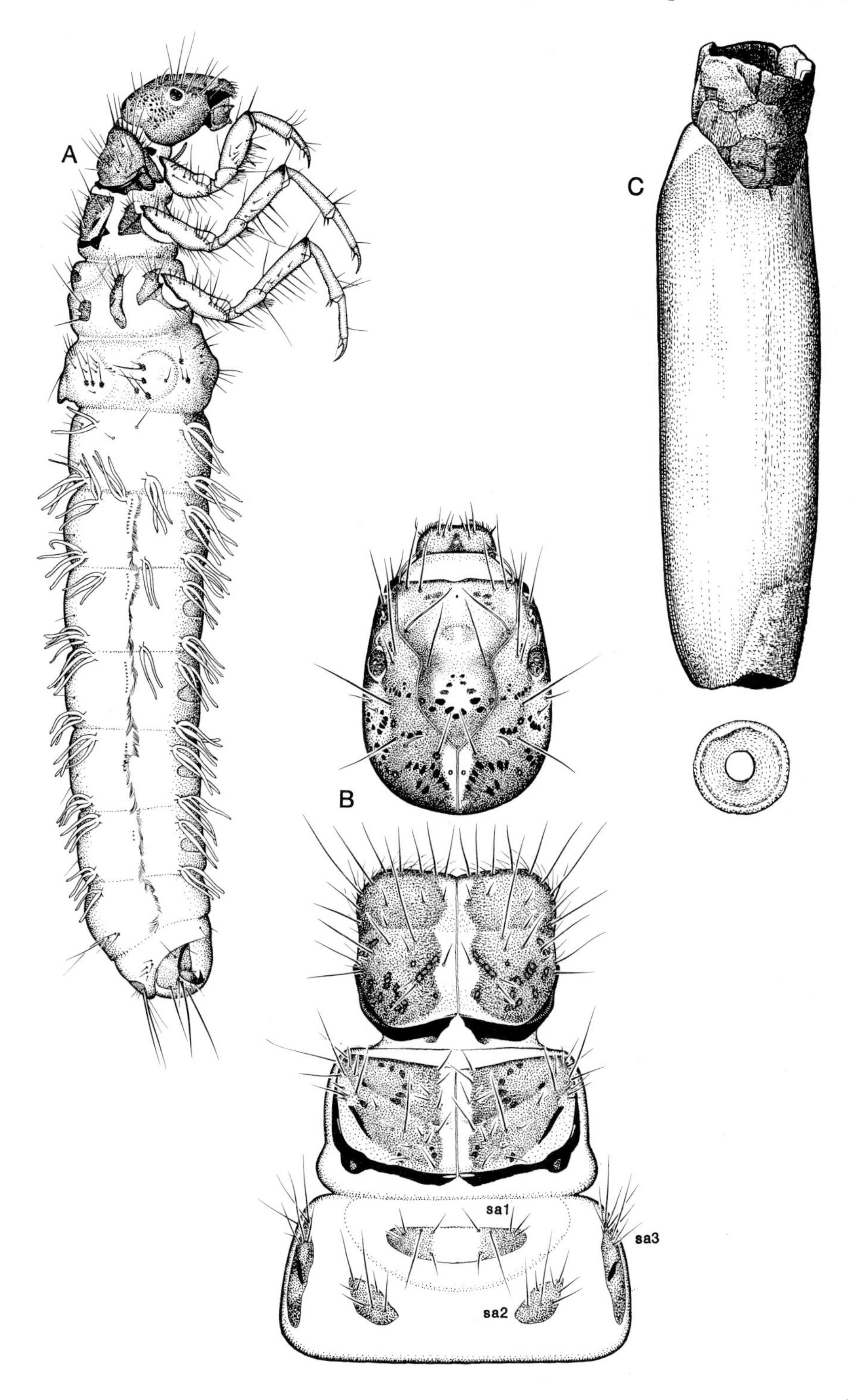
A
B
C
sa1
sa2
sa3

10.3 Genus **Anabolia**

DISTRIBUTION AND SPECIES Species of this genus are widely distributed in the Holarctic region (Schmid 1950a). Four are known from this continent: *A. bimaculata* (Walk.) and *A. consocia* (Walk.) are northern and transcontinental; *A. sordida* Hagen and *A. ozburni* (Milne) are northeastern.

Larvae have been described for all except *A. ozburni* (Flint 1960).

MORPHOLOGY Larvae have characteristic head markings of light brown with relatively large, dark spots coalescing, especially on the frontoclypeal apotome, into irregular blotches (B); spotted head markings may also occur in *Limnephilus, Grammotaulius*, and *Asynarchus* but the coalescence of spots in *Anabolia* appears to be diagnostic for that genus. Larvae of *Anabolia* can also be distinguished by a small patch of stout spines on the anterolateral corner of the pronotum (A,D). The hind femur bears only two major setae on the ventral edge (A). Chloride epithelia are present dorsally and laterally as well as ventrally. Abdominal gills are mostly two- or three-branched; and the anal claw usually bears three accessory hooks (A). Length of larva up to 29 mm.

CASE Larval cases are made largely of elongate pieces of twigs or stems arranged lengthwise (C); sometimes larvae of this genus construct three-sided cases of broader pieces of leaves (Flint 1960). Length of larval case up to 50 mm.

BIOLOGY Larvae inhabit marshes, slow-flowing streams, and temporary pools (Flint 1960; Wiggins 1973a). Gut contents of larvae (3) examined were largely pieces of vascular plant tissue; because of their widespread distribution and frequent local abundance these larvae are probably important detritivores in lentic communities. Diapause occurs in the last larval instar of the European species *A. furcata* Brauer (Novák 1960), but none is known in Nearctic species of *Anabolia.*

Anabolia bimaculata (Alberta, Blackfoot, 16 June 1962, ROM)
A, larva, lateral x7, anal claw enlarged; B, head and thorax, dorsal, mesonotal setal areas detailed; C, case, ventral x3; D, spines on anterolateral corner of pronotum, lateral

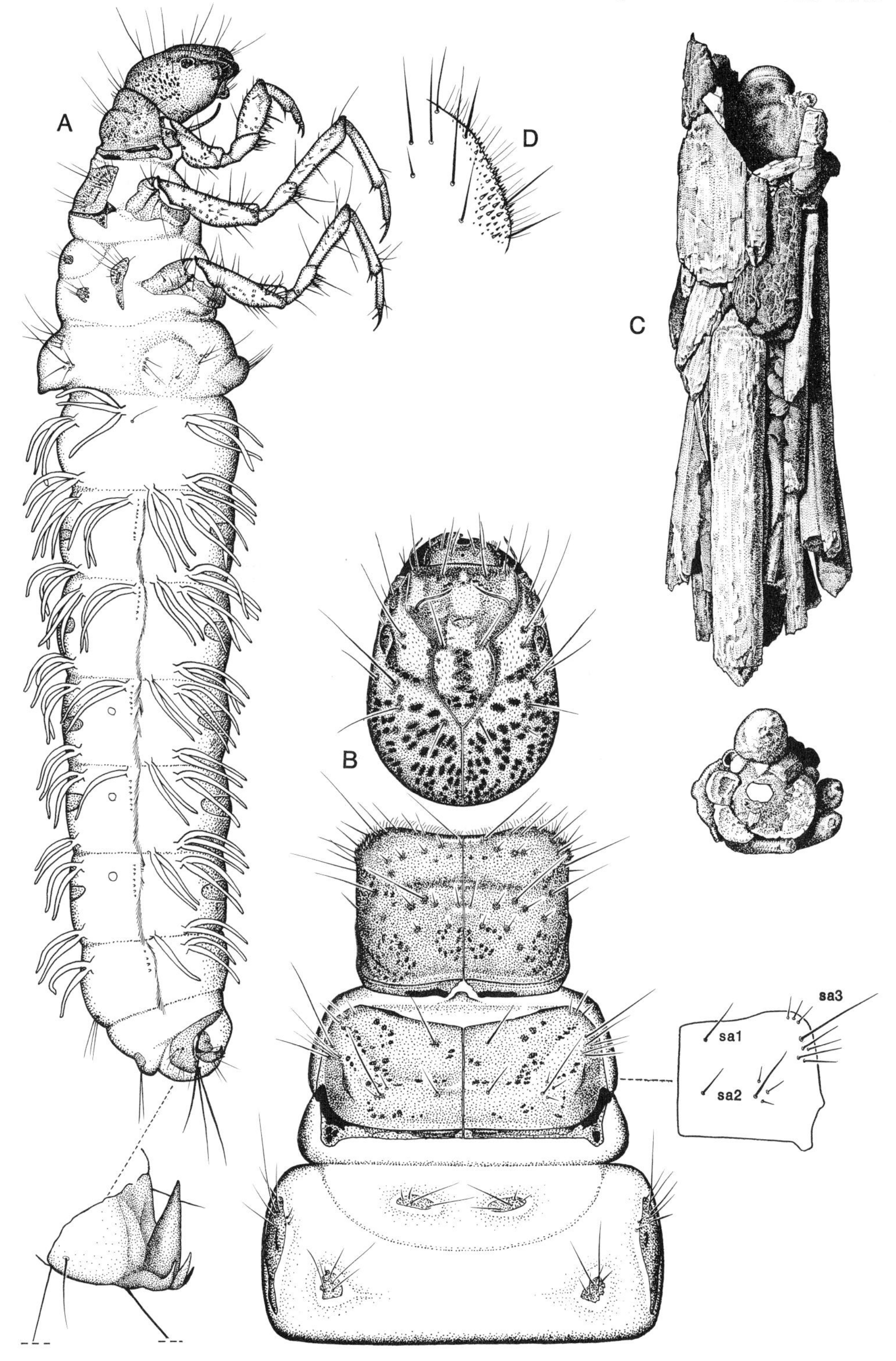
A
B
C
D
sa1
sa2
sa3

10.4 Genus Apatania

DISTRIBUTION AND SPECIES *Apatania* is a genus of some 50 species in the Holarctic and Oriental regions (Schmid 1953, 1954a). Some live in far northern arctic latitudes, and a few are circumboreal. Fifteen species are now recognized in North America where some range southward along the eastern and western mountain ranges to Georgia and Arizona respectively.

Larvae have been described (Flint 1960) only for *A. incerta* (Banks) in the east, and (Wiggins 1973c) for one western species, *A. arizona* Wiggins. Larvae are known for three northern species: *A. crymophila* McL. (Lepneva 1966); *A. stigmatella* (Zett.) (Flint 1960; Lepneva 1966); *A. zonella* (Zett.) (Lepneva 1966). We associated the larva for one other western species, *A. tavala* Denning.

MORPHOLOGY Larvae can be distinguished from those of any other limnephilid in North America by absence of the metanotal *sa*1 sclerites and arrangement of those setae in a continuous transverse band (B). Length of larva up to 7.5 mm.

CASE The larval case (C) is constructed of coarse rock fragments, slightly curved dorsoventrally and tapered strongly from front to rear; in some species larger pieces are arranged along each side. Final-instar larvae extend the dorsal edge of the anterior opening to overhang the ventral edge, resulting in the characteristic *Apatania* pupal case in which the anterior opening is on the ventral surface. Length of larval case up to 9.5 mm.

BIOLOGY Larvae are inhabitants of cool waters; eastern and western montane species usually occur in spring streams, but at more northerly latitudes larvae occur in deeper waters of lakes. The Holarctic species *Apatania zonella* was recorded from Lake Hazen, Ellesmere Island (71°18'W, 81°49'N), where females greatly outnumbered males and were believed to reproduce parthenogenetically for the most part (Corbet 1966). Similar observations on other species of *Apatania* in Denmark (Nielsen 1950) indicate that this phenomenon occurs at more southern latitudes as well. Food of *Apatania* larvae is largely diatoms and other algae (Nielsen 1942, 1943a; Lepneva 1966). Biology of the European *A. muliebris* McL. was studied by Elliott (1971).

REMARKS This group has often been treated under the name *Radema.*

Apatania arizona (Arizona, Coconino Co., 4 July 1966, ROM)
A, larva, lateral x14, claws of tarsus and anal proleg enlarged; B, head and thorax, dorsal; C, case, ventrolateral x10; D, mandible, ventral; E, labrum, dorsal; F, head, dorsal; G, head, ventral; H, segment IX and anal prolegs, dorsal

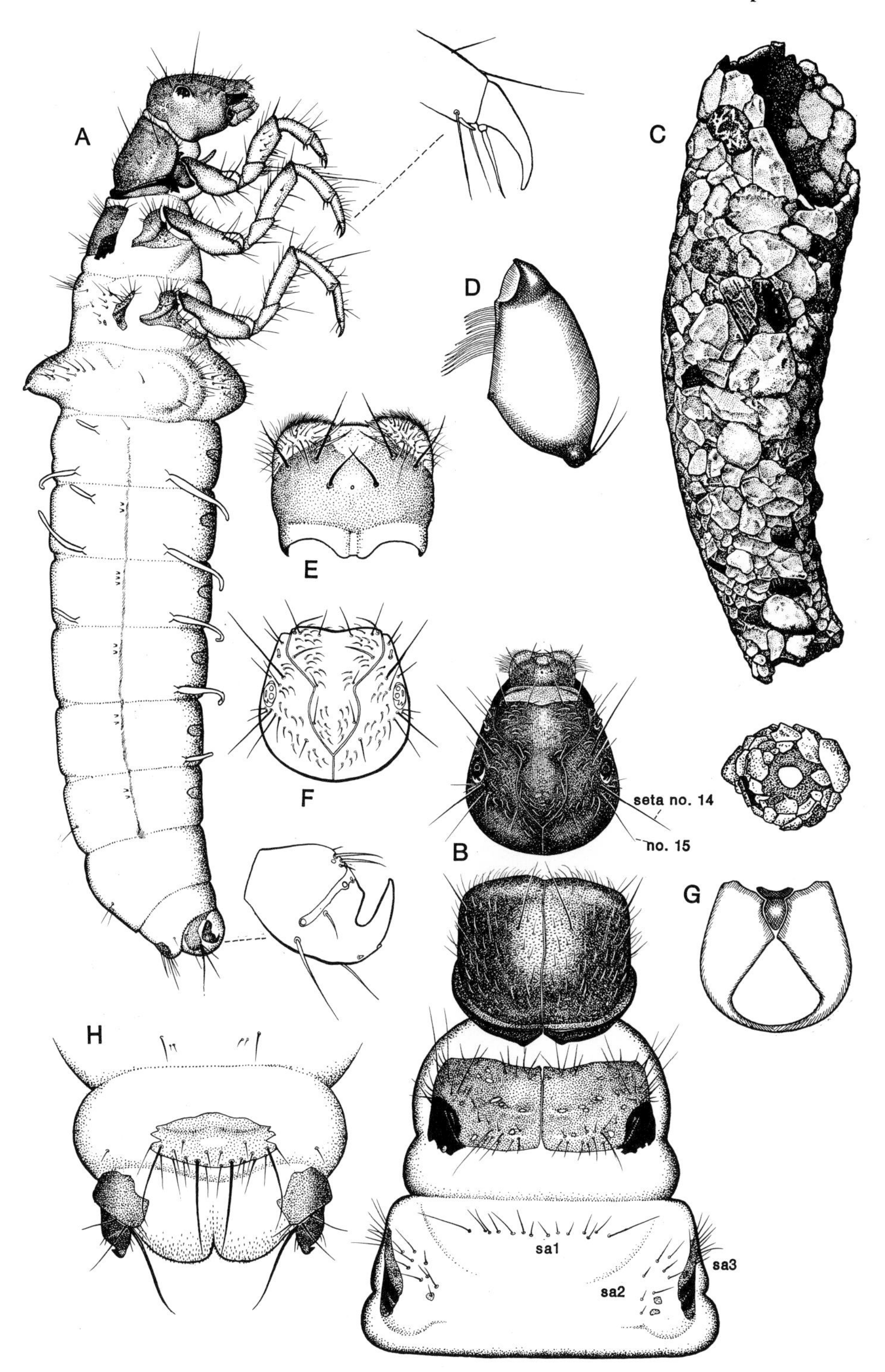
A
C
D
E
F
B
seta no. 14
no. 15
G
H
sa1
sa2
sa3

10.5 Genus **Arctopora**

DISTRIBUTION AND SPECIES *Arctopora* is a small Holarctic genus. There are two North American species: *A. pulchella* (Banks) from the northeast and *A. salmon* (Smith) from Idaho. A third species, *A trimaculata* (Zett.), occurs in northern Eurasia and Alaska.

Larvae of only *A. pulchella* are known, reared by us in Ontario, and described by Flint (1960).

MORPHOLOGY Meso- and meta-notal setae are reduced in number and confined to the primary areas; metanotal *sa*2 sclerites are lacking, and the *sa*1 sclerites very small (B). Setae of abdominal segment I are also reduced (A,D). Sclerotized parts of the head and thorax are light brown with few markings except faint muscle scars. Most gills have two or three branches; chloride epithelia are present on the venter of segments II–VII, and are apparent on the dorsum of some segments as well (A). Length of larva up to 15 mm.

CASE The larval case of *A. pulchella* is constructed of pieces of grass or sedge leaves arranged longitudinally to form a cylinder of rather regular outline. Length of larval case up to 27 mm.

BIOLOGY *A. pulchella* was reared from larvae collected in a temporary pool.

REMARKS These species have been treated under *Lenarchulus* (Schmid 1955) but it was pointed out by Fischer (1969) that *Arctopora* Thomson is the valid name for the genus. Taxonomy of adults was reviewed by Schmid (1952b) and Smith (1969).

Arctopora pulchella (Ontario, Algonquin Prov. Park, May 1959, ROM)
A, larva, lateral x12; B, head and thorax, dorsal; C, case, ventral x5; D, segment I, ventral

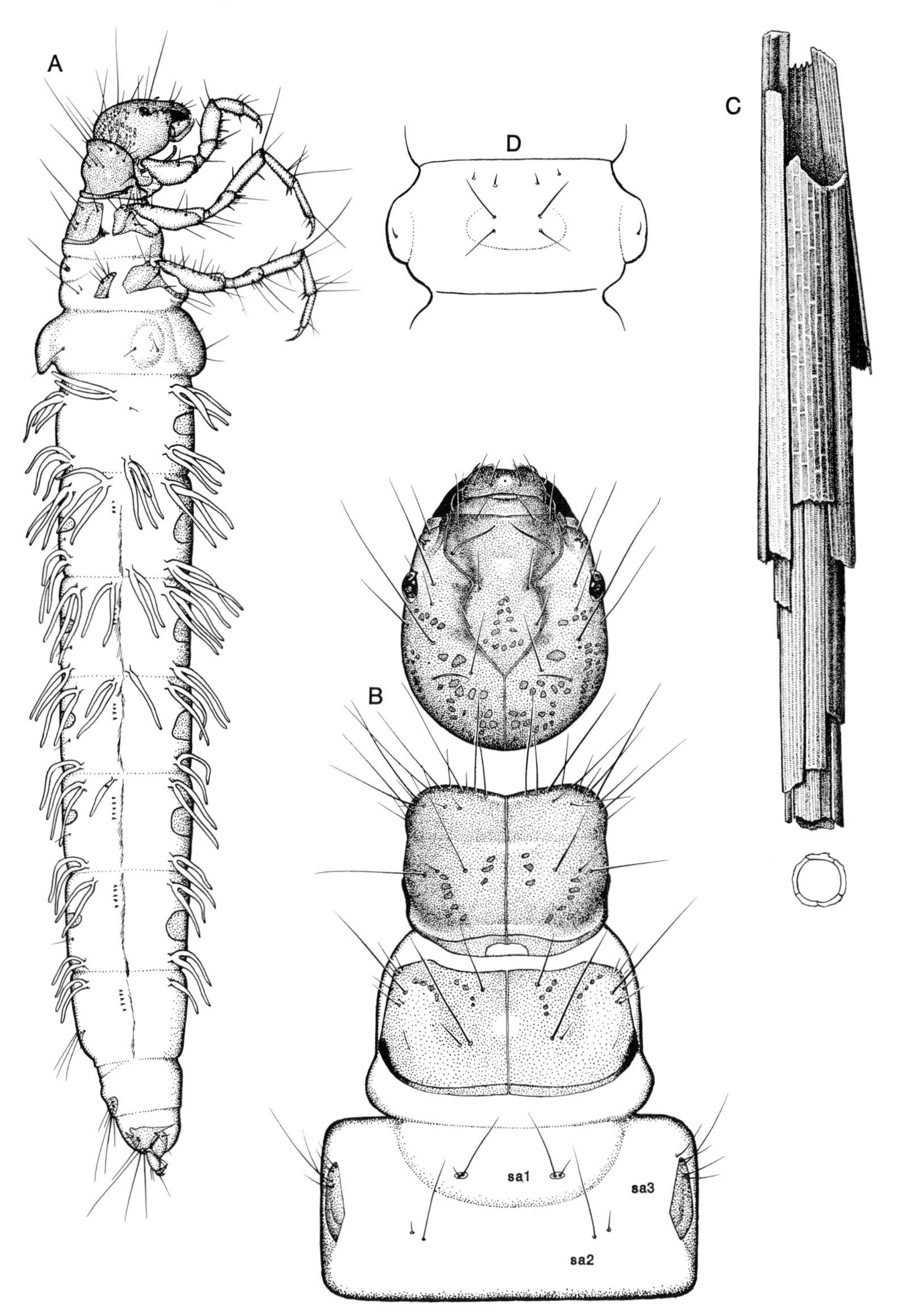
A
B
C
D
sa1
sa2
sa3

10.6 Genus Asynarchus

DISTRIBUTION AND SPECIES Ten of the 16 species assigned to this genus occur in North America where they are widespread across the northern half of the continent, extending south in the western mountains to Utah and Colorado. The remaining species occur in northern Eurasia; several species are Holarctic in distribution.

For the North American fauna, larvae have been described previously only for *A. montanus* (Banks) (Flint 1960, as *A. curtus*); we have associated larvae for four additional species.

MORPHOLOGY Larvae of *Asynarchus* usually have blotches or bands of contrasting colour on the dorsum of the head with three light areas in the narrowed part of the frontoclypeal apotome (B). Since there is a tendency for reduced contrast in these head marking, the genus can be reached at two places in the key. Mesonotal *sa*1 setae range from 2 to 8 and are rather variable within species, but evidently there is always more than one seta in this position. In most characters, larvae fit within the range of *Limnephilus*, and indeed the species of *Asynarchus* have been assigned to *Limnephilus* in the past. Chloride epithelia are, however, present dorsally on all Nearctic *Asynarchus* now known; dorsal chloride epithelia have not yet been found in Nearctic *Limnephilus*, but since they occur in some European larvae of this genus, this distinction may not hold for all species on this continent. Length of larva up to 23 mm.

CASE Larval cases in this genus are constructed either of small rock fragments (C) or of plant materials arranged lengthwise. Length of larval case up to 28 mm.

BIOLOGY We have collected larvae in streams, ponds, and temporary pools.

REMARKS Taxonomy of *Asynarchus* adults was reviewed by Schmid (1954b). The larva of *A. rossi* (Leonard and Leonard), which we have identified, will key to *Limnephilus* in the present work; I am advised by F. Schmid that based on characters of the adults, this species would probably be better placed in *Limnephilus*, and I am so treating it.

Asynarchus circopa (Montana, Beaverhead Co., 7-8 July 1964, ROM)
A, larva, lateral x8, detail of dorsal chloride epithelia; B, head and thorax, dorsal, detail of mesonotum; C, case x5; D, segment I, ventral

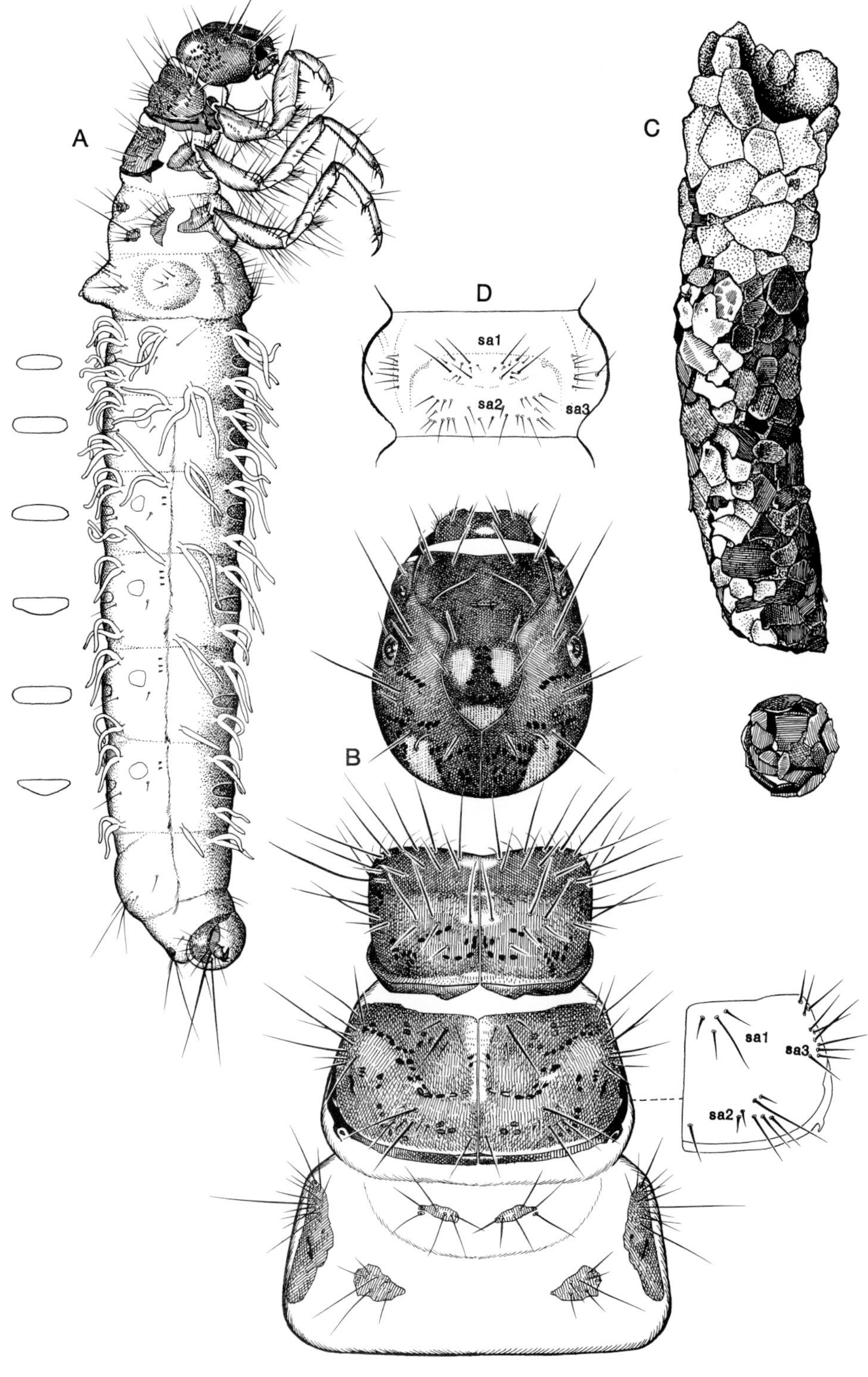
A
B
C
D
sa1
sa2
sa3
sa1
sa3
sa2

10.7 Genus **Chyranda**

DISTRIBUTION AND SPECIES *Chyranda centralis* (Banks) is the sole species recognized in the genus. Its range extends across most of Canada, north to Alaska and south to California, Utah, and Colorado in the western mountains; our collection records indicate that the species is much more abundant in the western part of the continent.

The larva and pupa of *C. centralis* were described by Wiggins (1963).

MORPHOLOGY The larva of *Chyranda* bears little resemblance to any other. The head (B) is of fairly uniform dark brown colour, smooth and shiny, round in circumference, and the dorsum somewhat flattened. The shape of the unusually large, irregular sclerite at the base of each lateral abdominal hump is diagnostic. On the venter of abdominal segment I a light sclerite in the *sa*2 position bears approximately eight setae (A); several setae on this segment have small basal sclerites. Abdominal gills are single and lateral tubercles are apparently absent; chloride epithelia are present ventrally on segments III–VII. Length of larva up to 18 mm.

CASE The larval case (C) consists of pieces of thin bark or stout leaves arranged to form a straight tube with a prominent flange-like seam along each side. In cross-section the case is broadly elliptical. Larval cases in only *Clostoeca* (Fig. 10.9C) are similar, but the lateral flanges are much wider in that genus and fewer leaf pieces are incorporated into the case. Length of larval case up to 35 mm.

BIOLOGY Larvae live in small spring streams, and are usually found in accumulations of leaves. Pieces of vascular plants and moss were the dominant items in guts of larvae (3) we examined.

Chyranda centralis (Oregon, Douglas Co., 9 June 1968, ROM)
A, larva, lateral x9, segment I enlarged; B, head and thorax, dorsal; C, case, ventrolateral x5

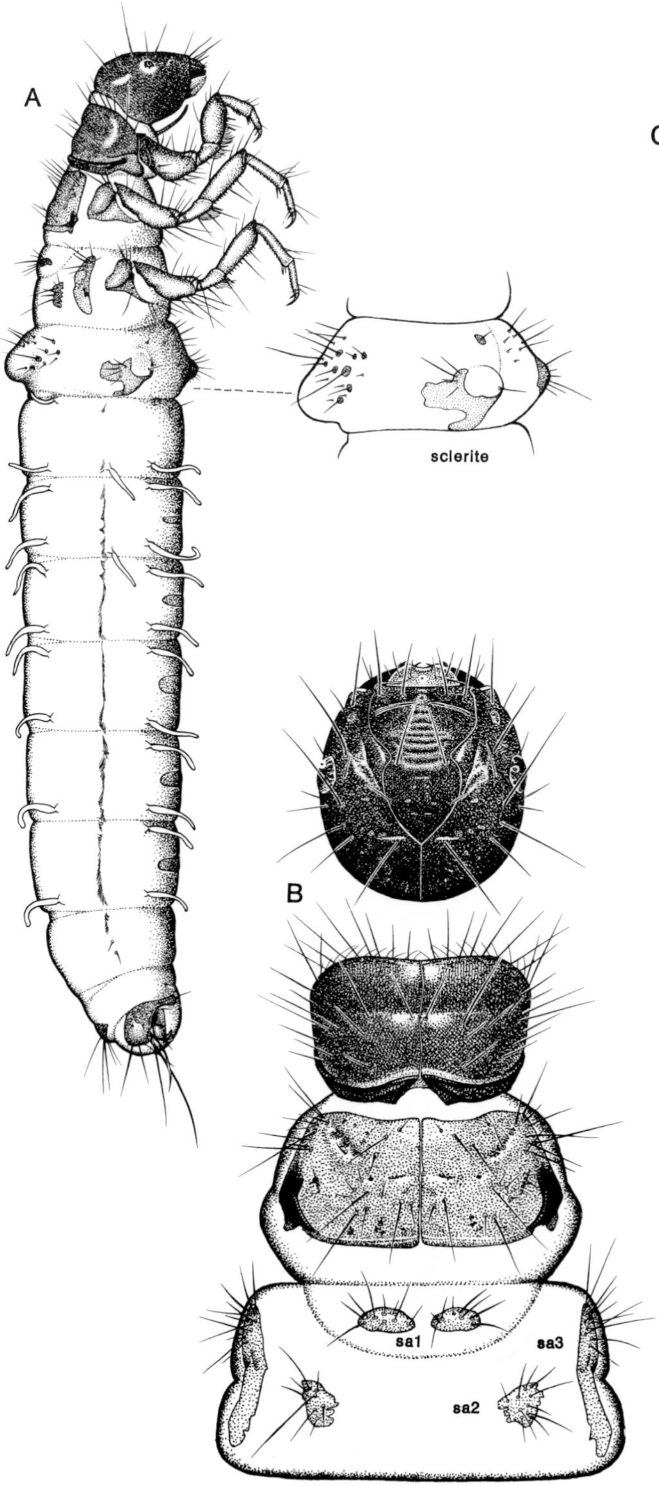

A

sclerite

B

sa1

sa3

sa2

C

10.8 Genus **Clistoronia**

DISTRIBUTION AND SPECIES This genus is confined to North America, with four species distributed over western montane areas from Alaska to Arizona. An additional species was described from Mexico at an elevation of 8,800 feet (approximately 2,700 m) (Denning and Sýkora 1966; Flint 1967c).

Larvae of *Clistoronia* have not been identified previously in the literature. We have reared material of *C. magnifica* (Banks) from localities in British Columbia, Oregon, and Washington.

MORPHOLOGY The larva of *C. magnifica* has head markings generally similar to those of related genera such as *Limnephilus* and *Asynarchus*, but the venter of abdominal segment I is unusually hairy, with more than 100 setae in all setal positions combined, merging from both sides (D). The head and pronotum are covered with small spines (B), and the lateral sclerite of the anal proleg bears several short, stout setae (A). Most gills are three-branched. Length of larva up to 28 mm.

CASE The larval case (C) is composed of small pieces of wood arranged irregularly to form a cylinder with little curvature or taper. In our associated series from Washington a new case of fine rock fragments was constructed before pupation to replace the original larval case of wood pieces. Length of larval case up to 34 mm.

BIOLOGY Species of *Clistoronia* generally occur at higher elevations, the larvae of at least *C. magnifica* in ponds and small lakes. According to Winterbourn (1971a), who studied the life history of *C. magnifica* in Marion Lake, British Columbia (elev. approx. 300 m), oviposition occurred in August and September, larvae fed primarily on detrital materials in bottom sediments and grew quickly to overwinter as final instars; adults emerged in May and June but apparently did not become sexually mature until late summer. In a population of *C. magnifica* that we sampled at a higher elevation, adults evidently emerged later (Washington, Mt Rainier National Park, elev. approx. 1,500 m, pupae 4 July).

Clistoronia magnifica (Oregon, Jefferson Co., 1 June 1968, ROM)
A, larva, lateral x7, anal proleg enlarged; B, head and thorax, dorsal; C, case x3; D, segment I, ventral

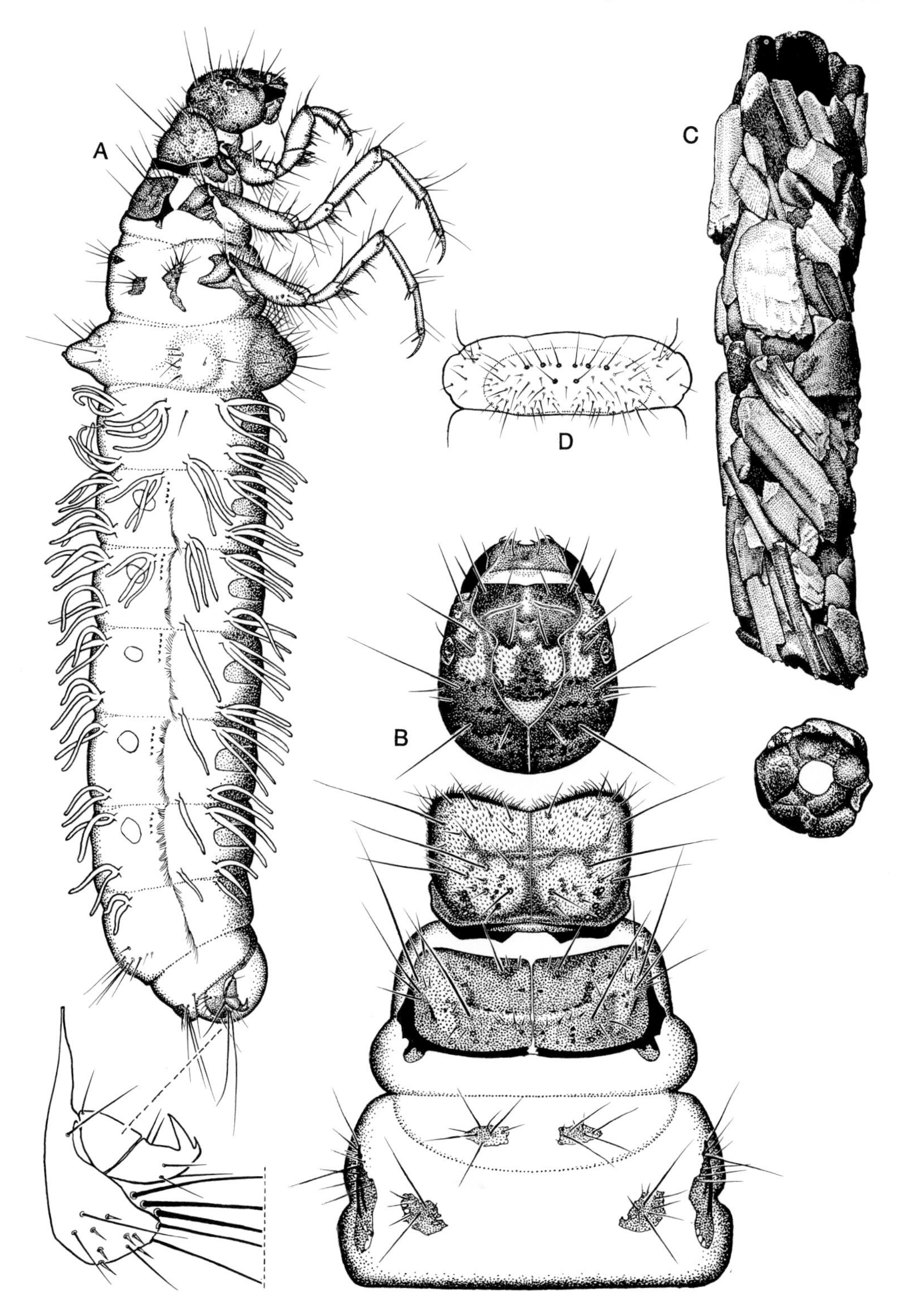
A
B
C
D

10.9 Genus **Clostoeca**

DISTRIBUTION AND SPECIES This genus occurs only in western North America, from California to British Columbia and possibly to Alaska (Flint 1960). All populations are currently recognized as a single, somewhat variable, species *C. disjuncta* (Banks).

The larva was described by Flint (1960); we reared larvae from Oregon.

MORPHOLOGY Sclerites of the head and thoracic nota are uniform reddish to yellowish brown, shiny and smooth, marked only by faint muscle scars (B); mesonotal setae are sparse, the primary setal areas widely separated. Lateral humps of abdominal segment I lack basal sclerites. Ventral chloride epithelia are present on segments II–VII, lateral tubercles are absent, and gills are single. Length of larva up to 14 mm.

CASE The larval case in *Clostoeca* is constructed of leaf pieces but is similar only to that in *Chyranda* because of the lateral flanges; the flanges are much wider in *Clostoeca* (C) and there are fewer individual pieces in the entire case. Length of larval case up to 18.5 mm.

BIOLOGY Larvae we reared in Oregon were collected under wet leaves in a spring seepage area.

Clostoeca disjuncta (Oregon, Benton Co., 8 April 1964, ROM)
A, larva, lateral x11; B, head and thorax, dorsal, detail of mesonotum; C, case, ventral x8

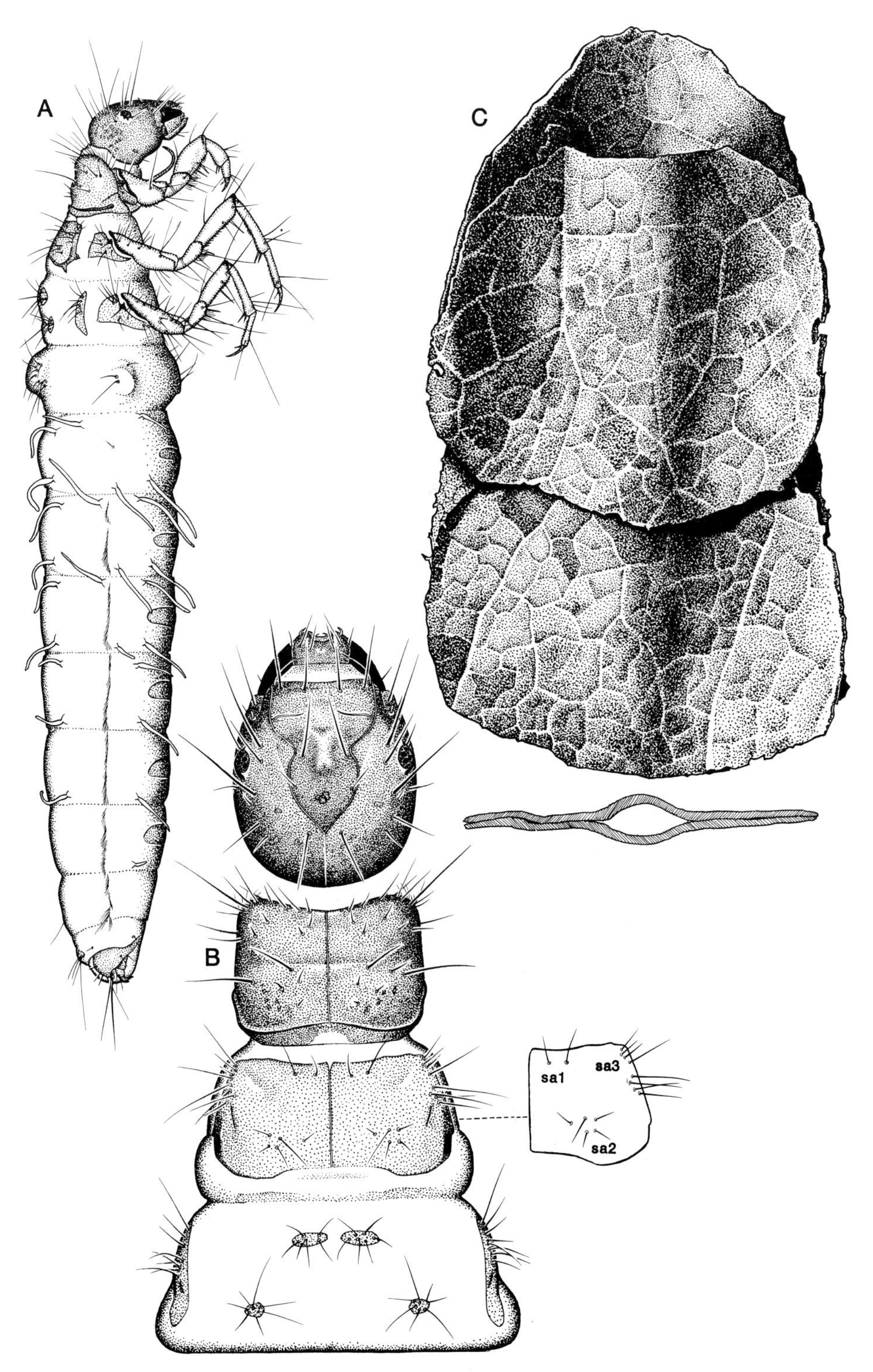
A
C
B
sa1
sa3
sa2

10.10 Genus **Cryptochia**

DISTRIBUTION AND SPECIES Seven species are known in this genus, all confined to western North America. We have collected larvae in mountainous sections of Alberta, British Columbia, California, Idaho, Montana, Oregon, Washington, and Wyoming.

Larvae of *Cryptochia* have not been previously identified in the literature, although the larva described as Dicosmoecinae 1 by Flint (1960) belongs to this genus; we associated larvae of *C. pilosa* (Banks) in Oregon.

MORPHOLOGY The anterior edge of the pronotum bears a dense fringe of very long setae (A), often heavily laden with sand and silt. The dorsum of the head is flattened and bears a peripheral carina thickly beset with setae, and the frontoclypeal apotome bears two bands of dense setae (B). The ventral apotome has concave lateral margins terminating posteriorly in a sharp point (E), and sclerites of the submentum and stipes are unusual in bearing many long setae (F); the galea is shorter than in other members of the Dicosmoecinae, but its ring-like sclerite is evident. Dorsal and ventral abdominal gills are of one or two filaments, but most of the lateral gills have four filaments. The legs bear small spines in comb-like clusters (D). Length of larva up to 8 mm.

CASE The unique larval cases of *Cryptochia* (C) are flattened, taper strongly from front to rear, and are made of plant materials arranged transversely. The smooth outer surface of the posterior part of the case suggests that the larva shapes the rough edges with its mandibles, although I have not observed this behaviour. Length of larval case up to 13 mm.

BIOLOGY We have found larvae in a wide range of small, cold, spring streams, usually in mountainous terrain. Some larvae have been found beneath the water surface in areas of little current, but more often they were collected above the surface on pieces of wet wood or in wet leaves at the water's edge. Pupae have been found in similar wet logs above the water surface in June. Vascular plant tissue and fine organic particles were dominant in the guts of larvae (3) we examined.

REMARKS A key to males of seven species was given by Denning (1975). One species in this genus, *C. pilosa* (Banks), was originally assigned to the European genus *Parachiona* (see Ross 1944: 297).

Cryptochia pilosa (Oregon, Douglas Co., 9 June 1968, ROM)
A, larva, lateral x25; B, head and thorax, anterolateral; C, case, anteroventral x14; D, middle leg, lateral, spines enlarged; E, head, ventral; F, maxilla and half of labium, ventral

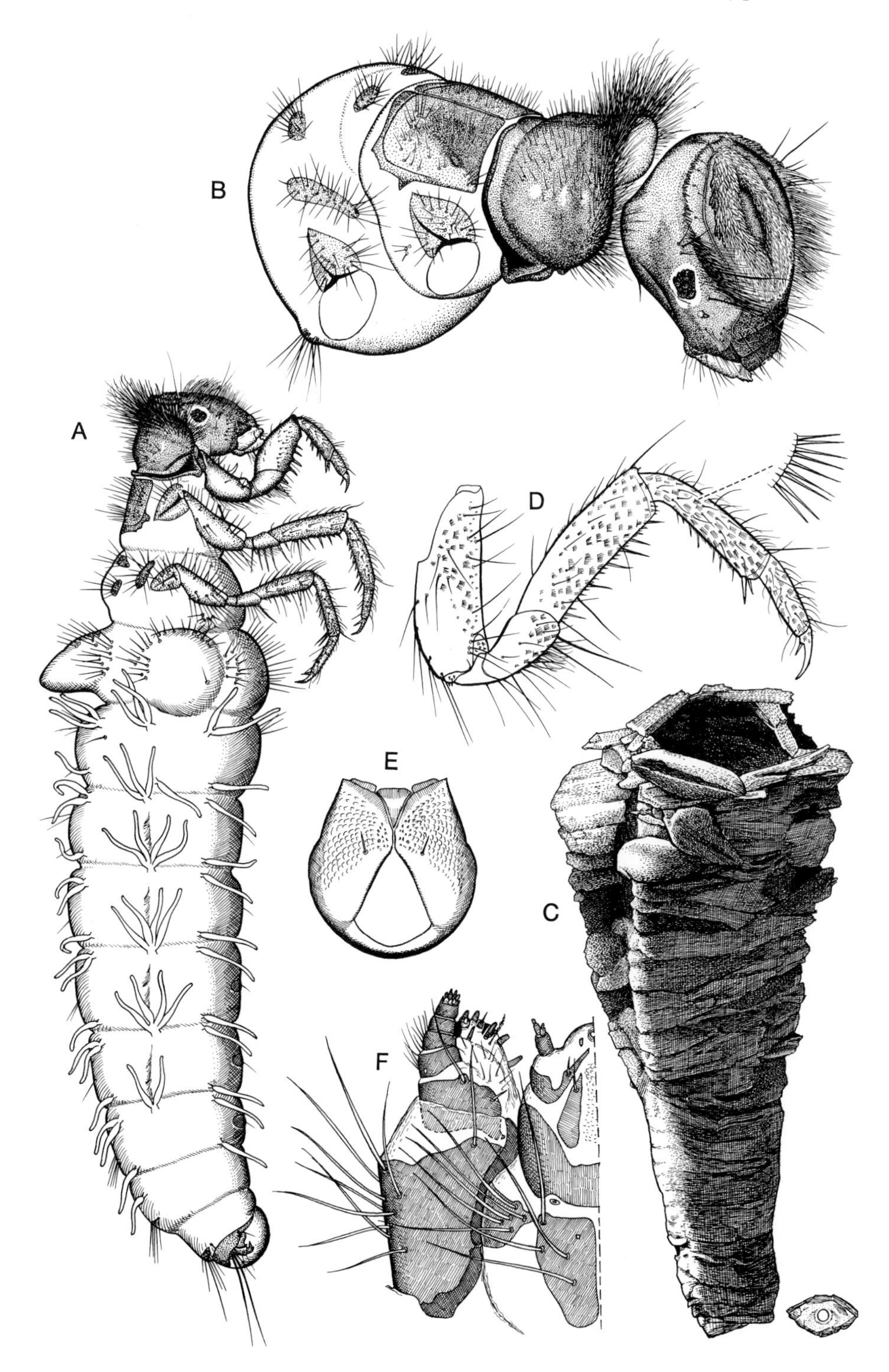
B
A
D
E
C
F

10.11 Genus Desmona

DISTRIBUTION AND SPECIES This genus consists of the single species *D. bethula* Denning from California.

Larvae have not been identified in the literature heretofore; we collected larvae and pharate adults from Sagehen Creek at the Research Station of the University of California in Nevada County.

MORPHOLOGY The diagnostic character by which larvae of *D. bethula* can be recognized is two small, ring-shaped sclerites on abdominal segment I at the base of each lateral hump (A). Sclerotized parts of the head and thorax are dark brown without prominent markings; there are small secondary setae on the dorsum of the head (B). The thoracic nota are heavily setate; some metanotal setae usually arise from the membrane between the primary sclerites, although the number of these setae is highly variable, with sometimes only a few between the *sa*2 sclerites. The prosternal horn is long, and coxae of the hind and middle legs have coarse spines scattered over their anterior surfaces. The sclerite on abdominal segment IX bears many setae (D). Abdominal gills are single; chloride epithelia are present ventrally on segments II–VII. Length of larva up to 15 mm.

CASE The larval case (C) consists mainly of small rock fragments, with some small pieces of wood added, the exterior regular in outline. Length of larval case up to 17 mm.

BIOLOGY Larvae we collected were in small spring streams, frequently buried in sand deposits. Final-instar larvae were collected in July, and adults emerge in October. Vascular plant pieces and fine organic particles were predominant in guts of larvae (3) we examined.

REMARKS Diagnostic characters of the male of *D. bethula* were given by Denning (1956).

Desmona bethula (California, Nevada Co., 20 July 1966, ROM)
A, larva, lateral x10, segment I and hind femur enlarged; B, head and thorax, dorsal; C, case x8; D, segments VIII and IX, dorsal

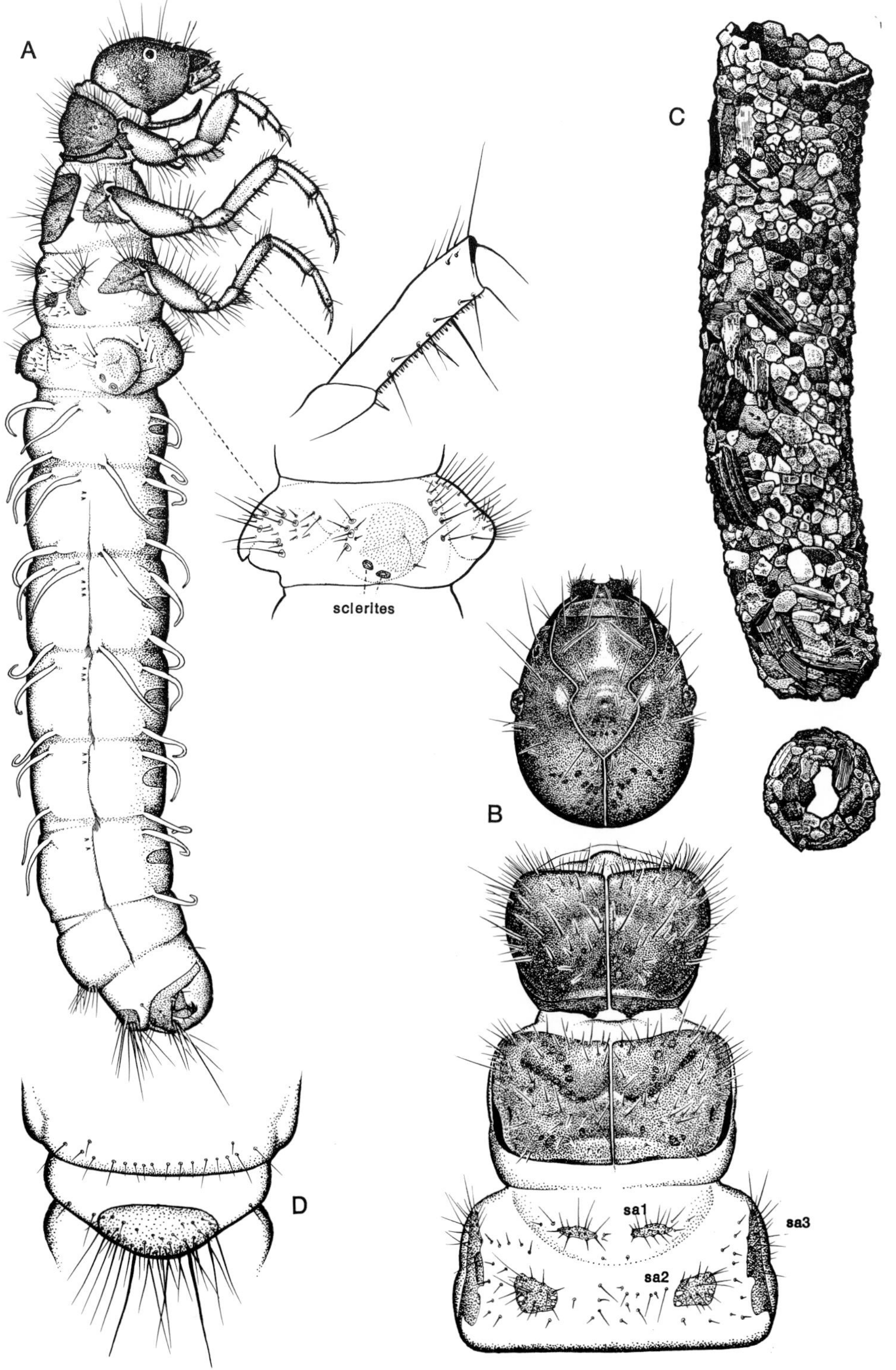
A
C
sclerites
B
D
sa1
sa3
sa2

10.12 Genus **Dicosmoecus**

DISTRIBUTION AND SPECIES *Dicosmoecus* is a Holarctic genus confined to western North America, Japan, and the eastern USSR; the Nearctic species occur in montane areas from Alaska to California. There has been some uncertainty about the actual number of species in North America. Four species remained from the available names listed by Schmid (1955) and variously synonymized by Flint (1966): *atripes* (Hagen), *gilvipes* (Hagen), *palatus* (McLachlan), and *pallicornis* Banks; but *jucundus* Banks was later distinguished from *atripes* (Nimmo 1971).

Larvae have been described for *D. gilvipes* (Flint 1960), and for the Palaearctic species, *D. palatus* (Lepneva 1966), recorded also from Alaska. We have associated larvae for these and for *D. atripes* and *pallicornis.*

MORPHOLOGY These larvae are large and stout-bodied; sclerotized parts of the head and thorax are mostly uniform dark brown to black. The tibiae of all legs have several pairs of stout spurs (A) and there are metanotal setae on the membrane between the primary sclerites (B); on segment IX a band of 20 to 40 setae extends ventrad from each side of the dorsal sclerite. Larvae we associated for two species heretofore assigned to *Dicosmoecus, frontalis* Banks (Flint 1966) and *schmidi* Wiggins (Wiggins 1975), do not share these diagnostic characters; and on the basis of evidence from the adults Dr F. Schmid will propose (Canadian Faunal Series, pers. comm.) that the same two species be transferred to the genus *Onocosmoecus.* That position is adopted here. Length of larva up to 35 mm.

CASE Final-instar larvae have a case of fine gravel, regular in outline, slightly curved, and often somewhat flattened (C). In the early instars cases are largely of plant materials, and transitional examples (Lepneva 1966, fig. 117) are often collected, demonstrating an abrupt change to gravel. Length of case up to 41 mm.

BIOLOGY *Dicosmoecus* larvae live in streams of various sizes. Gut contents in each of 3 larvae of *D. gilvipes* were largely fine organic particles and fine sand, generally an indication that rock surfaces are grazed; smaller proportions of filamentous algae, animal remains, and vascular plant pieces were also found. Larvae fix their cases to the underside of rocks in mid-summer and remain in diapause for several weeks before metamorphosis takes place, adults emerging in late summer and early autumn.

Dicosmoecus gilvipes (Oregon, Jefferson Co., 19-23 Sept. 1966, ROM)
A, larva, lateral x5, middle tibia enlarged; B, head and thorax, dorsal, detail of mesonotum; C, case, ventrolateral x3; D, segment I, dorsal; E, segments I and II, ventral

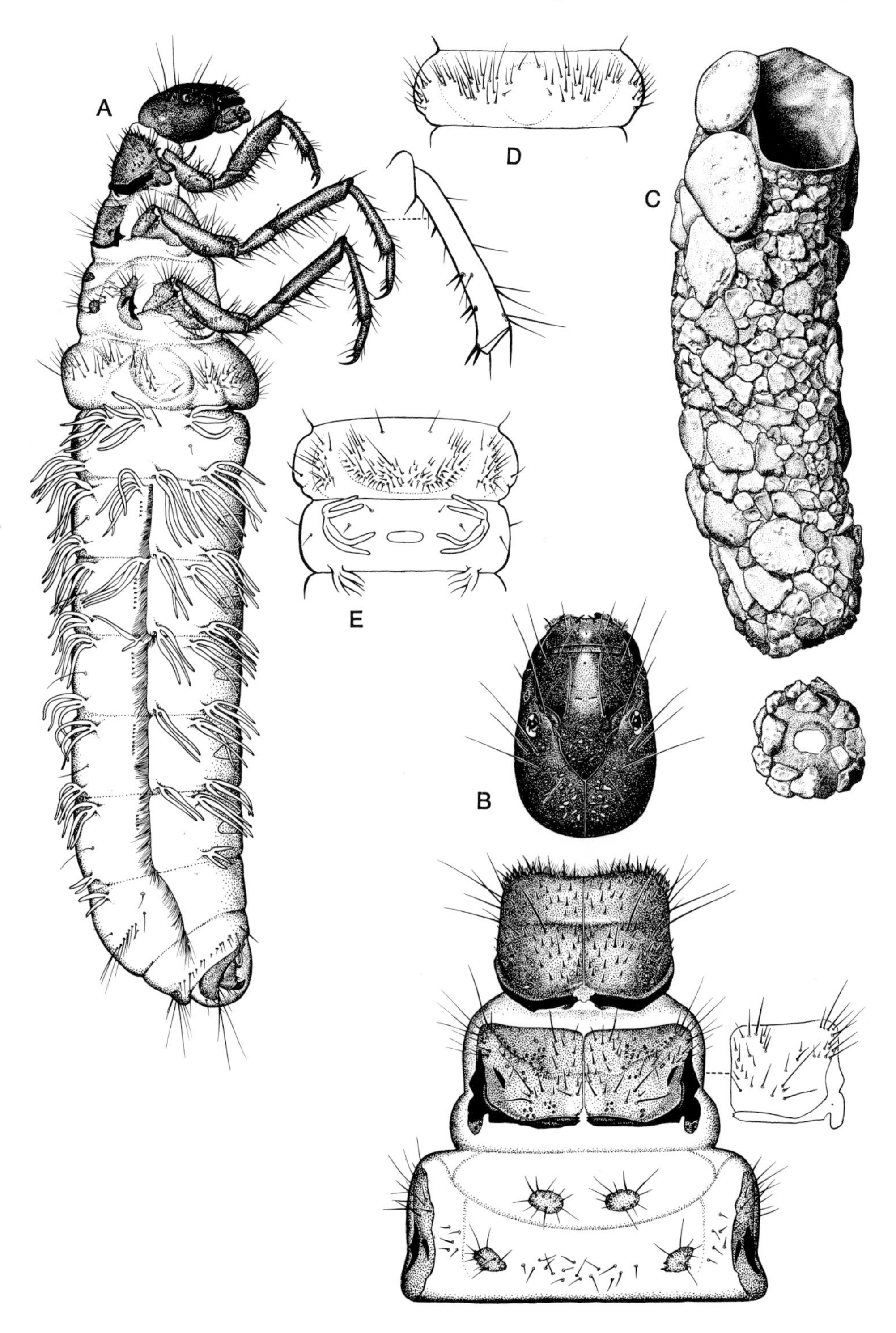
A
D
C
E
B

10.13 Genus Ecclisocosmoecus

DISTRIBUTION AND SPECIES This is a genus of two species: the type species from Sakhalin (Schmid 1964a) and *E. scylla* (Milne) from western North America. For many years the latter was assigned to *Ecclisomyia*, but when the larva was discovered it was evident that *E. scylla* was not congeneric with *Ecclisomyia* spp. (Wiggins 1975). We have collected larvae in British Columbia, Oregon, and Washington.

Larvae of *Ecclisocosmoecus* have not been characterized heretofore; we reared pupae taken with larvae at Mt Hood, Oregon.

MORPHOLOGY The head and pronotum are bright, uniform, brownish red in colour, and have a finely pebbled texture; both are rotund in lateral aspect and wide in dorsal aspect, the pronotum strongly constricted at its posterior margin (A,B). Some metanotal setae arise from the membrane between the *sa*2 sclerites (B); mesonotal setae are largely continuous from *sa*1 to *sa*2 to *sa*3. The lateral abdominal fringe does not extend across segment VIII; abdominal gills are single, and ventral chloride epithelia are evident on segments III–VII. Length of larva up to 17 mm.

CASE Larval cases of *E. scylla* (C) are made of sand grains closely fitted into a smooth exterior surface; the case has a pronounced taper and curvature. Length of larval case up to 19 mm.

BIOLOGY Larvae of *Ecclisocosmoecus* occur in small, cold mountain streams; most of those we collected were concealed in sand and gravel. Pupal cases are fastened to rocks, adults emerging in June and July; the range of larval instars in some collections indicates that more than one year may be required for a generation. Guts of larvae (3) we examined contained mostly pieces of vascular plants and moss with fine organic particles.

REMARKS Diagnostic characters of the male of *E. scylla* were given by Ross (1950, under *Ecclisomyia*).

Ecclisocosmoecus scylla (Oregon, Clackamas Co., 23–29 Sept. 1966, ROM)
A, larva, lateral x10; B, head and thorax, dorsal, detail of mesonotum; C, case, x7

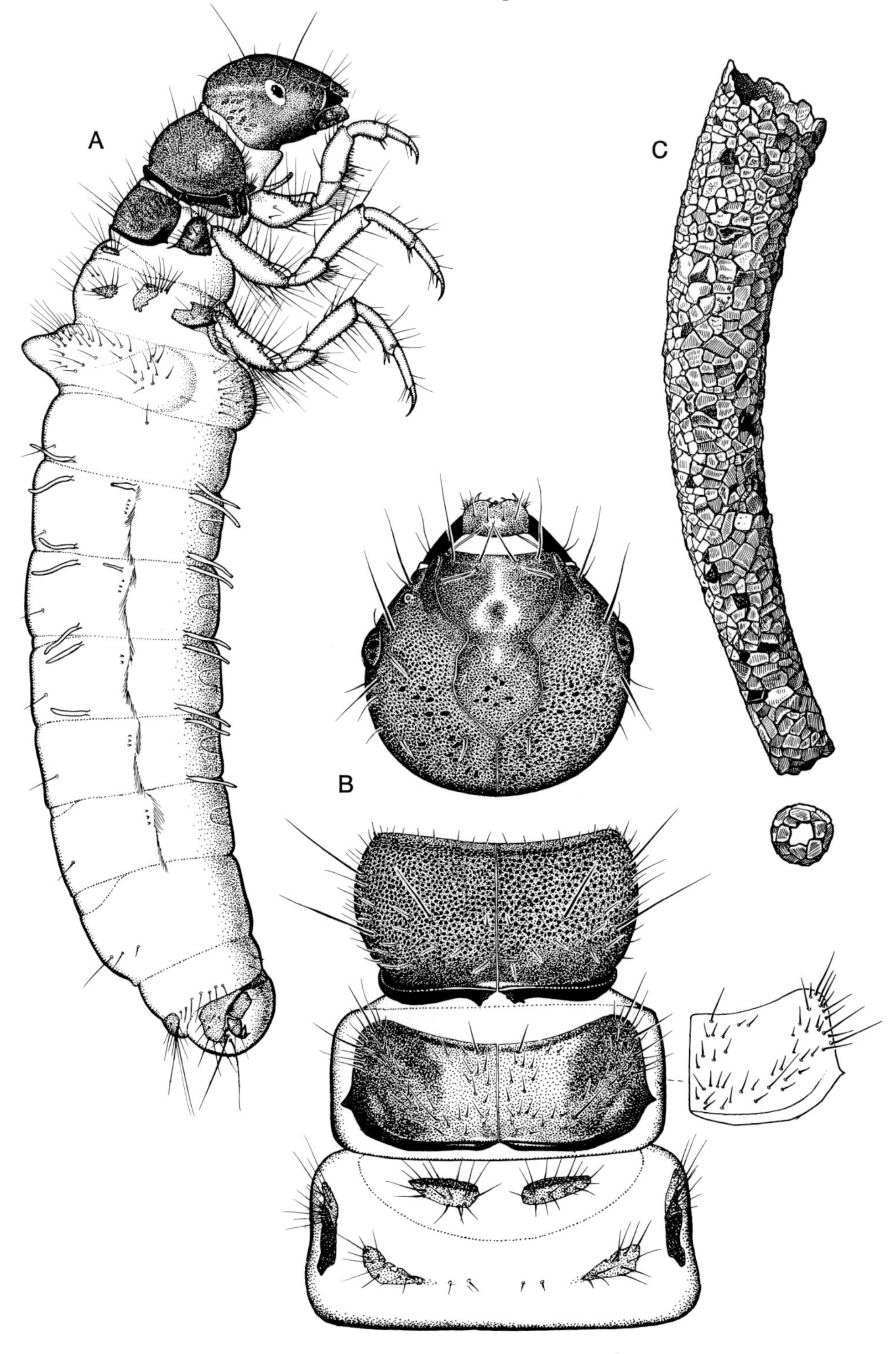
A
C
B

10.14 Genus **Ecclisomyia**

DISTRIBUTION AND SPECIES *Ecclisomyia* is a small genus of the Nearctic and Asian Palaearctic regions. The three North American species are confined to western montane areas, ranging as a group from Alaska to California.

Larvae have been described previously for *E. conspersa* Banks (Flint 1960) and for a Siberian species *E. digitata* (Martynov) (Lepneva 1966); we have associated larvae for all three North American species.

MORPHOLOGY Larvae of *Ecclisomyia* can be recognized by the large size of the metanotal *sa*1 and *sa*2 sclerites in relation to the area of the metanotum (B); the distance between the *sa*2 sclerites is never more than twice the largest dimension of one *sa*2 sclerite. The light-coloured stripe on the mid-dorsal line of the pro- and meso-nota and the coronal suture in the illustration occurs in both *E. maculosa* Banks and *conspersa* Banks but not in *E. bilera* Denning. Mesonotal setal areas are continuous from *sa*1 to *sa*2 to *sa*3. As in many of the Dicosmoecinae, the middle and hind femora bear more than two major setae ventrally (A). Abdominal segment I bears many setae, often with small sclerites incorporating the bases of several of them; gills are single. Length of larva up to 19 mm.

CASE Larval cases in *Ecclisomyia*, often more slender than in most other limnephilids and straight with little taper, are constructed of rather coarse rock fragments; some long pieces of plant materials are often incorporated. Length of case up to 31 mm.

BIOLOGY Larvae of *Ecclisomyia* live in cool mountain streams, among rocks and gravel. In guts of larvae (3) we examined, diatoms were the dominant part of the contents with small amounts of vascular plant pieces and fine organic particles. Fine organic particles were found to be the major component in the food of *E. maculosa*, with a small proportion of vascular plant fragments (Mecom 1972a).

REMARKS Diagnostic characters of males of the North American species were reviewed by Ross (1950).

Ecclisomyia conspersa (Oregon, Linn Co., 1 June 1968, ROM)
A, larva, lateral x11; B, head and thorax, dorsal; C, case x5

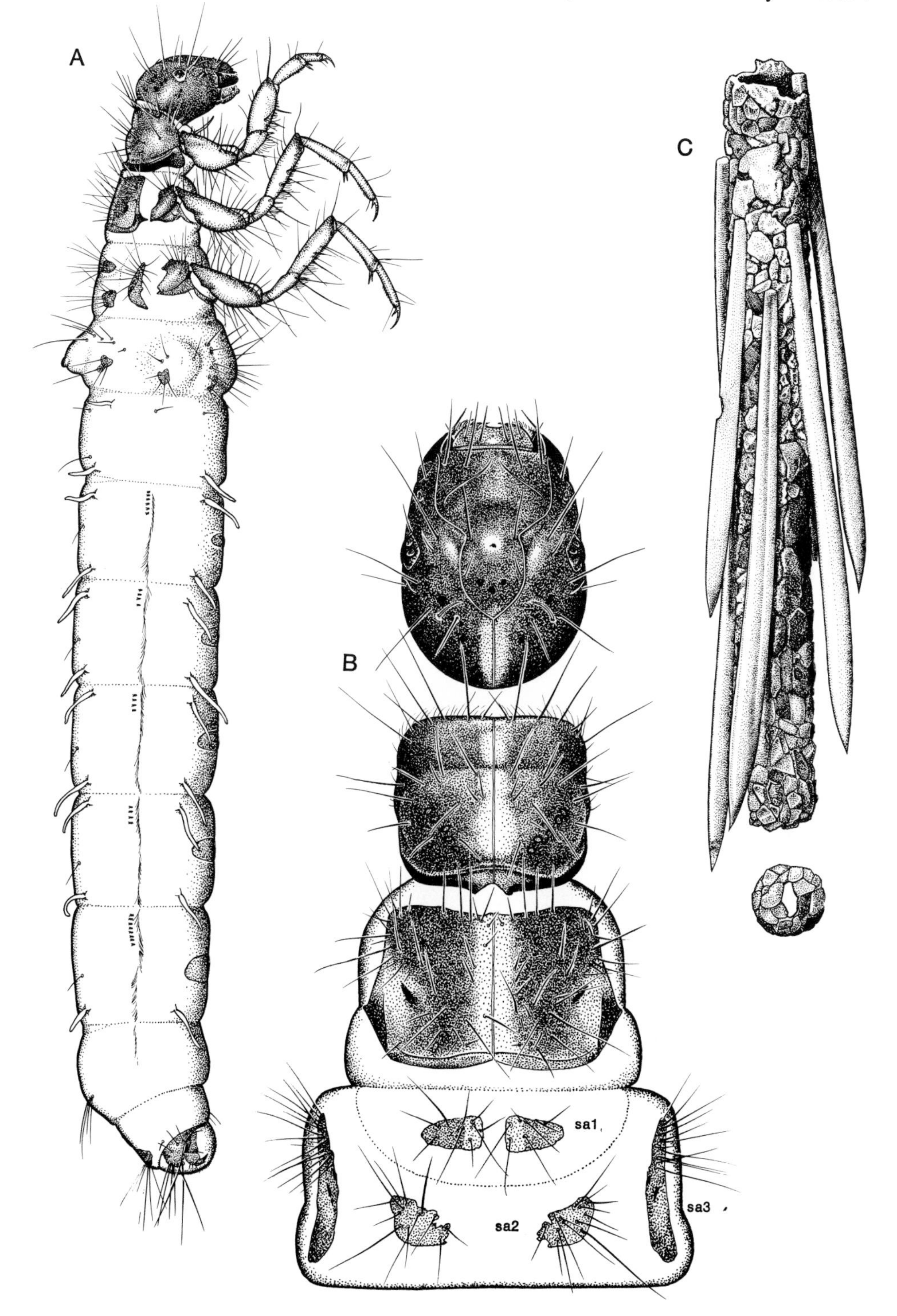
A
B
C
sa1
sa2
sa3

10.15 Genus **Farula**

DISTRIBUTION AND SPECIES *Farula* is confined to western North America where it is represented by seven species recorded from California, Oregon, and Washington.

The larva of *F. malkini* Ross was described by Denning (1973); we have associated materials for two species and have collections of larvae from much of the range of the genus.

MORPHOLOGY Larvae of *Farula* are very much like those of *Neothremma* and probably have often been so identified in the past. Structurally, the two genera are quite unlike any others in North America. Larvae are very long and slender; sclerotized parts are mostly uniform dark brown or black. The dorsum of the head bears secondary setae (B); the mandibles (D) are scraper-like without separate teeth, and the labrum has a membranous anterior fringe. The pronotum is longer than wide, covered with short setae, and the prosternal horn is prominent. The mesonotal plates bear an anteromesial excision, and metanotal *sa*1 is a single seta. Larvae are without abdominal gills and lateral tubercles. In addition to diagnostic characters in the key, larvae of *Farula* lack a prothoracic sternellum and have chloride epithelium on segment V alone (A). Length of larva up to 8.5 mm.

CASE Larval cases in both *Farula* and *Neothremma* are so exceedingly slender that they could be mistaken for conifer needles, but *Farula* cases are the more slender. The case (C) is constructed of fine sand grains fitted neatly together, the exterior surface as well as the interior covered with a thin layer of silk; frequently much of the case is of silk alone. Although the posterior end of the larval case is not reduced with silk, a subterminal perforate silken membrane is spun across the inside by final-instar larvae in making a pupation chamber (C). Length of case up to 14.5 mm.

BIOLOGY Species of *Farula* inhabit small, cold streams in mountainous areas. Larvae crawl over rock surfaces, and frequently pupate in clusters on the underside of rocks. Gut contents of larvae (3) we examined were mostly fine organic particles with some filamentous algae and fine rock fragments.

REMARKS Taxonomy of males in this genus was reviewed by Denning (1958b).

Farula jewetti (Oregon, Clackamas Co., 20 April 1964, ROM)
A, larva, lateral x26, anterior edge segment II enlarged; B, head and thorax, dorsal; C, case, lateral x11, portion cut away to show position of subterminal membrane; D, mandible, ventral

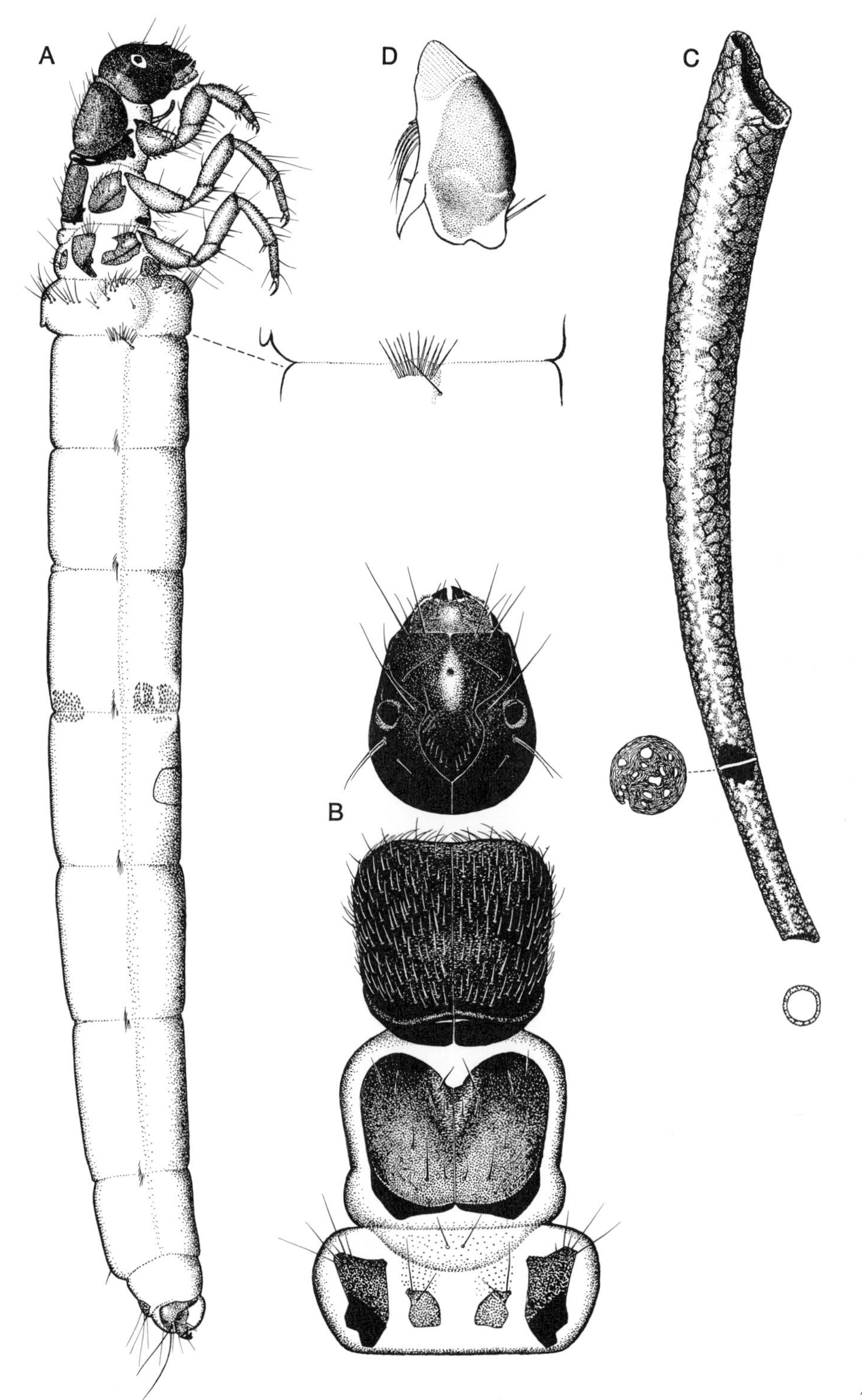
A
D
C
B

10.16 Genus Frenesia

DISTRIBUTION AND SPECIES *Frenesia* is a Nearctic genus, its two species *F. difficilis* (Walker) and *F. missa* (Milne) confined to the northeastern part of the continent, roughly within the area bounded by Nova Scotia, Minnesota, and Virginia.

The larva of *F. difficilis* was described by Lloyd (1921, under *Chilostigma*), and of *F. missa* by Ross (1944). Larvae of both species were described and distinguished by Flint (1956); we have associated larvae for both.

MORPHOLOGY Sclerites of the head and thorax in these larvae are light brown with few markings other than light muscle scars on the posterior part of the head. The head is covered with tiny spines; the pronotum bears stout, spine-like setae, especially along the anterior margin (B); and in contrast to *Glyphopsyche* the legs are not banded. On the venter of abdominal segment I, *sa*1 and *sa*2 are conjoined (D). Most abdominal gills are two- or three-branched, a few consisting of a single filament; chloride epithelia are present on the venter of segments II–VII. The lateral sclerite of the anal proleg bears stout spine-like setae similar to those of *Glyphopsyche*. Length of larva up to 17 mm.

CASE Larval cases in *Frenesia* are mainly of rock fragments, often with small bits of wood added (C). Length of case up to 21 mm.

BIOLOGY An extensive account of the biology of both species has been given by Flint (1956). Larvae live in cold streams and spring seepage areas, where they feed on leaves and decaying wood (Lloyd 1921). Larvae reach the final instar in September and October, pupate, and emerge as adults in November.

REMARKS Taxonomy of adults was reviewed by Schmid (1952c).

Frenesia difficilis (Ontario, York Co., 11 Sept. 1953, ROM)
A, larva, lateral x10; B, head and thorax, dorsal, detail of mesonotum; C, case x5; D, segment I, ventral

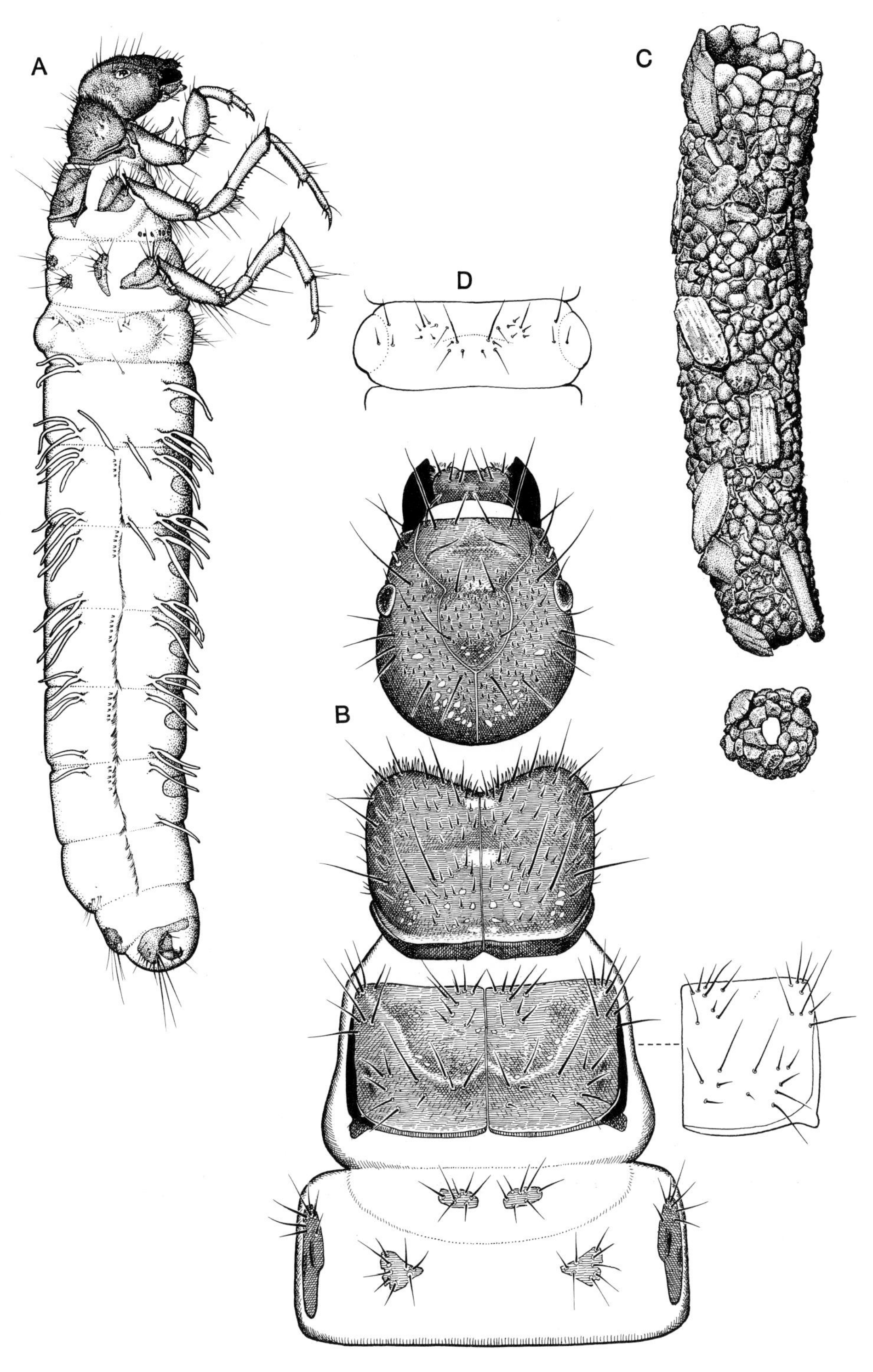
A
B
C
D

10.17 Genus **Glyphopsyche**

DISTRIBUTION AND SPECIES This is another of the genera confined to the Nearctic region. Two species are known: *G. irrorata* Fab., northern and transcontinental, in the western mountains from the Yukon to California; and *G. missouri* Ross, recorded only from Missouri.

Larvae have been described and assigned to both species: *G. missouri* by Ross (1944) and Flint (1960); and *G. irrorata* by Flint (1960) although no positive evidence of association was available for the latter species. We have reared larvae of *G. irrorata* and have collected them at many localities from Ontario to British Columbia.

MORPHOLOGY Few of the Limnephilinae larvae in North America have dark-coloured bands on the legs. Among those with single gills are only certain species of *Psychoglypha*, and of those with branched gills, there are only species of *Glyphopsyche.* Larvae of this genus have stout setae on the lateral sclerite of the anal proleg (A) and on the anterior edge of the pronotum (B). Length of larva up to 21 mm.

CASE Larval cases range from a rather even cylinder of rock and plant pieces (C) to very uneven cases incorporating larger wood pieces arranged longitudinally. Length of case up to 34 mm.

BIOLOGY *G. irrorata* is largely a species of marshes and the edges of slow streams; larvae are often associated with an abundance of decaying vegetation. Data from our collections indicate that larvae are in the final instar by late summer; adults have been collected from September through most months to May.

REMARKS Schmid (1952c) has reviewed taxonomy of the adults.

Glyphopsyche irrorata (California, Lake Co., 6 Sept. 1946, ROM)
A, larva, lateral x7, anal proleg enlarged; B, head and thorax, dorsal; C, case, lateral x4

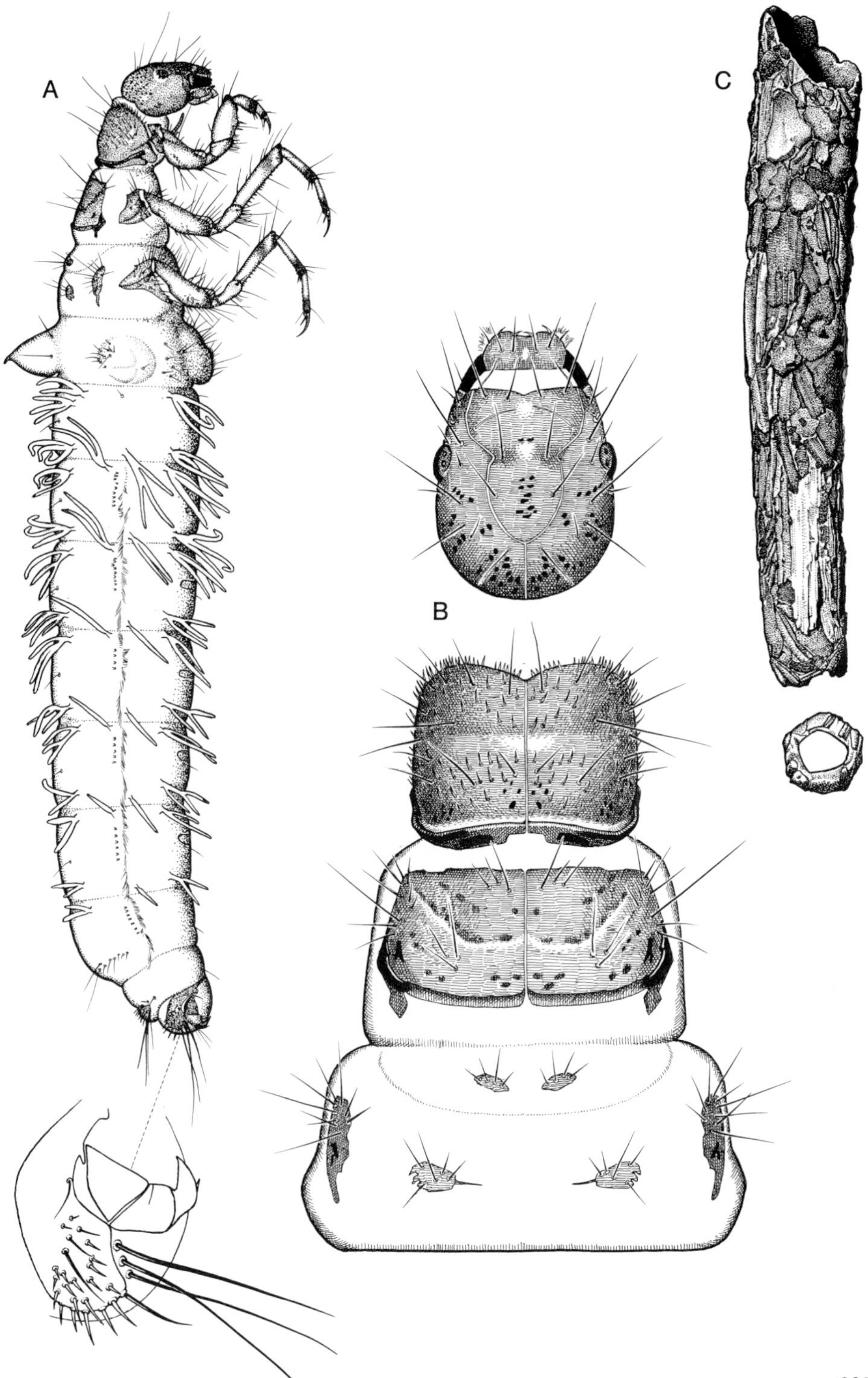
A
B
C

10.18 Genus **Goera**

DISTRIBUTION AND SPECIES Species of *Goera* are distributed over much of the Holarctic and Oriental regions. Six species are known in North America, five within the eastern half of the continent and *G. archaon* Ross in Oregon.

Larvae have been described for *G. calcarata* Banks, *G. fuscula* Banks, and *G. stylata* Ross by Flint (1960); we have associated larvae for three species, including *G. archaon*, illustrated here but not previously identified in the literature.

MORPHOLOGY Among Nearctic goerine larvae, species of *Goera* alone have gills of more than one filament (A). The head bears secondary setae (D) and usually has ridges of some sort; the pronotum is thickened laterally (A), bears a pair of anterolateral processes and frequently a median ridge (B). Each mesonotal plate is subdivided into separate sclerites (B). Ventral as well as dorsal chloride epithelia are present (A). Length of larva up to 10.5 mm.

CASE With a row of larger pebbles along each side of the central tube of small rock fragments, larval cases of *Goera* (C) are similar only to those of *Neophylax* in North America. Generally, *Goera* cases have fewer and larger ballast stones on each side, usually two; *Neophylax* cases usually have smaller ballast stones, thus more than two. Larvae of *Goera* have a tendency to cover irregularities on the outside of the case with silk. The posterior opening of the larval case in *Goera* is restricted with silk to a very small central hole. Length of larval case normally up to 14 mm.

Larval cases almost twice as long as the larva occur infrequently in collections, and demonstrate the way in which *Goera* larvae enlarge their case by building a new one on the end of the existing case and finally cutting it free. Since most caddisfly larvae increase the size of their cases by more gradual addition of new pieces at the anterior end, case-building behaviour in *Goera* seems to have been modified to accommodate the larger ballast stones. Similar behaviour was studied in the related European genus *Silo* by Hansell (1968b).

BIOLOGY These are larvae of running waters where they live in the current on larger rocks and feed by scraping periphytic algae and fine detrital particles from these surfaces (Coffman et al. 1971).

Goera archaon (Oregon, Benton Co., 16 April 1964, ROM)
A, larva, lateral x16; B, head and thorax, dorsal; C, case x11; D, head, dorsal

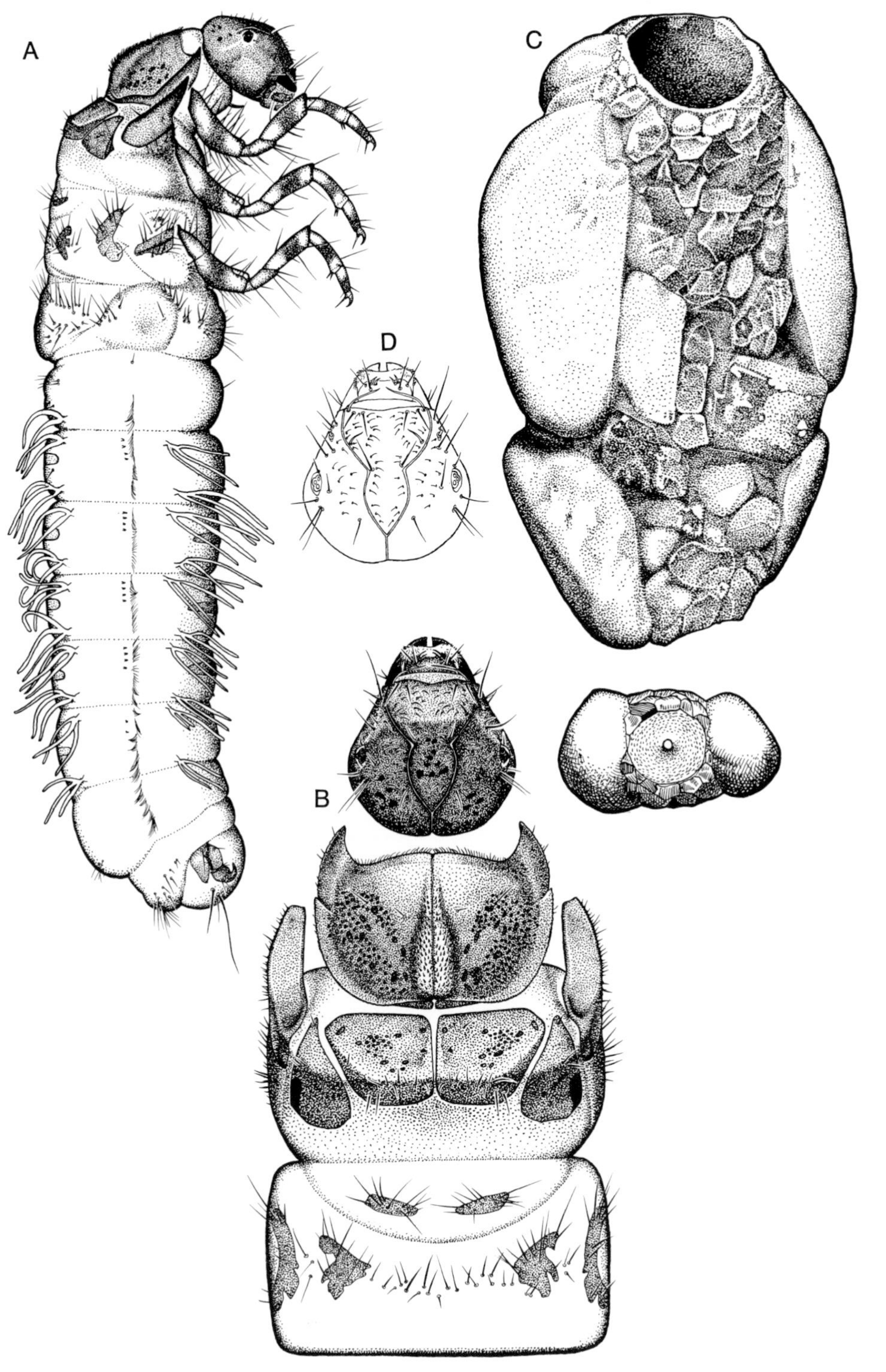
A
B
C
D

10.19 Genus Goeracea

DISTRIBUTION AND SPECIES *Goeracea* is a Nearctic genus recorded from montane areas of western North America: British Columbia, California, Idaho, Montana, and Oregon. Two species are known, *G. genota* (Ross) and *G. oregona* Denning.

We have associated larvae for both species, and data for larvae and adults were presented elsewhere (Wiggins 1973b).

MORPHOLOGY Larvae of *Goeracea* are unusual in appearance and unlikely to be mistaken for any other genus. The pronotum is flat and rounded in dorsal outline, heavily thickened laterally; the mesepisternum is laterally compressed and very long (A). Each metanotal *sa*1 consists of one or two setae without a sclerite or with a very small sclerite. Gills are single and confined to segments III and IV or to III alone. Ventral chloride epithelia are present; and the two stout setae comprising the basal tuft of the anal proleg arise from a small sclerite, the dorsal plate (G). Typically for the Goerini, mandibles have scraping edges rather than distinct teeth (D), the basal seta of each tarsal claw extends nearly to the tip of the claw (A), and the anterior part of the labrum is membranous (F). Length of larva up to 6.6 mm.

CASE The larval case is of rock fragments, curved and tapered, with a row of larger pebbles along each side; frequently a mid-dorsal ridge of small stones is also attached to the central tube. The silken membrane reducing the posterior opening of the case has an eccentric, relatively small opening. Length of larval case up to 8 mm.

BIOLOGY Larvae of *Goeracea* inhabit small, cold streams in mountainous areas where they are usually found on rocks. Gut contents of larvae (6) we examined were largely fine organic and inorganic particles with a small proportion of vascular plant pieces.

Goeracea genota (Oregon, Benton Co., 13 April 1964, ROM)
A, larva, lateral x26, tarsal claw enlarged; B, head and thorax, dorsal; C, case, dorsolateral x14; D, mandible, dorsal; E, head, ventral; F, labrum, dorsal; G, segment IX and anal prolegs, dorsal

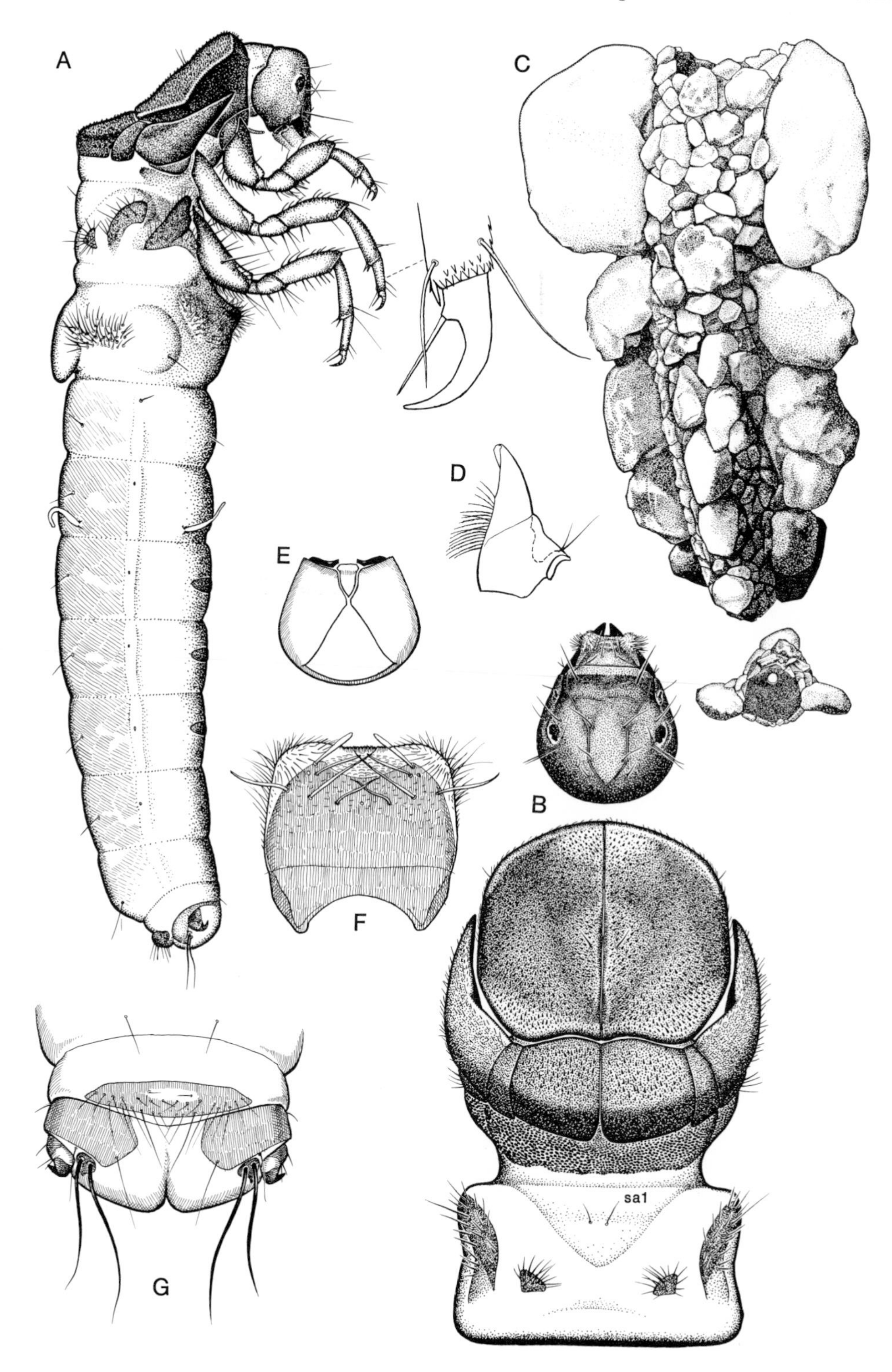
A
C
D
E
B
F
sa1
G

10.20 Genus **Goereilla**

DISTRIBUTION AND SPECIES This is another of the remarkable goerine genera found only in western North America. The single species *G. baumanni* Denning is known from Montana and Idaho.

We have reared larvae from Montana; a more extensive account of the larva, pupa, adults, and life history of this species appears elsewhere (Wiggins 1976).

MORPHOLOGY In all other Nearctic larvae of the Goerinae the mesepisternum is conspicuously elongated as in *Goera*, but in *Goereilla* the mesepisternum is modified to form only a rounded, spinose lobe (A,G); mesonotal sclerites are subdivided as in other members of the subfamily. The pronotum bears a few coarse teeth on the anterior margin (A), but is not thickened laterally as in other goerines. As in *Lepania*, the mandibles have separate tooth-like points (E), the labrum is not membranous anteriorly (D), and the basal seta of the tarsal claws is short (A); all of these are characters discordant with other goerines. *Goereilla* is also distinctive among the Goerinae in having a well-developed accessory hook on the anal claw (A). Setae of the basal tuft of the anal proleg arise from the dorsal plate (F). Abdominal gills are single and chloride epithelia are lacking. Length of larva up to 9 mm.

CASE Larvae of *G. baumanni* have smoothly contoured cases of sand grains with a few small pieces of wood. The posterior opening is reduced with silk to a central hole with a crenulate circumference (C). Length of larval case up to 11 mm.

BIOLOGY Larvae live in the watery organic ooze and muck of spring seepage areas; they are almost impossible to see in these habitats, but can be collected by washing samples of the material in a fine-mesh sieve in flowing water. Gut contents of larvae (3) examined were largely pieces of vascular plants with fine organic particles. We have collected larvae sealed up in pupal cases in August; adults were recorded in May.

Goereilla baumanni (Montana, Missoula Co., 17-18 August 1973, ROM)
A, larva, lateral x21, anal claw enlarged; B, head and thorax, dorsal; C, case, lateral x11; D, labrum, dorsal; E, mandible, ventral; F, segment IX and anal prolegs, dorsal; G, mesopleuron, lateral

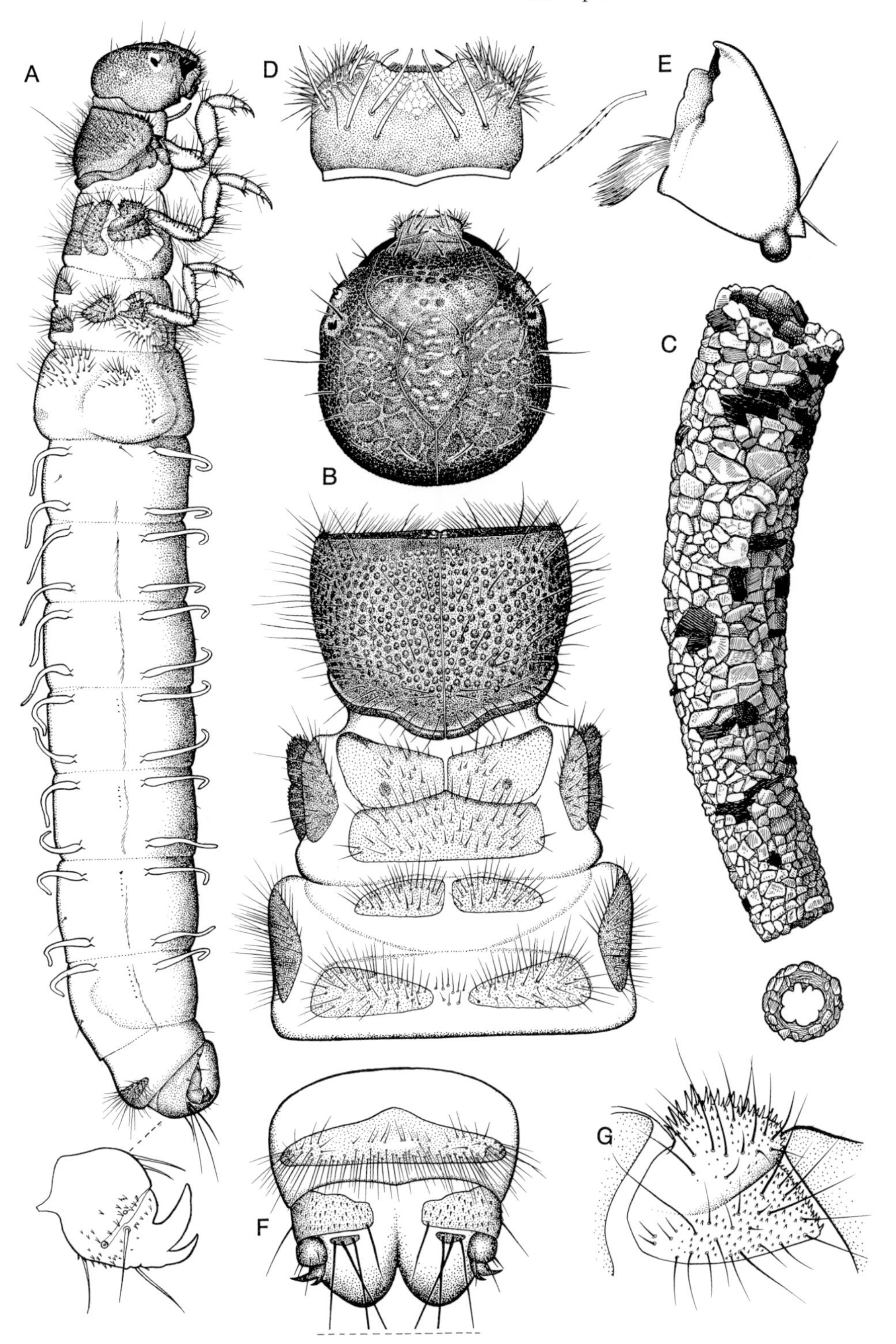
A
D
E
C
B
F
G

10.21 Genus **Goerita**

DISTRIBUTION AND SPECIES The genus is confined to North America and is known only from the eastern part of the continent; there are two species, *G. semata* Ross and *G. betteni* Ross.

Larvae identified circumstantially as *G. semata* were described by Flint (1960) and by Wiggins (1973b); larvae of *G. betteni* were identified by Wiggins (1973b), and taxonomic data for adults of both species were also provided.

MORPHOLOGY These are the only North American goerine larvae known with no abdominal gills. The prominent median hump on the pronotum is distinctive, the anterolateral processes and lateral thickening similar to *Goera* (A,B). The antennae are set in depressions (B) as in several other goerine genera. Sclerotized parts are reddish brown, the head and thoracic nota with a pebbled surface. Ventral chloride epithelia are present on segments IV-VI; the two stout setae comprising the basal tuft of the anal proleg arise from the dorsal plate (D) as in several other genera of the Goerinae. Length of larva up to 6 mm.

CASE Larvae construct a smooth-sided case of sand grains; those of *G. betteni* incorporate larger pieces in the lateral walls of the case (C), behaviour suggestive of that in *Goeracea* where a row of larger stones is added to each side of the case. The posterior opening is restricted with silk to a small hole dorsad of centre. Length of larval case up to 6.5 mm.

BIOLOGY Larvae of both species occur on rocks in small, cold spring runs in mountainous areas. Gut contents of larvae (3) we examined were mainly fine organic particles with some fine mineral particles, as would be expected from larvae grazing on rock surfaces. Colonies appear to be exceedingly local in distribution for they are not often found.

Goerita betteni (Tennessee, Franklin Co., 14-15 May 1970, ROM 700337)
A, larva, lateral x22; B, head and thorax, dorsal; C, case, ventrolateral x16; D, segment IX and anal prolegs, dorsal

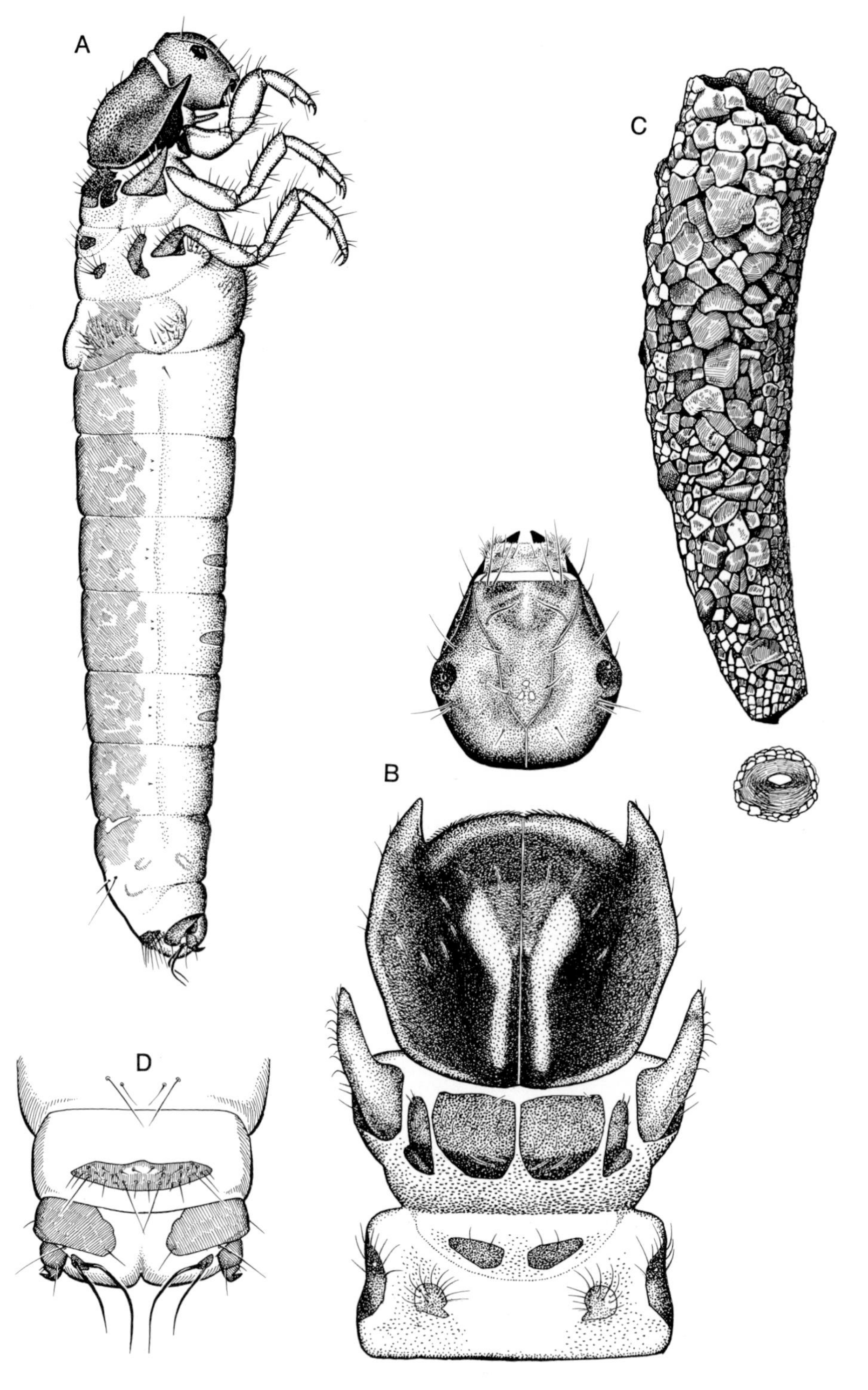
A
B
C
D

10.22 Genus **Grammotaulius**

DISTRIBUTION AND SPECIES *Grammotaulius* is a northern group widespread over much of the Holarctic region. Five species are known in North America and these are largely western and montane from Alaska to Colorado; one species, *G. interrogationis* Zett., extends across the northern part of the continent as far south as Minnesota and occurs also in Greenland.

Larvae have been identified for only one North American species, *G. lorettae* Denning from Colorado (Flint 1960). Larvae of four Eurasian species have been described (Lepneva 1966).

MORPHOLOGY Although there is still much to learn about the North American species of *Grammotaulius*, the consensus from the Holarctic fauna is that larvae have a light, yellowish brown head, bearing numerous small, dark spots (B). North American larvae I have examined, apparently of this genus, have all three setal areas separate on both dorsum and venter of abdominal segment I (D,E), with 1-3 setae at ventral *sa*1. Length of larva up to 28 mm.

CASE Larval cases in *Grammotaulius* (C) are made of lengths of sedge or similar leaves arranged lengthwise, often in a spiral, although this is difficult to interpret because of the long ends. Length of larval case up to 41 mm.

BIOLOGY The larvae of *G. lorettae* described by Flint (1960) were collected in a small weed-filled pond near tree-line in Colorado. The habitat of *G. betteni* H.-G. was described by Hill-Griffin (1912) as small ponds and slow streams at much lower elevations around Corvallis, Oregon; the same author observed that although adults of this species emerged in March and April, adult specimens were all taken in September, October, and November. It seems likely, then, that in at least some of the North American species of *Grammotaulius*, adults emerging in spring are in diapause, their sexual maturity delayed until the onset of a shorter photoperiod in late summer; and in fact the European *G. atomarius* Fab. was one of the species studied in the elucidation of this type of life cycle in Trichoptera (Novák and Sehnal 1963, 1965).

REMARKS Schmid (1950b, 1964b) has reviewed taxonomic data for adults of *Grammotaulius.*

Grammotaulius sp. (Utah, Summit Co., 12 June 1961, ROM)
A, larva, lateral x5, segment I enlarged; B, head and thorax, dorsal, detail of mesonotum; C, case, ventral x3; D, segment I, dorsal; E, segment I, ventral; F, segments VIII and IX, dorsal

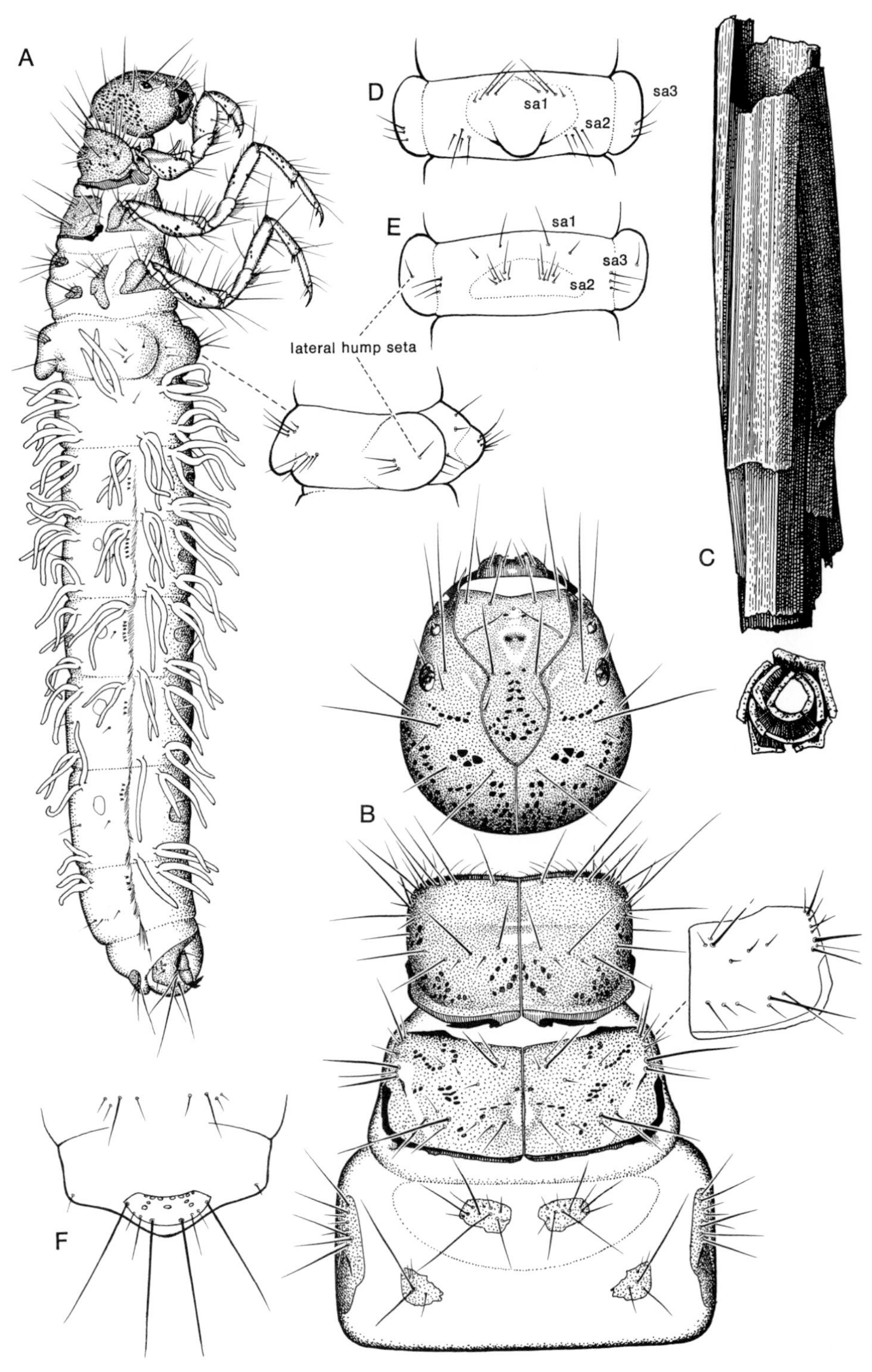
A
D
sa1
sa2
sa3
E
sa1
sa3
sa2
lateral hump seta
C
B
F

10.23 Genus **Grensia**

DISTRIBUTION AND SPECIES The single boreal species, *G. praeterita* (Walker), is common to both Palaearctic and Nearctic regions. In North America the few records available indicate that the species occurs in arctic tundra areas of the Northwest Territories and Alaska, extending northward at least to Victoria Island.

No information on larval stages has been available previously; we have an associated series of larvae and pharate adults from the Northwest Territories.

MORPHOLOGY The pronotal plates and the head are covered with fine spines (B), the head is dark brown in colour with some muscle scars; mesonotal *sa*1 and *sa*2 are separated by a gap without setae. The lateral hump of abdominal segment I lacks basal sclerites, and gills are single. The lateral sclerites of the anal prolegs bear many stout, clear setae (D). Length of larva up to 14 mm.

CASE For their cases larvae use small pieces of woody plant tissue, some rock fragments, and some organic materials with concentric rings that appear to be opercula of prosobranch snails (C). Length of larval case up to 16 mm.

BIOLOGY Our material of this species indicates that it is largely an inhabitant of lakes located beyond the northern limit of trees; but we have one Alaskan series of larvae, apparently of this genus, with the habitat note 'seeps in cliffs.'

REMARKS Most of our larval and pupal specimens to date have come from stomachs of Arctic char collected by the Arctic Survey of the Fisheries Research Board of Canada. Diagnostic characters of adults were illustrated by Betten and Mosely (1940, under *Frenesia*).

Grensia praeterita (Northwest Territories, Drink Lake, 12 Sept. 1956, ROM)
A, larva, lateral x10; B, head and thorax, dorsal, portion of pronotum enlarged; C, case, lateral x7; D, segment IX and anal prolegs, dorsal

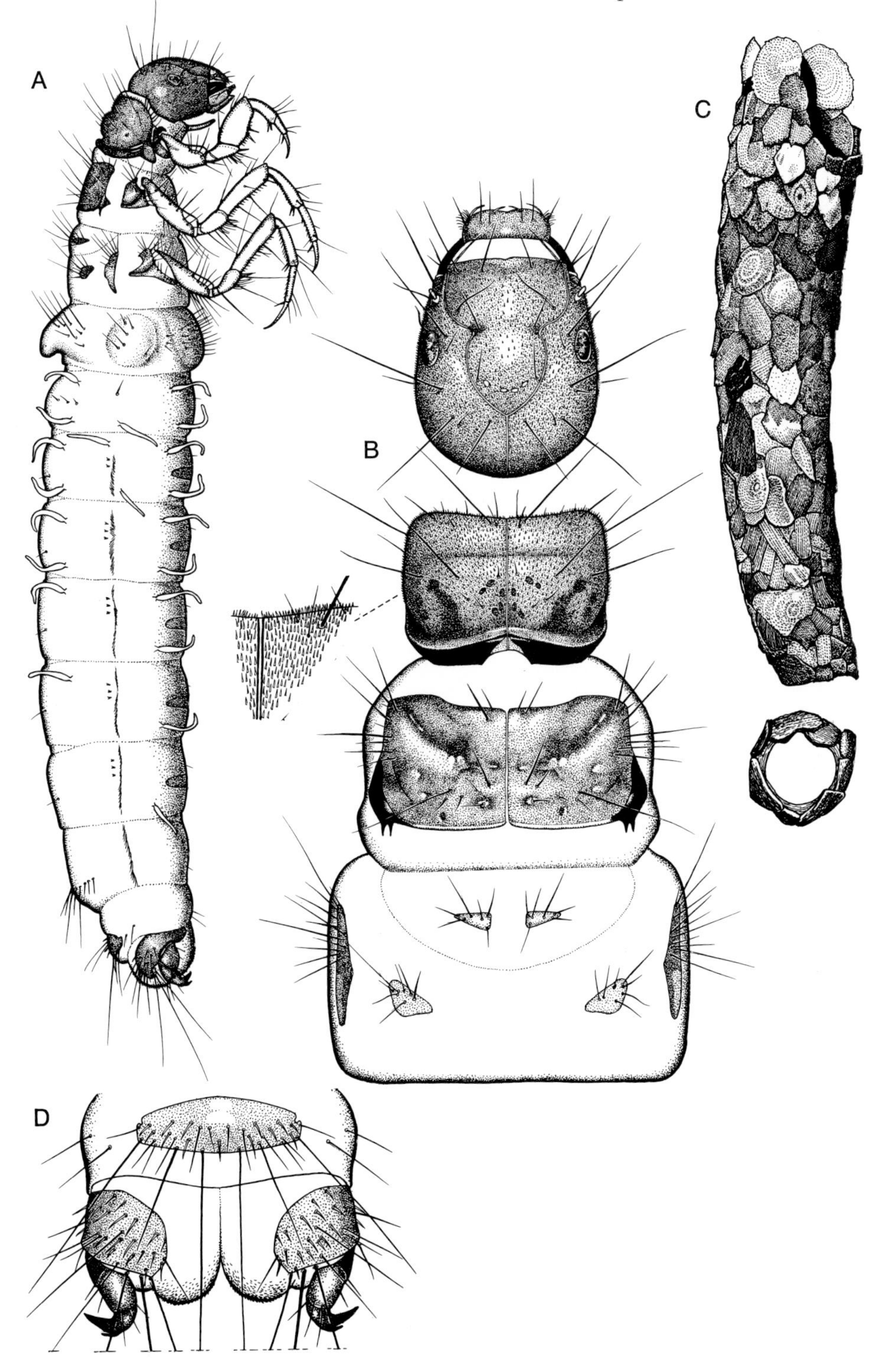
A
B
C
D

10.24 Genus **Halesochila**

DISTRIBUTION AND SPECIES The genus is confined to western North America where a single species, *H. taylori* (Banks), is known.

Larvae have not been identified previously in the literature; we have a series reared in British Columbia, and have collected larvae evidently congeneric with them from many localities in British Columbia, Oregon, and Washington.

MORPHOLOGY Three dark and often indistinct bands on the head are characteristic of *H. taylori*, as is a single seta in the mesonotal *sa* 1 position (B). Dorsal chloride epithelia are present in addition to lateral and ventral series. The venter of abdominal segment I is heavily setate in all three primary positions (D). Most abdominal gills are three-branched. Length of larva up to 24 mm.

CASE Most larval cases are constructed of leaf pieces or wood fragments laid flat to form a cylinder of rough exterior (C), although not always as wide at the anterior end as in the illustration; some larvae in our collection, apparently of this genus, have three-sided cases of large pieces of leaves. Final instars construct cases of fine gravel prior to pupation. Length of larval case up to 35 mm.

BIOLOGY Larvae live in small lakes and ponds, some of the latter reputed to be temporary. Life history data compiled by Winterbourn (1971a) show that food of larvae in a small British Columbia lake was largely detritus and that larvae of this species grew rapidly in spring and early summer to reach the final instar by June, final-instar larvae occurring from June to September; pupae were found partially buried in bottom sediments of the larval habitat. Adults are active in October.

REMARKS Taxonomic data for adults of *H. taylori* were given by Schmid (1950c).

Halesochila taylori (British Columbia, Vancouver Is., May 1965, ROM)
A, larva, lateral x7; B, head and thorax, dorsal, detail of mesonotum; C, case, lateral x4; D, segment I, ventral

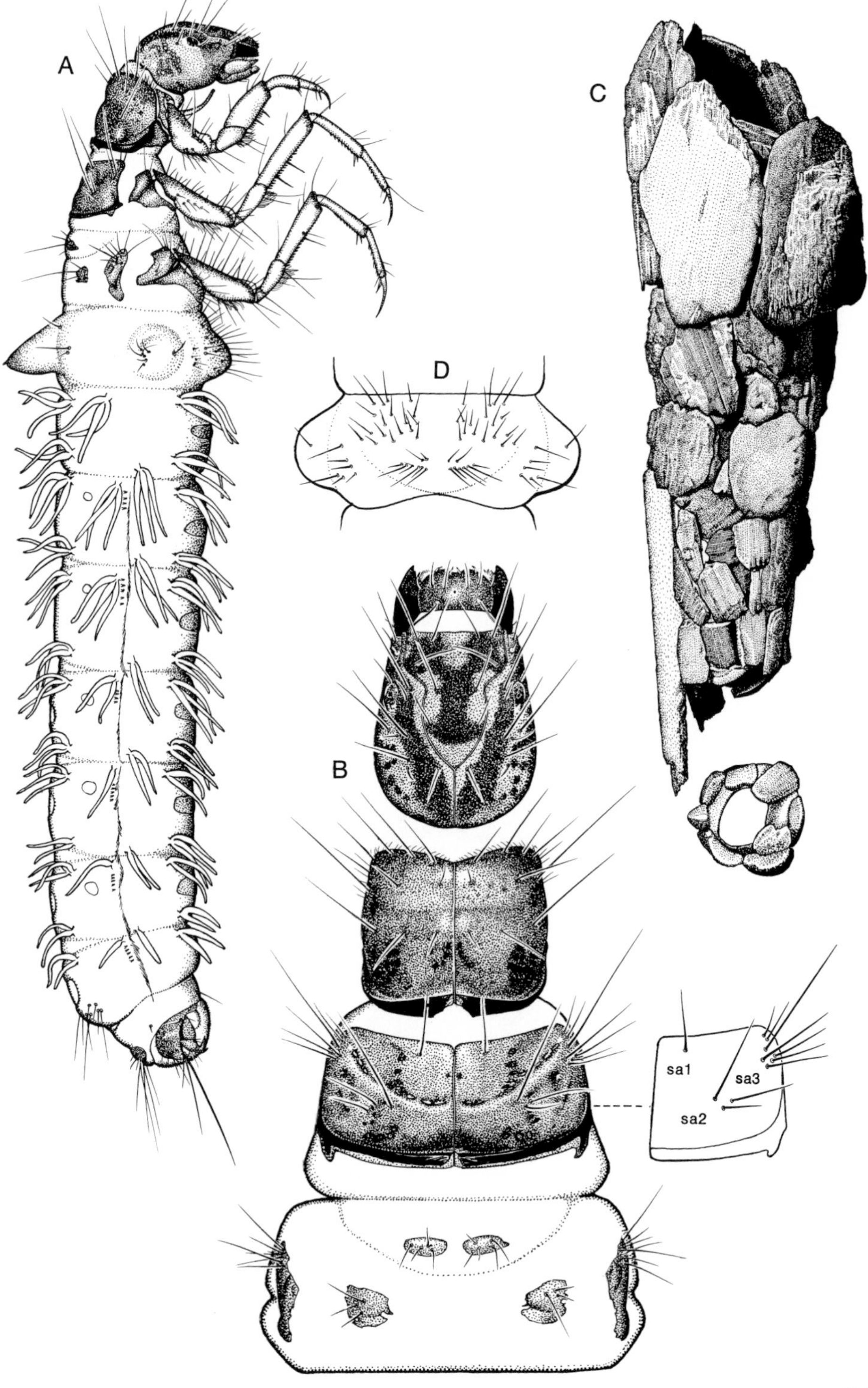
A
B
C
D
sa1
sa3
sa2

10.25 Genus **Hesperophylax**

DISTRIBUTION AND SPECIES *Hesperophylax* is exclusively a Nearctic genus with species widely distributed and common throughout much of the continent. Six species have been described: five in the west and one, *H. designatus* (Walker), in eastern and central areas.

Larvae of *H. designatus* have long been known (Vorhies 1905, 1909 and Lloyd 1921, as *Platyphylax*; Ross 1944; Flint 1960). We have a large amount of associated material in this genus from the west. Generic concordance is good, but species differences among the larvae appear subtle, at best; and the problem of larval identification to species in *Hesperophylax* is compounded by variations in species characters of the adults.

MORPHOLOGY This is one of the few limnephilid genera in which most gills of the dorsal and ventral series have four or more branches (A). Mesonotal *sa*1 has several setae, and setae are numerous on the metanotal membrane between the primary sclerites (B). Femora of the second and third legs have only two major setae along the ventral edge (A). The head is medium brown with few markings save for some dark muscle scars. Length of larva up to 33 mm.

CASE Most cases consist entirely of rock fragments (C), but in western localities larvae sometimes incorporate small pieces of wood; cases are cylindrical, often much more irregular than illustrated when larger pieces are used. Length of larval case up to 40 mm.

BIOLOGY For the most part, larvae live in small streams and, as a genus, appear to have wide temperature tolerance. *Hesperophylax* larvae also occur in temporary streams and, at higher elevations and latitudes, in lentic habitats. Young instars of *H. designatus* feed mainly on diatoms, older ones ingest vascular plant materials as well (Vorhies 1905, 1909; Lloyd 1921). Extended observations on biology and behaviour of *H. designatus* were given by Vorhies (1905, 1909). Collections indicate that adults emerge from spring through to late summer.

Hesperophylax designatus (Ontario, Durham Co., 13 March 1966, ROM)
A, larva, lateral x7, middle femur enlarged; B, head and thorax, dorsal, detail of mesonotum; C, case, lateral x5

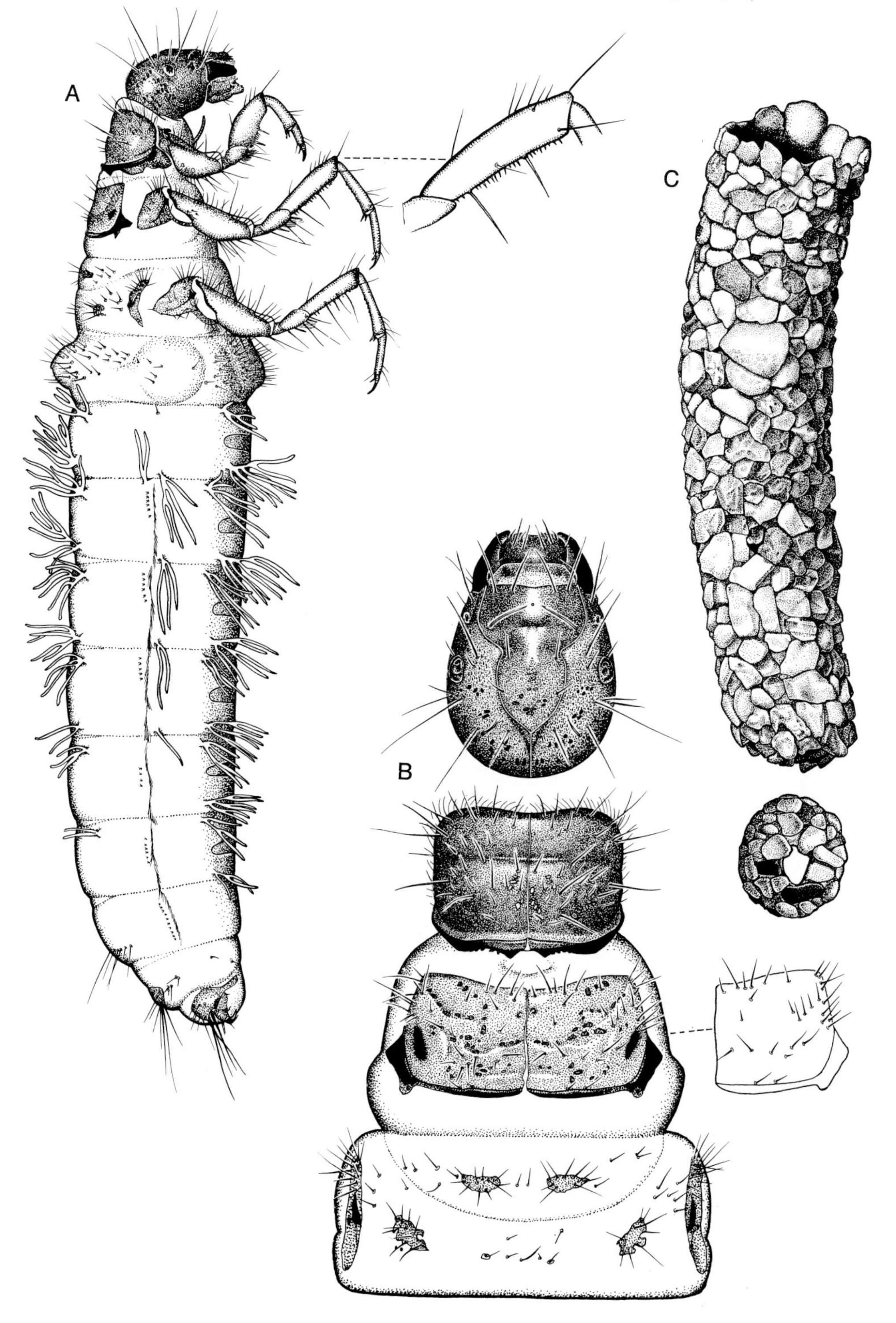
A
B
C

10.26 Genus **Homophylax**

DISTRIBUTION AND SPECIES This genus is exclusively Nearctic, represented throughout western montane parts of the continent from Alaska to California. Ten species are known.

Larvae of *Homophylax* have not been identified previously. We collected associated material for three species, and have taken larvae generally similar to these at many other localities.

MORPHOLOGY Larvae of *Homophylax* are most similar to those of *Psychoglypha* but material available indicates that the two can be separated by the form of the sclerite at the base of each lateral hump of abdominal segment I; in *Homophylax* the sclerite is a short bar along the posterior edge of the hump (A). Sclerotized parts of the head and thorax in *Homophylax* larvae are medium brown without contrasting markings; muscle scars are ring-like with open centres. The anterior border of the labrum is unpigmented but not especially membranous (D). Femora have only two major setae on the ventral edge. Abdominal gills are single; the dorsal sclerite of segment IX bears many setae (E). Length of larva up to 21 mm.

CASE Larval cases of late-instar *Homophylax* are usually constructed of thin pieces of bark arranged irregularly into a smooth-walled cylinder with relatively little taper or curvature; the thin walls are often flexible. Cases of this type (C) are usually diagnostic for *Homophylax*, although occasionally larvae have a three-sided case. Flat rock pieces are sometimes utilized as well, making distinction from *Psychoglypha* difficult. Length of larval case up to 25 mm.

BIOLOGY Larvae we collected were restricted to small, cold streams of mountain slopes, where they are most often concealed in fine gravel or accumulations of leaves. Adults of different species occur at various times through spring and summer months.

REMARKS A taxonomic review of adults was given by Denning (1964). Although *Homophylax* was provisionally assigned to the subfamily Pseudostenophylacinae (Schmid 1955), I follow here a more recent view (F. Schmid, pers. comm.) that on the basis of characters of the adults *Homophylax* is more closely allied to the Chilostigmini of the Limnephilinae; larval data are largely consistent with this position.

Homophylax andax (Washington, Mt Rainier National Park, 3 July 1969, ROM 690164)
A, larva, lateral x8, segment I enlarged; B, head and thorax, dorsal; C, case, lateral x5; D, labrum, dorsal; E, segment IX and anal prolegs, dorsal, lateral seta of basal tuft enlarged

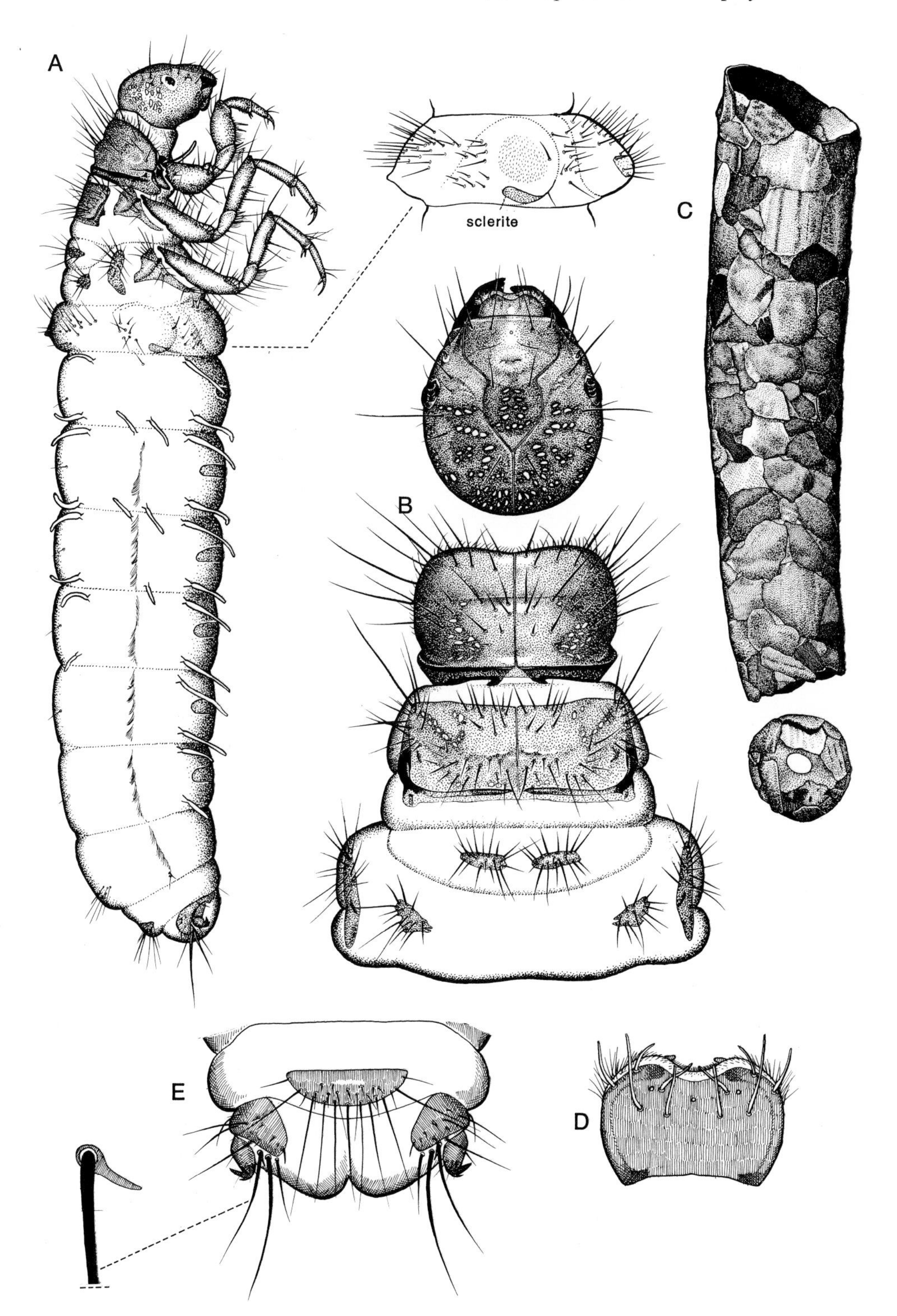
A
sclerite
C
B
E
D

10.27 Genus **Hydatophylax**

DISTRIBUTION AND SPECIES *Hydatophylax* is a genus of the Holarctic region, with four North American species. In the east, *H. argus* (Harr.) is widespread from South Carolina to Quebec and west to Minnesota; *H. victor* Banks is known only from the northeast. In the west, *H. hesperus* (Banks) occurs from Alaska to California; *H. variabilis* (Mart.), a species widespread in northern Europe and Siberia, has been recorded from Alaska.

Larvae have been described for *H. argus* (Lloyd 1915a, under *Astenophylax*; Flint 1960); we have associated larvae for *H. hesperus*, illustrated here, in Oregon and Washington, and for *H. argus*.

MORPHOLOGY The general similarity between *Hydatophylax* and some *Pycnopsyche* larvae complicates generic identification as is evident in the key. *Hydatophylax* larvae have fused metanotal *sa*1 sclerites (B,D) and one species (*H. argus*) has chloride epithelium on the venter of abdominal segment II; larvae of the western *H. hesperus* (A) differ from *H. argus* in lacking chloride epithelium on segment II, and in having a tuft of 3-6 setae on segment IX laterad of the dorsal sclerite. In both species, sclerotized parts of the head and thorax are light to medium brown and have diffuse spots somewhat like those in *Anabolia*. As in *Pycnopsyche*, an elongate sclerite lies along the posterior margin of each lateral hump, and gills are single. Length of larva up to 35 mm.

CASE Typical larval cases in *Hydatophylax* are constructed of bulky pieces of wood or leaves. The cases are irregular, *H. argus* probably having a larger case than any other North American caddisfly. Occasionally a larva in a flattened case of leaves is found. Length of larval case up to 76 mm.

BIOLOGY Larvae of *Hydatophylax* live in streams, often small but not exclusively so. They are largely confined to accumulations of plant debris, or to vegetation along the edge; *H. argus* feeds on dead wood and bark (Lloyd 1921).

REMARKS A taxonomic review of adults was given by Schmid (1950d). The genus was previously known as *Astenophylax*.

Hydatophylax hesperus (Oregon, Lincoln Co., 10 July 1963, ROM)
A, larva, lateral x6, segments I and IX enlarged; B, head and thorax, dorsal; C, case lateral x2; D, metanotal *sa*1 sclerite enlarged

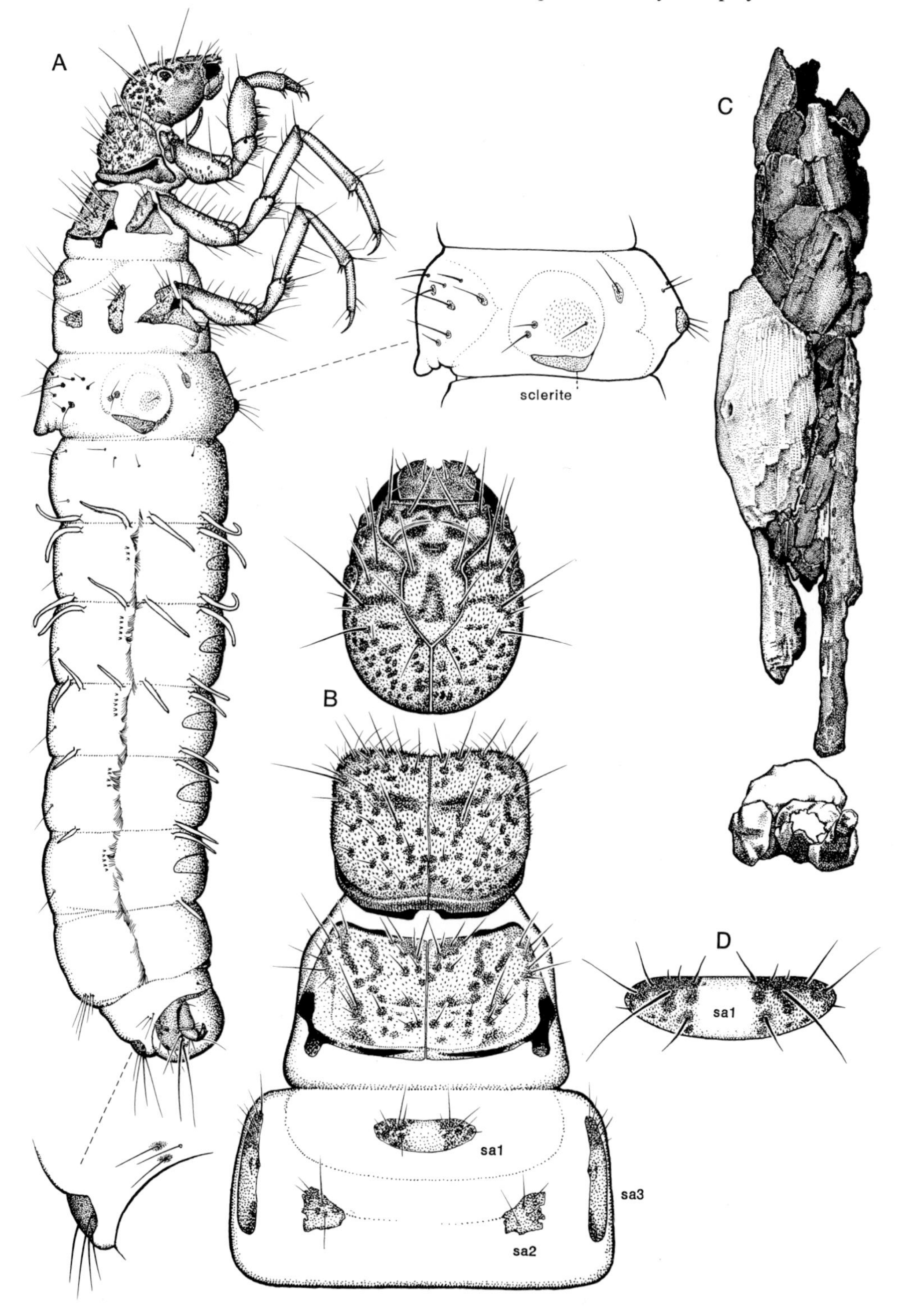

A
C
sclerite
B
D
sa1
sa1
sa3
sa2

10.28 Genus **Imania**

DISTRIBUTION AND SPECIES This genus contains 11 species known in mountainous areas of western North America from Alaska to Colorado and Nevada; other species occur in eastern Siberia.

Larvae referred to Limnephilid Genus D by Ross (1959) and Flint (1960) are almost certainly *Imania.* We identified larvae for four species (Wiggins 1973c) and have collected series belonging to this genus from many western localities. The larva of a Siberian species, *I. sajanensis* Levanidova, has also been described (Levanidova 1967).

MORPHOLOGY The head in larvae of *Imania* is flattened dorsally, the area frequently concave and bounded by a sharp, semicircular carina (B); posterodorsal horns on the head (A,B) are known only in *I. scotti* Wiggins. The head and pronotum have a pebbled texture. Other characters of the head include mandibles with apical edges entire and not subdivided into teeth (D), a T-shaped ventral apotome (F), and a labrum with a membranous anterior margin (E). The pronotum is strongly convex, the mesonotal plates shorter than in most other genera, and the basal seta of the tarsal claws extends almost to the tip of the claw (A,B). Abdominal gills are lacking. Length of larva up to 11.5 mm.

CASE Larvae build a tapered, cylindrical case of small, rather coarse rock fragments. The case of *I. scotti* (C) is unique in having the basal quarter sharply constricted from the remainder. In cases of *I. cidoipes* Schmid there is a ridge of small stones along each side. Length of case up to 13 mm.

BIOLOGY Larvae live in small, cold mountain streams, often at high elevations; frequently they are found on vertical rock faces in a thin layer of flowing water, but also occur on rocks in turbulent water. Gut contents of larvae (3) we examined were largely fine organic and mineral particles, which is consistent with the interpretation that *Imania* larvae scrape the upper surfaces of rocks. It is evident that species require more than one year to complete their life cycle because for several species we have collected adults and larvae of the last two or three instars at the same time.

REMARKS No taxonomic review of adults has been published since that of Ross (1950). Some species of this genus have been treated under the name *Allomyia* (see Ross 1944: 297).

Imania scotti (Oregon, Clackamas Co., 11-12 June 1967, ROM)
A, larva, lateral x19, tarsal claw enlarged; B, head and thorax, frontolateral; C, case, lateral x12; D, mandible, ventral; E, labrum, dorsal; F, head, ventral; G, segment IX and anal prolegs, dorsal

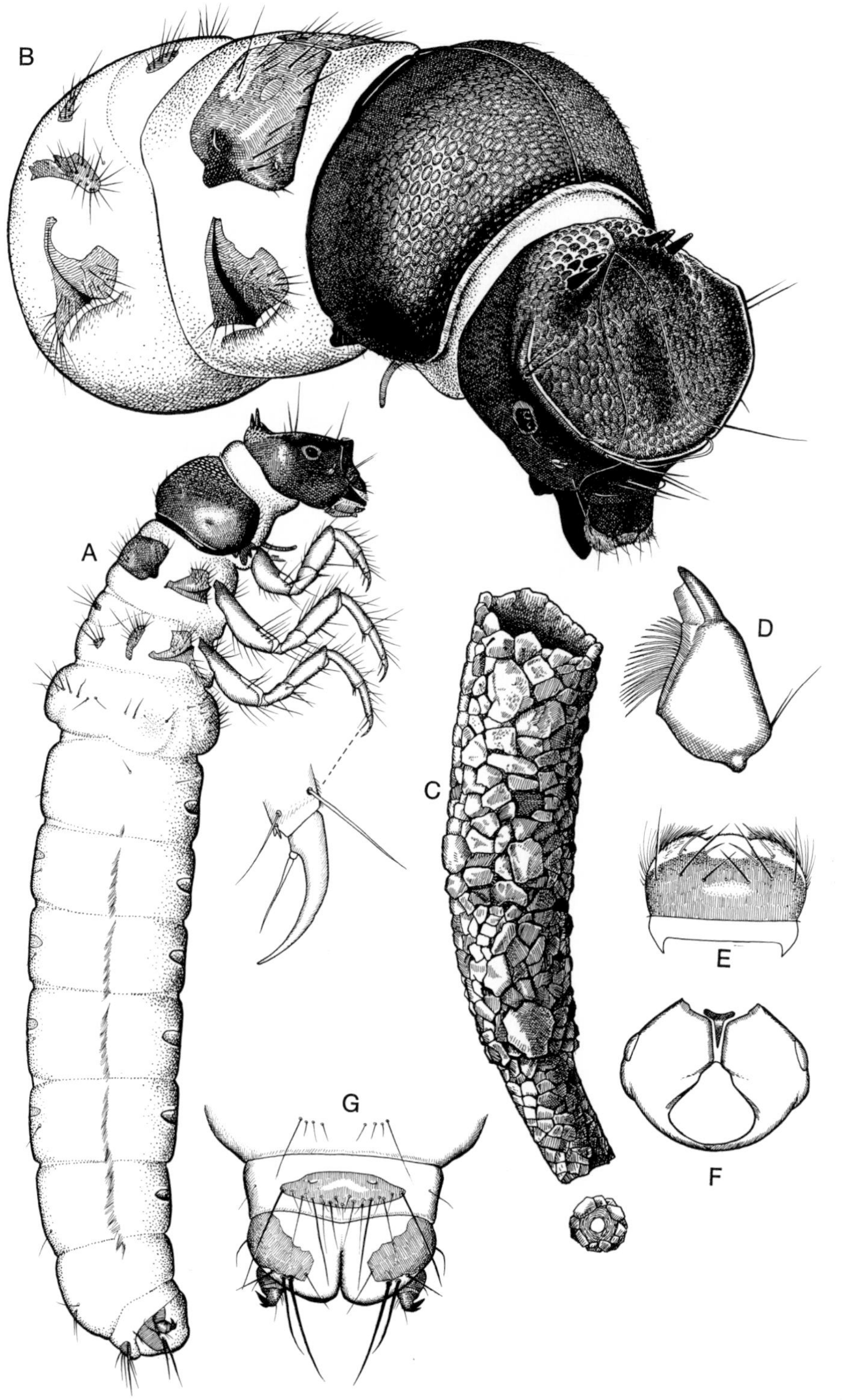

B
A
C
D
E
F
G

10.29 Genus **Ironoquia**

DISTRIBUTION AND SPECIES *Ironoquia* is a Holarctic genus. In North America four species are known, all within the eastern part of the continent from North Carolina to Nova Scotia and Wisconsin; a fifth species is widely distributed in Europe.

Larvae were described for *I. punctatissima* (Walker) and *I. parvula* (Banks) by Flint (1960). There is associated material for *I. lyrata* (Ross) in the ROM collection from work by Mackay (1969) in Quebec.

MORPHOLOGY *Ironoquia*, one of four North American limnephilid genera in which most abdominal gills of the dorsal and ventral rows have more than four branches, is distinct from the others in having more than two (usually five) major setae along the ventral edge of the middle and hind femora (A). Many metanotal setae arise from the integument between the primary sclerites (B). Length of larva up to 22 mm.

CASE Two types of larval case are known. In *I. punctatissima* and *I. lyrata* cases are made of bark and leaves, curved but little tapered; the case of *I. parvula* is made of sand grains (Flint 1960, fig. 48). Length of larval case up to 27 mm.

BIOLOGY Although North American larvae of the subfamily Dicosmoecinae are largely restricted to cool, running waters, species of *Ironoquia* are the sole exception, living in temporary pools and streams (Ross 1944; Flint 1958, 1960; Wiggins 1973a); these habitats are characterized by extremes of temperature and by a drought period of several months. Observations by Flint (1958) and Williams and Williams (1975) have demonstrated that larvae avoid drought by aestivating in leaf litter around the margin of the receding water from late spring until pupation in early autumn. But sexually mature females of *I. lyrata* emerged in June from larvae reared in Quebec by Mackay (1969) and there was no evidence of autumn activity by adults around the temporary-stream habitat. Filamentous algae and vascular plant pieces were the dominant items in guts of larvae (3) we examined.

REMARKS Taxonomy of *Ironoquia* adults has been reviewed by Schmid (1951). Species have been assigned to *Caborius* in the past.

Ironoquia sp. (Missouri, Franklin Co., 12 May 1961, ROM)
A, larva, lateral x8, middle femur enlarged; B, head and thorax, dorsal, detail of mesonotum; C, case x5

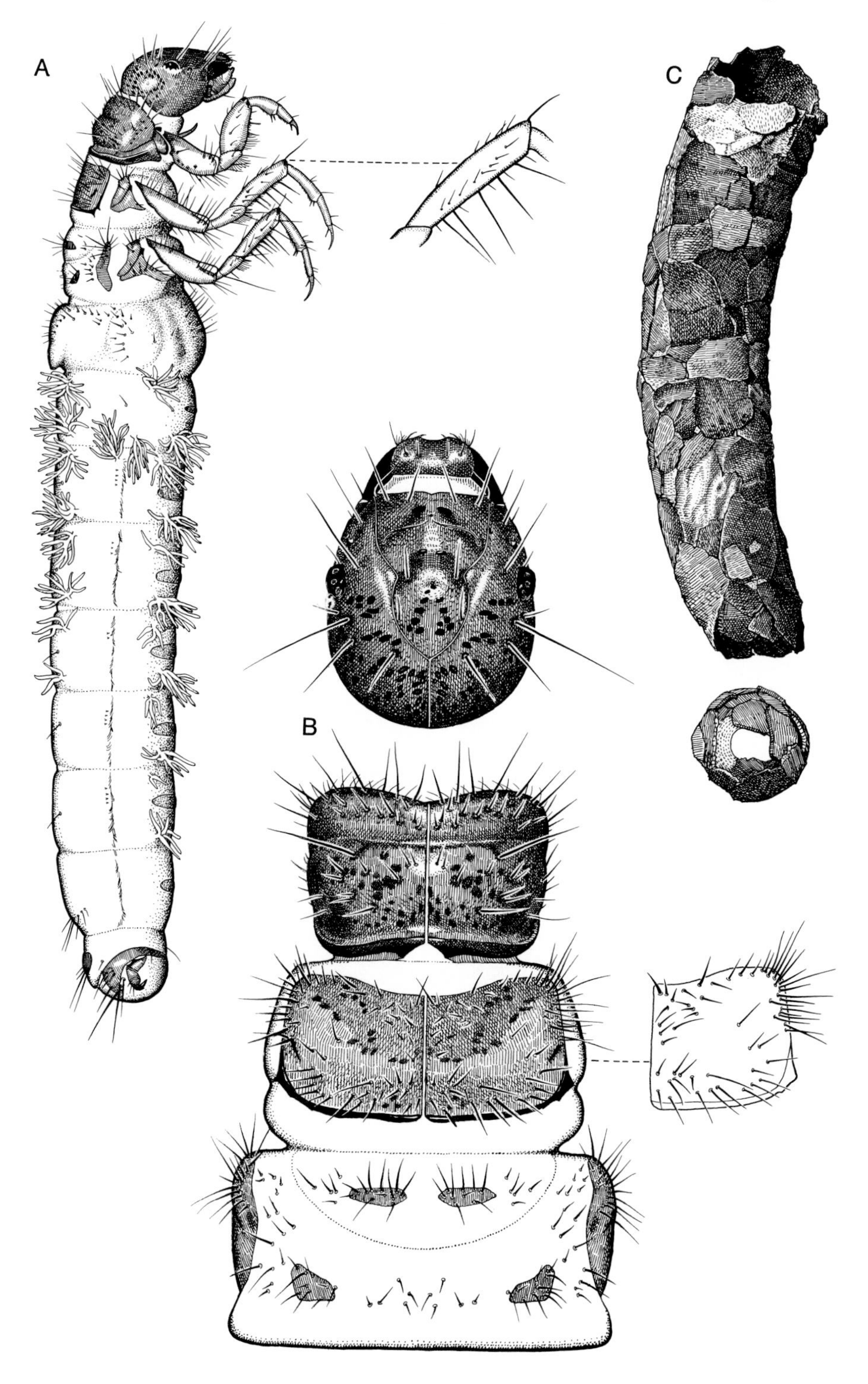
A
B
C

10.30 Genus **Lenarchus**

DISTRIBUTION AND SPECIES Species of *Lenarchus* are widely distributed throughout the northern part of the Holarctic region. Of nine species known in North America most are western, occurring at higher elevations from Alaska to California; eastern species have been recorded from Michigan to Massachusetts.

Larvae have not previously been identified for *Lenarchus*; we have associated larvae for *L. vastus* (Hagen) from Oregon and *L. rho* (Milne) from British Columbia. Larvae described as Limnephilini 1 by Flint (1960) are probably of this genus.

MORPHOLOGY *Lenarchus* is one of the few North American genera in which most abdominal gills of the dorsal and ventral rows have more than four branches. Larvae of *Lenarchus* can be recognized because they have only two major setae on the ventral edge of the second and third femora (A), have all metanotal setae confined to the primary sclerites, and have separate mesonotal setal areas (B). Larvae have a dark brown head and thorax; in some species there is a light line running obliquely through the eyes (A,B), but not in all. Length of larva up to 30 mm.

CASE Larval cases in this genus are somewhat variable. Sometimes lengths of sedge leaves are arranged longitudinally (C). We have collected other larvae that probably are *Lenarchus* in cases of irregular fragments of bark and leaves. Length of larval case up to 55 mm.

BIOLOGY These are species of standing waters - edges of small lakes, ponds, marshes, and temporary pools, especially at higher elevations and latitudes. Winterbourn (1971a) found final-instar larvae of *L. vastus* in Marion Lake, British Columbia, from October through April; the larvae fed on organic sediments.

REMARKS A review of adults of *Lenarchus* was provided by Schmid (1952b); a key to adults of North American species was given by Ross and Merkley (1952).

Lenarchus vastus (Oregon, Jefferson Co., 17-18 June 1968, ROM)
A, larva, lateral x8; B, head and thorax, dorsal, detail of mesonotum; C, case, lateral x2

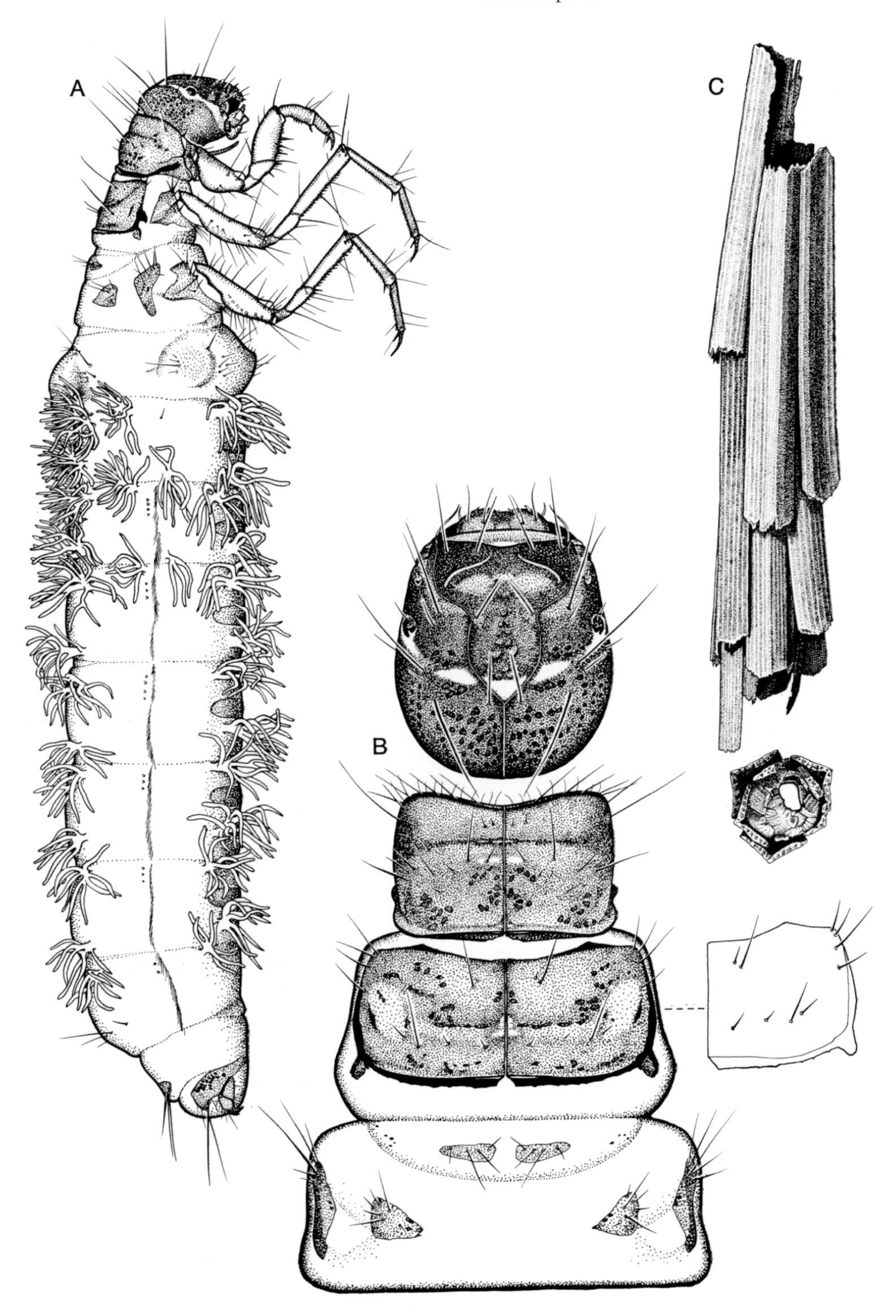
A
B
C

10.31 Genus **Lepania**

DISTRIBUTION AND SPECIES *Lepania* is an unusual Nearctic genus with a single species, *L. cascada* Ross, known from Oregon and Washington. Morphology of all stages has been described and assessed by Wiggins (1973b).

MORPHOLOGY Subdivided plates of the mesonotum and an elongate mesepisternum (B), along with thickened lateral edges on the stout pronotum, establish the larva as a member of the subfamily Goerinae; within that group the larva of *Lepania* is most readily distinguished by the dorsally depressed mesepisternum and the well-developed metanotal *sa*1 sclerites. The larva of *L. cascada* also has mandibles with separate teeth (D) and short basal setae on the tarsal claws (A); because of this combination of characters, shared only with *Goereilla, Lepania* has been interpreted as one of the most primitive of the Goerinae (Wiggins 1973b). The head bears many secondary setae dorsally (B), and the antennae are set in lateral concavities (H); the ventral apotome has concave lateral margins (F). Abdominal gills are reduced to two pairs of single filaments on segment III (A), a small dorsal plate lies beside the basal tuft of the anal proleg (G), and the anal claw lacks an accessory hook. Length of larva up to 5.3 mm.

CASE Larvae of *L. cascada* construct a case of small rock pieces, curved and strongly tapered (C); rock pieces on the venter are smaller than those placed dorsally and laterally. Length of larval case up to 5.3 mm.

BIOLOGY This exceedingly local species occurs in spring seepage sites in mountainous country. From our field observations on Marys Peak, Oregon, it is evident that larvae are restricted to the water-saturated organic muck at the head of springs. Gut contents of larvae (3) we examined were mainly pieces of vascular plants with some algae. Adults fly in June, but the presence of early-instar larvae at the same time suggests that more than one year is required for completion of the life cycle.

Lepania cascada (Oregon, Benton Co., 14–15 June 1968, ROM)
A, larva, lateral x23, tarsal claw enlarged; B, head and thorax, dorsal, detail of mesonotum; C, case, lateral x18; D, mandible, ventral; E, labrum, dorsal; F, head, ventral; G, segments VIII, IX and anal prolegs, dorsal; H, right side of head, dorsal

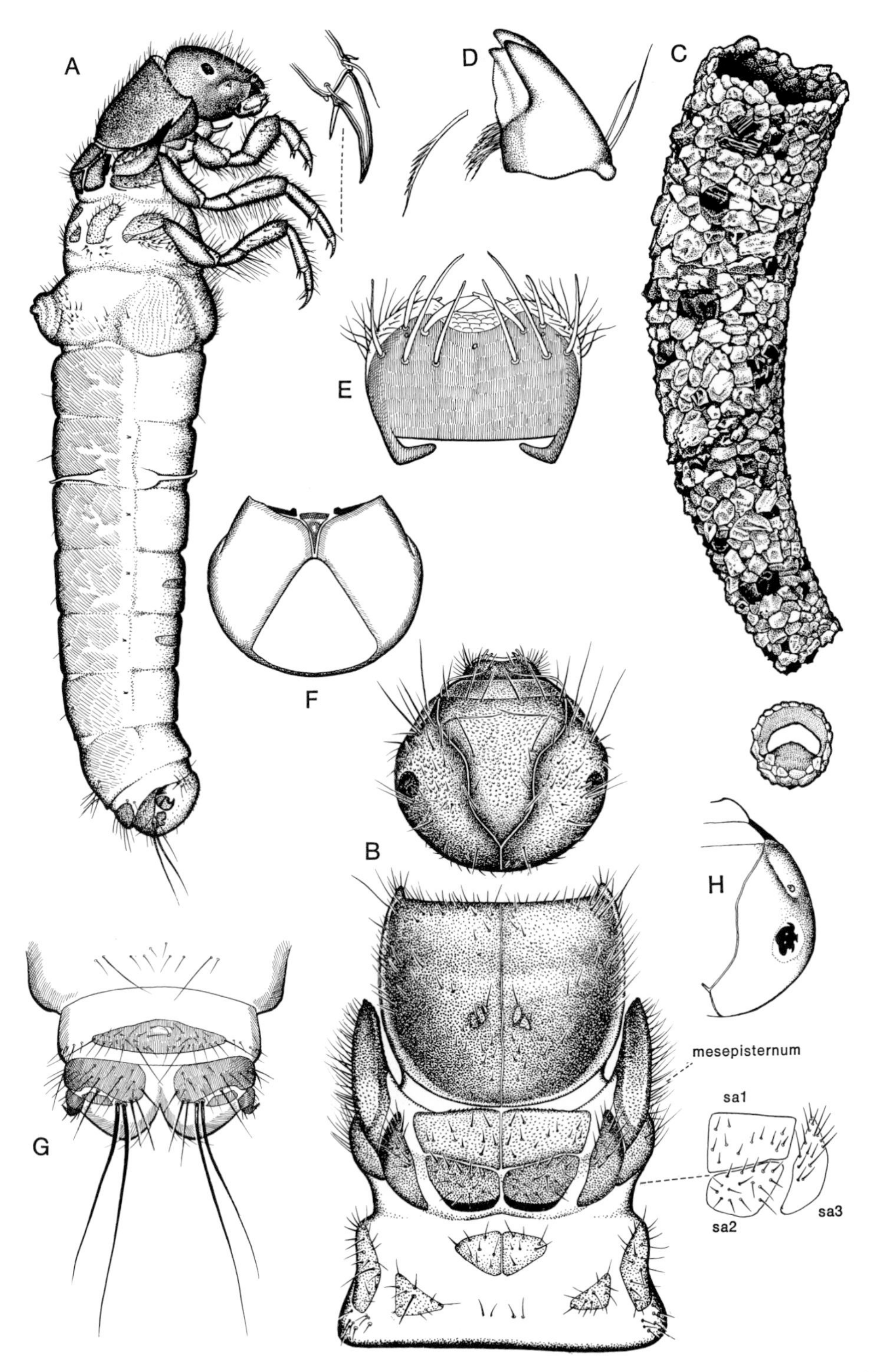
A
B
C
D
E
F
G
H
mesepisternum
sa1
sa2
sa3

10.32 Genus **Limnephilus**

DISTRIBUTION AND SPECIES *Limnephilus* is a large genus of the Holarctic region. Approximately 95 species are now known in North America and they occur over much of the continent.

Larval descriptions are available in the literature for only five species (Ross 1944; Flint 1960; Nimmo 1965); to date, we have associated material for 34 species in our field work.

MORPHOLOGY Colour patterns of head and thorax are of two basic types, and the genus emerges from the key accordingly in two places. In one type, the dorsum of the head bears bands or patches of contrasting colour, and within the narrowed posterior portion of the frontoclypeal apotome are three light areas - one along each side and one at the posterior extremity as in B and as in *Philarctus* (Fig. 10.41B). In the second type of *Limnephilus* the head lacks bands or other well-defined areas of contrasting colour, and may have prominent spots, much as illustrated for *L. frijole* Ross (G). Separation of *Asynarchus* (q.v.) and *Philarctus* (q.v.) larvae from *Limnephilus* in North America is questionable. Length of larva up to 29 mm.

CASE The variety of larval cases for *Limnephilus* spp. known from our associations is extremely broad: sand grains, pebbles (E), bark (F), wood, and leaves arranged lengthwise (D) or transversely (C). Length of case up to 51 mm.

BIOLOGY *Limnephilus* larvae are predominantly members of lentic communities in ponds, lake margins, and marshes, although a few have been collected in streams and cold springs. Some species are typical of temporary pools and streams, marginal habitats they can exploit because diapause delays sexual maturity until late summer when the height of the drought is over; and also because after hatching from the eggs a few weeks later, larvae can remain within the thick gelatinous egg-matrix for several months until the basin is flooded again (Wiggins 1973a). *Limnephilus* larvae feed mainly on detritus (Lloyd 1921; Hodkinson 1975).

REMARKS Ross and Merkley (1952) provided a key to males of North American *Limnephilus* spp. The northern species *L. pallens* (Banks) was originally assigned to *Zaporata* (see Ross 1944: 299).

Limnephilus indivisus (Ontario, Durham Co., May 1969, ROM)
A, larva, lateral x6; B, head and thorax, dorsal; C, case x3
D, *L. submonilifer* case x1.7
E, *L. canadensis* case x2.4
F, *L. atercus* case
G, *L. frijole* head, dorsal

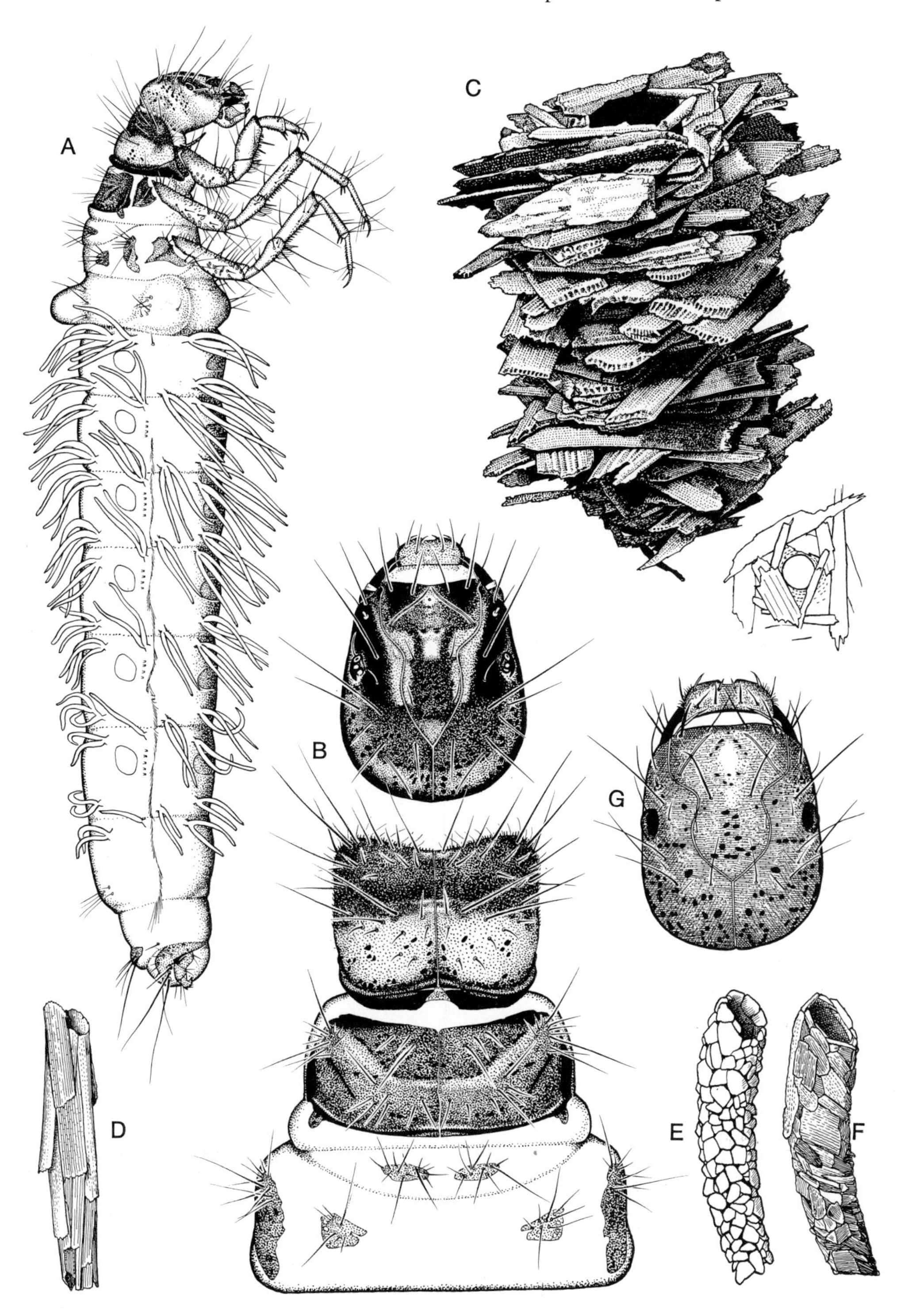
A
B
C
D
E
F
G

10.33 Genus **Manophylax**

DISTRIBUTION AND SPECIES This genus was established for a single species *M. annulatus* Wiggins collected in mountainous country in Idaho close to the Montana border. A fully associated series was collected; larvae, pupae, and adults have been described elsewhere (Wiggins 1973c).

MORPHOLOGY The larva of *M. annulatus* has a general similarity to *Imania* and *Apatania* larvae, all three having mandibles with a continuous scraping edge (D), tarsal claws with elongate basal setae (A), and a labrum with the anterior margin membranous (F). The larva of *M. annulatus* is distinctive in having two well-developed metanotal *sa*1 sclerites (B), and on the venter of abdominal segment I an elliptical sclerite with a central unsclerotized area (E). The pronotum in *Manophylax* is enlarged and bears stout setae around the periphery (B). Abdominal gills are single and chloride epithelia are present on the venter of segments III–VII; the anal claw has a well-developed accessory hook (A). Length of larva up to 8 mm.

CASE The larval case of *M. annulatus* is constructed of rock fragments, strongly tapered and slightly curved (C); small twigs and leaf fragments are attached to the dorsolateral surfaces of the case, and filamentous algae were growing on the cases collected. Cases in *Apatania* and *Imania* are made of rock fragments alone. Length of larval case up to 10 mm.

BIOLOGY Our single collection of *M. annulatus* was made in a small run, maximum width approximately 45 cm, cascading over boulders down a steep hillside. Larvae were crawling on flat rock surfaces in a thin film of flowing water. Pupal cases were found on damp rocks along the stream's edge, and emergence of adults was in progress in June.

Manophylax annulatus (Idaho, Idaho Co., 24 June 1968, ROM)
A, larva, lateral x18, claws of tarsus and anal proleg enlarged; B, head and thorax, dorsal; C, case, ventrolateral x9; D, mandible, ventral; E, segment I, ventral; F, labrum, dorsal; G, head, ventral; H, segment IX and anal prolegs, dorsal

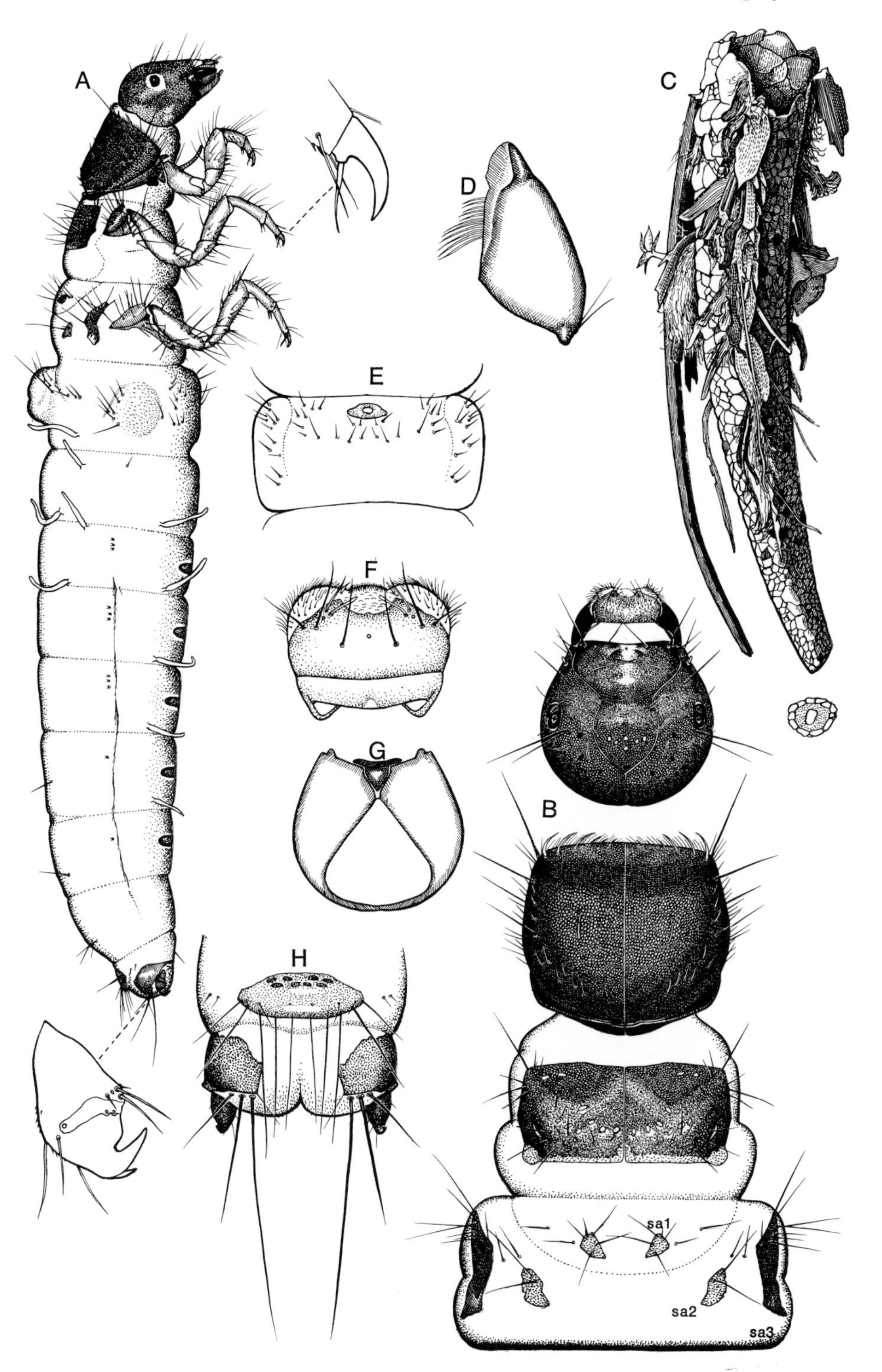
A
B
C
D
E
F
G
H
sa1
sa2
sa3

10.34 Genus Moselyana

DISTRIBUTION AND SPECIES *Moselyana* is another of the unusual Nearctic Limnephilidae. There is but a single species, *M. comosa* Denning, known from Oregon.

We collected an associated series of larvae on Marys Peak in Oregon; description and analysis of characters of larvae, pupae, and adults are available elsewhere (Wiggins 1973c).

MORPHOLOGY The head is rounded in dorsal aspect and bears secondary setae (B); the labrum is entirely sclerotized (E), and the mandibles toothed (D). The pronotum is robust, the mesonotum unusually wide, and the metanotum has many setae arising between the primary sclerites (B); basal setae of the tarsal claws are short (A). Sclerotized parts of the head and thorax are uniform brownish red. Abdominal gills are lacking as is the lateral fringe except for a tuft of setae on segment VIII; chloride epithelia are apparent, though faint, both dorsally and ventrally on some segments; the dorsal sclerite of segment IX and the basal segment of the anal prolegs are heavily setate (G). Length of larva up to 7mm.

CASE The larval case (C) is constructed of fine rock fragments, strongly tapered and curved, the posterior end obstructed with marginal silken points radiating toward the centre; a silken layer covers the exterior surface of the case. Length of larval case up to 8.5 mm.

BIOLOGY Larvae were found in the water-saturated organic muck of a spring seepage area; careful searching failed to produce any larvae in a small spring stream 2 m away from the seepage area. Gut contents of larvae (3) were largely vascular plant pieces, algae, and fine particles of both organic and inorganic origin. Pupae and adults were taken in June; many larvae of late instars were active at the same time.

Moselyana comosa (Oregon, Benton Co., 14-15 June 1968, ROM)
A, larva, lateral x21, claws of tarsus and anal proleg enlarged; B, head and thorax, dorsal; C, case, lateral x13; D, mandible, ventral; E, labrum, dorsal; F, head, ventral; G, segment IX and anal prolegs, dorsal

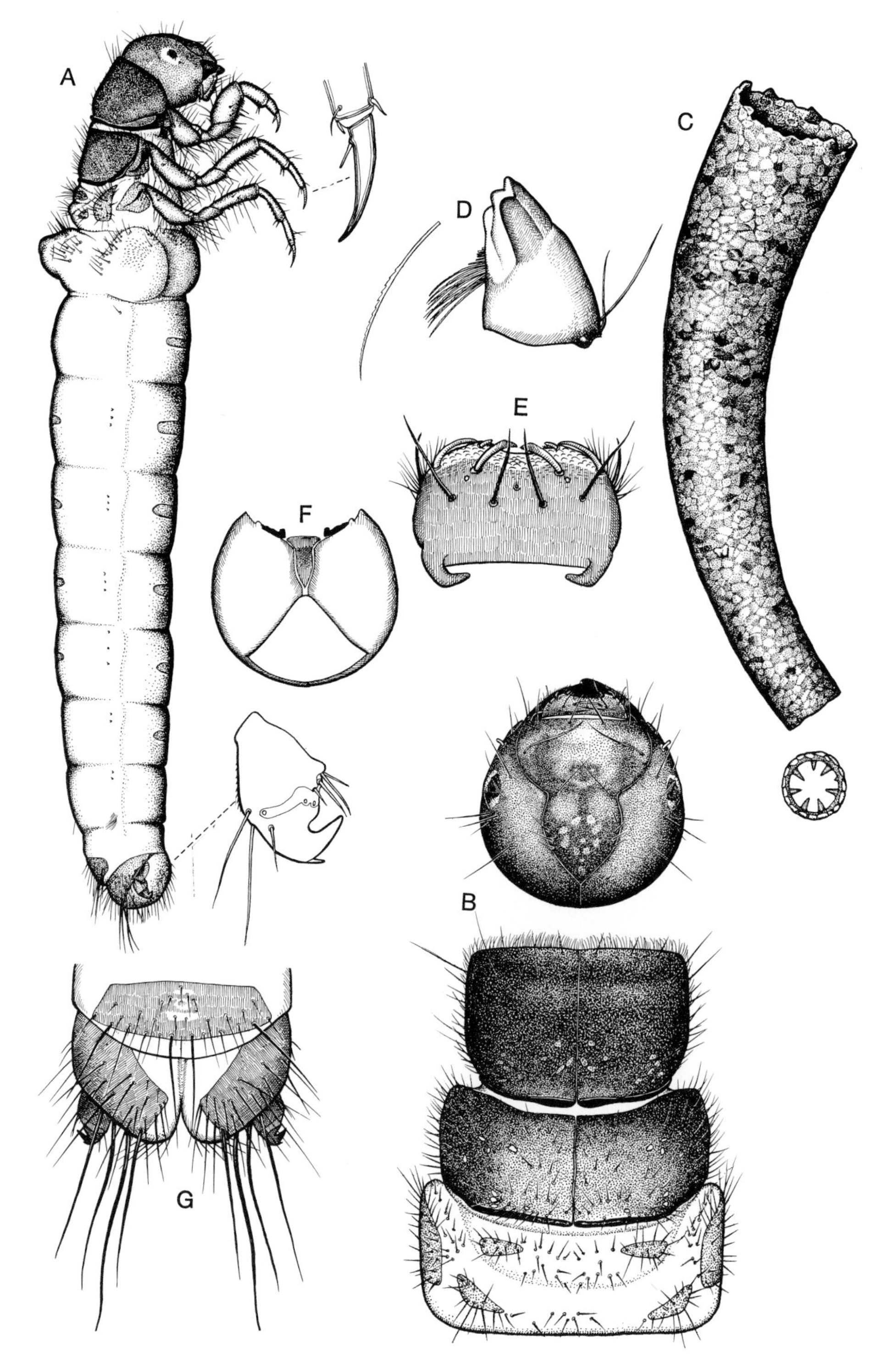
A
B
C
D
E
F
G

10.35 Genus **Nemotaulius**

DISTRIBUTION AND SPECIES *Nemotaulius* is a small Holarctic genus with four species in eastern Asia, one in Europe, and one in North America (Schmid 1952a). The single Nearctic species, *N. hostilis* (Hagen), occurs across the continent from British Columbia and Oregon to Newfoundland, south to the New England States and Michigan.

Larvae have been described by Lloyd (1921) and Ross (1944) under *Glyphotaelius*, and also by Flint (1960); we reared several series of larvae from Ontario.

MORPHOLOGY *Nemotaulius* larvae have distinctive head markings (B); on a base colour of light yellowish brown a dark U-shaped band extends through each eye, and a median band extends over most of the frontoclypeal apotome; the pronotum bears a central, transverse, dark band. There are spines on the head and pronotum. Abdominal segment I has a group of prominent setae in each of the three primary positions (D). The anal claw bears several accessory hooks (A). Length of larva up to 28 mm.

CASE From midsummer through autumn, *Nemotaulius* cases are constructed of pieces of leaves and twigs fastened more or less transversely; in the later instars rather large pieces are arranged in dorsal and ventral series, giving the case a flattened appearance (C). But when collected in May and early June just before pupation, cases are of leaf pieces fastened longitudinally much as shown for *Grammotaulius*. Thus it appears that case architecture is changed during the final instar and transitional cases are sometimes found (Flint 1960, fig. 60); but evidently this does not always occur (Bernhardt 1966). Length of case up to 70 mm.

BIOLOGY Larvae are typical inhabitants of standing waters, especially small ponds with a dense growth of aquatic plants.

Nemotaulius hostilis (New York, Tompkins Co., 12 Nov. 1957, ROM)
A, larva, lateral x7, anal claw enlarged; B, head and thorax, dorsal; C, case, ventral x2; D, segment I, ventral

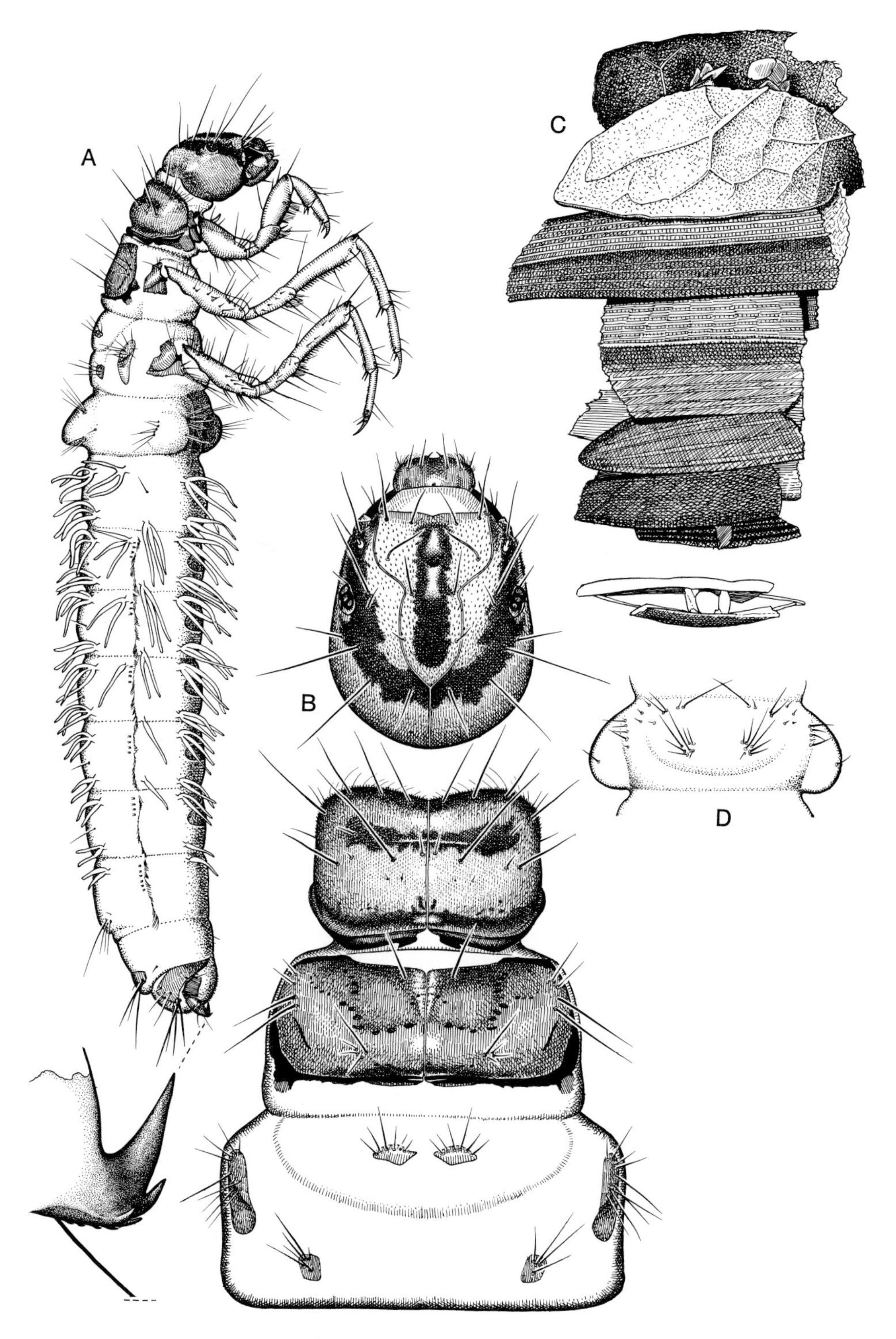
A
B
C
D

10.36 Genus **Neophylax**

DISTRIBUTION AND SPECIES This is a genus of the Nearctic, Asian Palaearctic, and Oriental regions. Fifteen species of *Neophylax* are known in North America: eastern and central species occur roughly within the area from Newfoundland to Minnesota and south through Missouri to Georgia; western species are widespread in the mountains from British Columbia to California.

Larvae of North American species have been identified in the literature by several workers. A key to eight eastern species was given by Flint (1960); we have larval material for 12 species.

MORPHOLOGY *Neophylax* larvae have several unusual characters: the head is long; the labrum has a membranous anterior edge (E); the ventral apotome is distinctly T-shaped (F); the mandibles have entire scraping edges (D); the anterolateral margin of the pronotum is broadly rounded; the prosternal horn is reduced; the mesonotum bears an anteromedian notch (B); each metanotal *sa*1 sclerite is reduced to a small area around the base of one seta (B); the basal seta of each tarsal claw extends to the tip of the claw (A); in some species abdominal segment I bears a rounded, ventral gill-like lobe (A); and the claw of the anal proleg bears an accessory hook. Length of larva up to 16.5 mm.

CASE The larval case (C) is relatively short and thick, constructed of coarse rock fragments with several larger stones along each side; it resembles the case of only one other genus in North America - *Goera.* Length of larval case up to 15 mm.

BIOLOGY Larvae are confined to flowing waters, where different species are characteristic of particular types or sections of streams. Food is largely diatoms and fine detrital particles from rock surfaces. For the most part, larvae grow during autumn and winter months; in spring and early summer, final instars fasten their cases in clusters to rocks, seal off the openings, but remain quiescent in larval diapause for several weeks before initiation of metamorphosis. In most species, adults emerge in the later part of summer; there are exceptions where adults fly in spring and early summer as in the eastern *N. ornatus* Banks and the western *N. occidentis* Banks, and as in certain populations within a species, e.g. *N. concinnus* McL. (Ross 1944, as *N. autumnus*).

Neophylax rickeri (Oregon, Benton Co., 11 April 1964, ROM)
A, larva, lateral x14, tarsal claw enlarged; B, head and thorax, dorsal; C, case, ventral x8; D, mandible, dorsal; E, labrum, dorsal; F, head, ventral

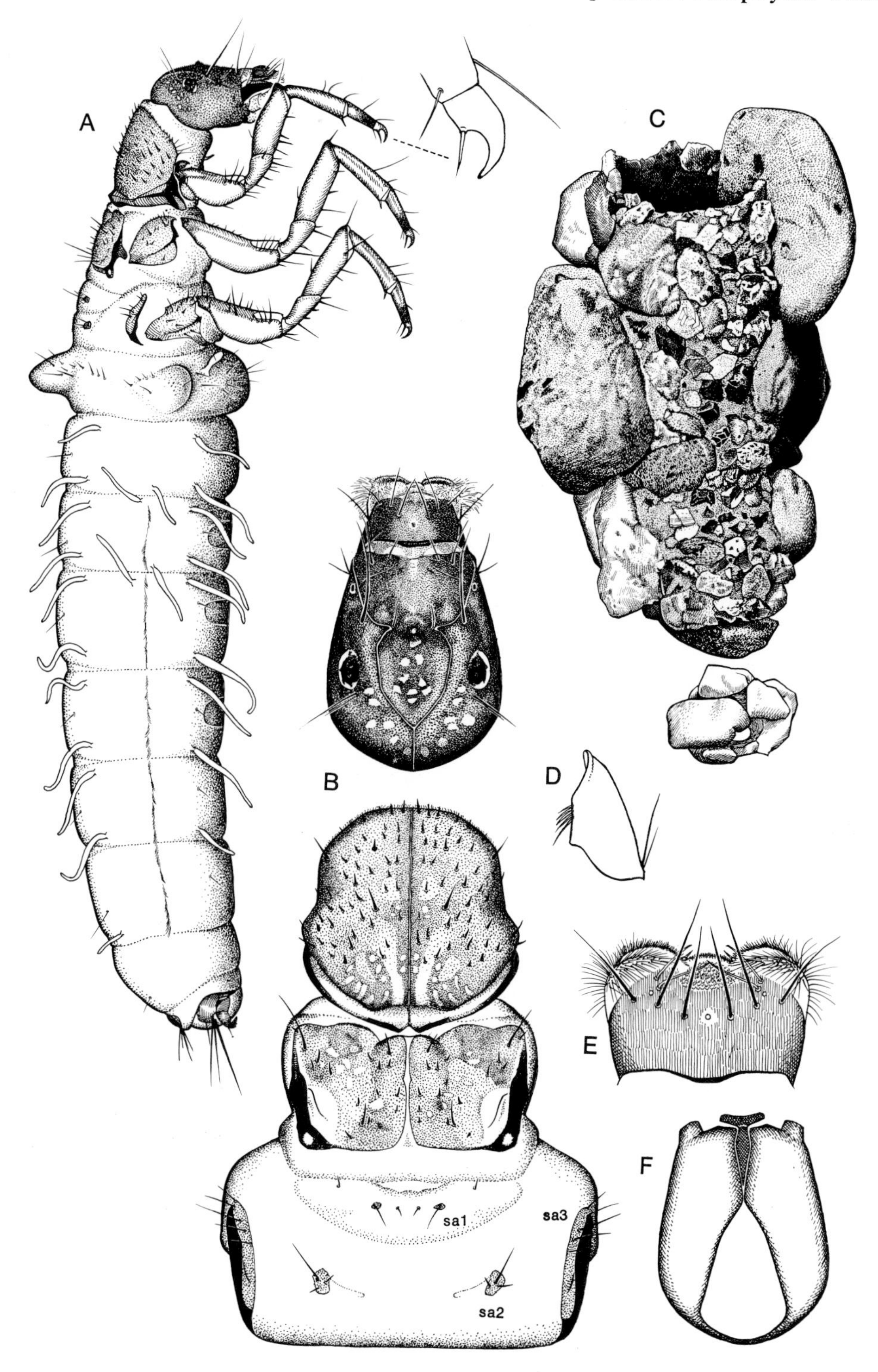
A
C
B
D
E
F
sa1
sa3
sa2

10.37 Genus **Neothremma**

DISTRIBUTION AND SPECIES Species of *Neothremma* are confined to western North America, but are widespread throughout mountainous areas from British Columbia and Alberta south to Colorado and California; six species are known (Denning 1975).

The larva of only *N. alicia* Banks has been identified in the literature (Flint 1960); we have associated larvae for three species.

MORPHOLOGY *Neothremma* larvae share the general characters outlined under *Farula*. In both genera the triangular ventral apotome (D) barely separates the genae which for the most part are closely appressed along the ventral ecdysial line. There are paired sternal sclerites on both the meso- and meta-thorax (E), and the basal seta of the tarsal claws is long, reaching nearly to the tip of the claw (A). Small spines occur dorsally and ventrally in the intersegmental areas between segments IV and V (A). In addition to the diagnostic characters given in the key, *Neothremma* larvae have a prothoracic sternellum (E) and ventral chloride epithelia on segments V and VI (A). Length of larva up to 9 mm.

CASE Larval cases in *Neothremma* (C) are only slightly stouter than those in *Farula*. Constructed of sand grains, the cases are lined inside as well as outside with a thin sheet of silk. As in *Farula*, final-instar larvae spin a posterior perforate sieve membrane inside the case in constructing a pupation chamber. Length of larval case up to 14 mm.

BIOLOGY Larvae occur in mountain springs and streams but, judging by collections we have made, species of *Neothremma* generally inhabit larger streams than *Farula* do. *Neothremma* larvae feed mainly on fine organic particles, with small proportions of diatoms and other algae (Muttkowski and Smith 1929; Mecom 1972a).

Neothremma alicia (Idaho, Idaho Co., 24 June 1968, ROM)
A, larva, lateral x27, spines of segments IV and V enlarged; B, head and thorax, dorsal, anterior border of pronotum enlarged; C, case x14; D, head, ventral; E, thorax, ventral

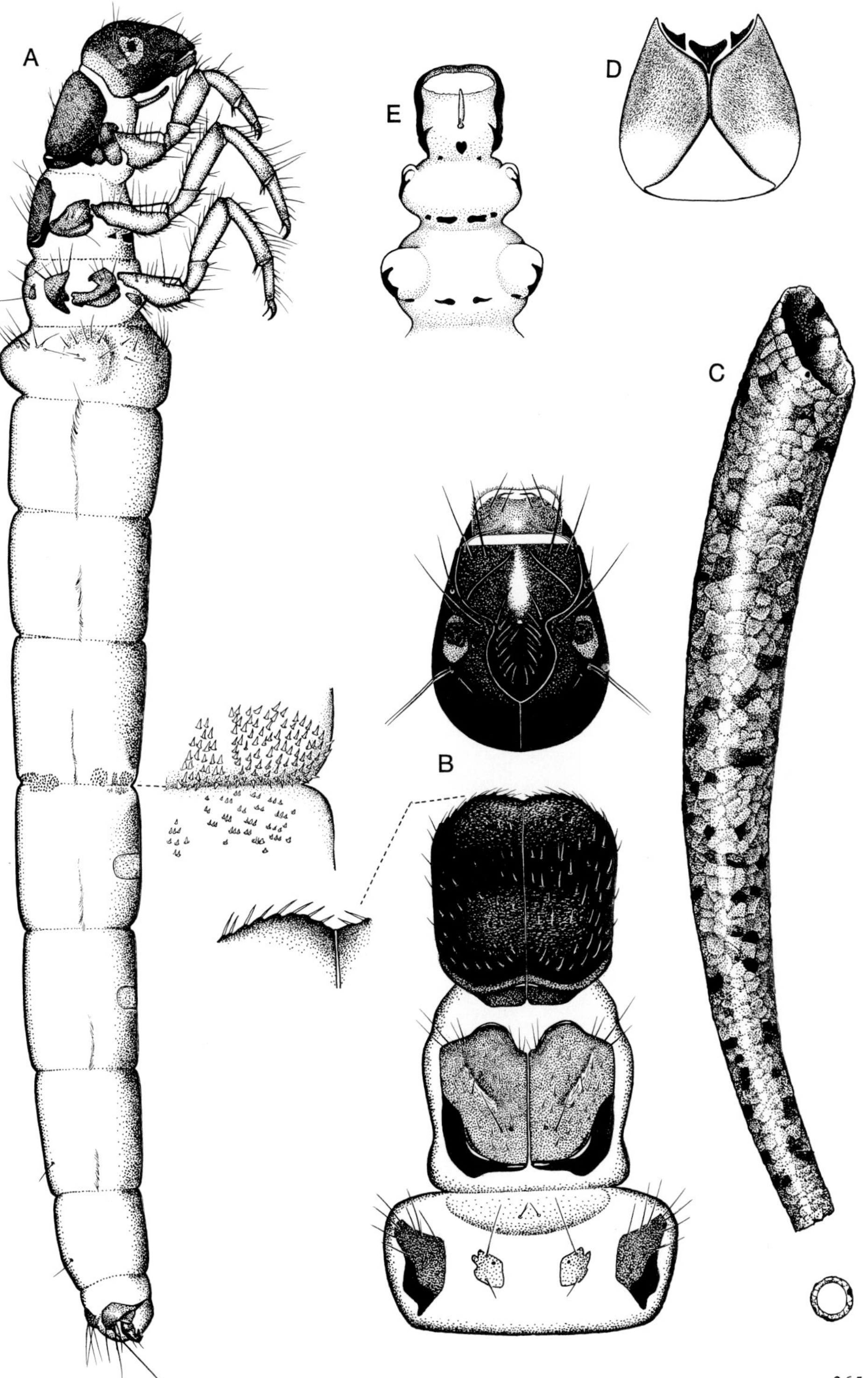
A
E
D
C
B

10.38 Genus **Oligophlebodes**

DISTRIBUTION AND SPECIES *Oligophlebodes* is exclusively a Nearctic genus, the seven species now known all being confined to mountainous areas of the west. Records range from British Columbia and Alberta through South Dakota to California and New Mexico.

Larvae have been identified in the literature for *O. minuta* (Banks) and *O. sierra* Ross (Flint 1960); we have associated material for *O. ruthae* Ross.

MORPHOLOGY Larvae are readily recognized because the pronotum is variously modified by longitudinal ridges and depressions, and the dorsum of the head bears a marginal carina (B). Sclerites of the head and first two thoracic segments are uniform dark brown in colour and coarsely pebbled. Mandibles lack distinct teeth; the ventral apotome is T-shaped as in *Neophylax* (D). Metanotal *sa*1 sclerites are lacking, represented only by a few setae. The basal seta of each tarsal claw extends almost to the tip of the claw. Abdominal gills are single, and the lateral fringe is lacking. Length of larva up to 7.5 mm.

CASE Larval cases are constructed of coarse rock fragments, strongly tapered and curved (C); the exterior is often irregular because relatively large pieces are incorporated. The posterior silken membrane has an unusual eccentric opening. Length of larval case up to 8.5 mm.

BIOLOGY Larvae of *Oligophlebodes* are characteristic of mountain streams, sometimes the most rapid sections. The only species for which life cycle observations are available is *O. sigma* Milne (Pearson and Kramer 1972): larvae developed from late autumn through to summer; within sealed cases a quiescent period, larval diapause similar to *Neophylax*, was followed by metamorphosis with maximum emergence of adults in August and September. Larvae fed actively upon upper rock surfaces during the day, but returned to the under surfaces at night; digestive tracts contained green and brown algae, diatoms, and unidentifiable plant fragments. An unusual tendency for downstream drift during the day was recorded for *O. sigma* (Waters 1968).

Oligophlebodes sp. (California, Shasta Co., 19 June 1967, ROM)
A, larva, lateral x19, tarsal claw enlarged; B, head and thorax, dorsal; C, case, lateral x13; D, ventral apotome of head, ventral

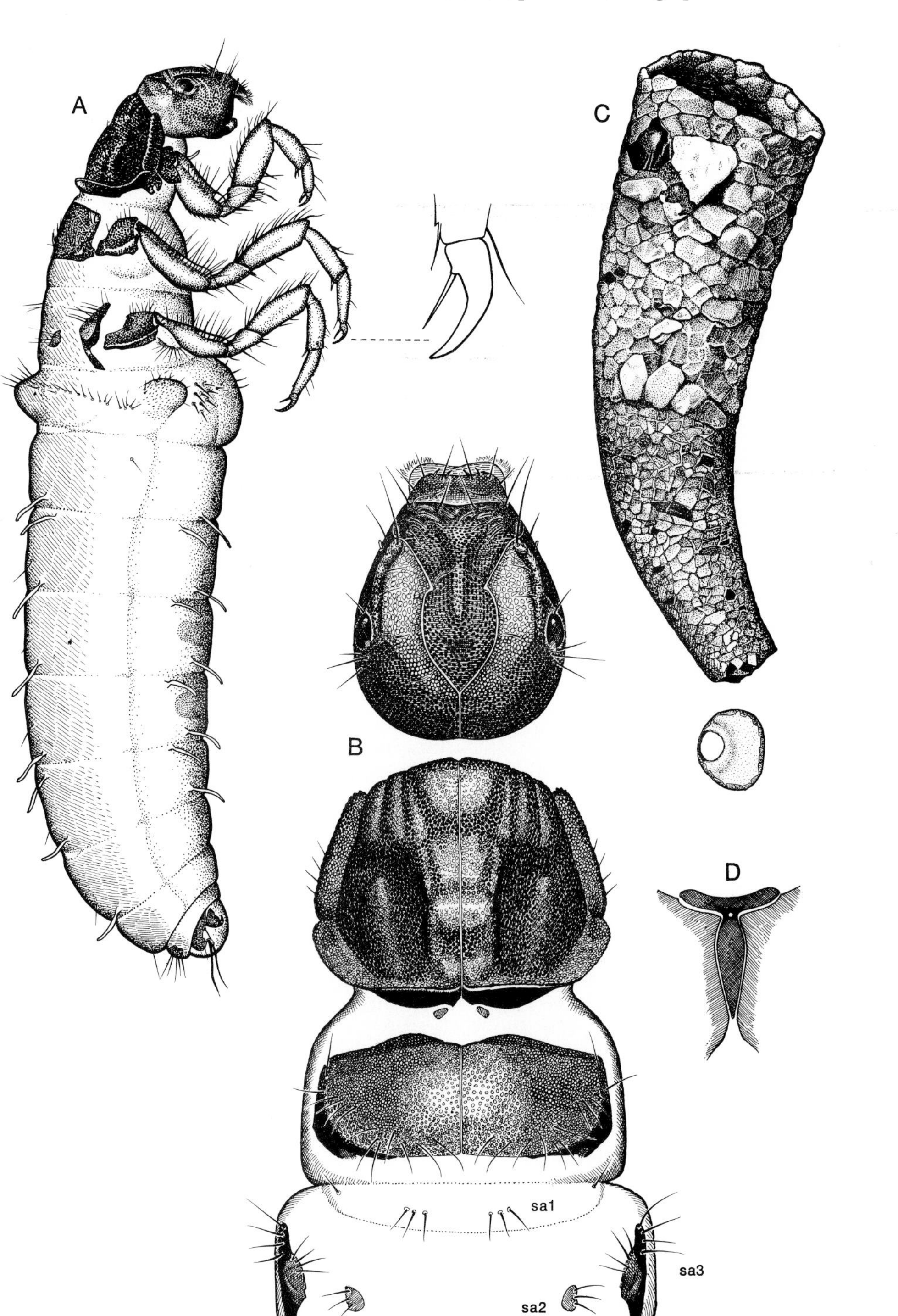
A
B
C
D
sa1
sa2
sa3

10.39 Genus **Onocosmoecus**

DISTRIBUTION AND SPECIES This is a genus of the Nearctic and eastern Palaearctic regions. The number of species actually occurring in North America is uncertain because consistent characters for separating many of the adults have not yet been found. There is probably a single species ranging from Newfoundland to Wisconsin, and perhaps also to Saskatchewan where we collected larvae. *Onocosmoecus* is widespread through most of the western montane area from California to Alaska where six other species have been proposed (Schmid 1955); and as indicated under *Dicosmoecus*, two of the western species heretofore assigned to that genus, *frontalis* Banks and *schmidi* Wiggins, are assigned to *Onocosmoecus.*

The larva of the eastern *O. quadrinotatus* Banks has been described (Flint 1960); we have larvae for this species, for other series collected in the west, and for *O. frontalis* and *O. schmidi.*

MORPHOLOGY *Onocosmoecus* larvae usually have a light median band on the pro- and meso-nota and on the coronal suture of the head (B), as do several other dicosmoecines. Only apical tibial spurs are present (A), and metanotal setae are confined to the primary sclerites; segment IX bears a row of 4 to 6 setae on each side ventrad from the dorsal sclerite. Length of larva up to 26 mm.

CASE Most larvae of this genus construct tubular cases with flat pieces of bark and wood (C). The case of *O. frontalis* is more irregular because it is made of stouter pieces of wood, but that of its closest relative, *O. schmidi*, is made entirely of gravel. Length of larval case up to 29 mm.

BIOLOGY Larvae live in cool waters, both lotic and lentic, but generally not in currents where *Dicosmoecus* occurs. Often they are found abundantly in accumulations of plant debris in pools of cool streams (Flint 1960). Larvae usually burrow in bottom materials before pupating. The life cycle of a western species was studied in a British Columbia lake by Winterbourn (1971a); guts of larvae contained mostly sediment with some animal fragments; developing from eggs deposited in September, most larvae overwintered in the third or fourth instar with final instars active until July. Larvae of a Siberian species, *O. flavus* (Martynov), fed mainly on leaves and other allochthonous plant materials (Levanidova 1975). *Onocosmoecus* larvae may therefore function more as detritivorous shredders than others in the Dicosmoecinae.

Onocosmoecus sp. (British Columbia, Marion Lake, 1969, ROM)
A, larva, lateral x6, middle tibia enlarged; B, head and thorax, dorsal; C, case x4

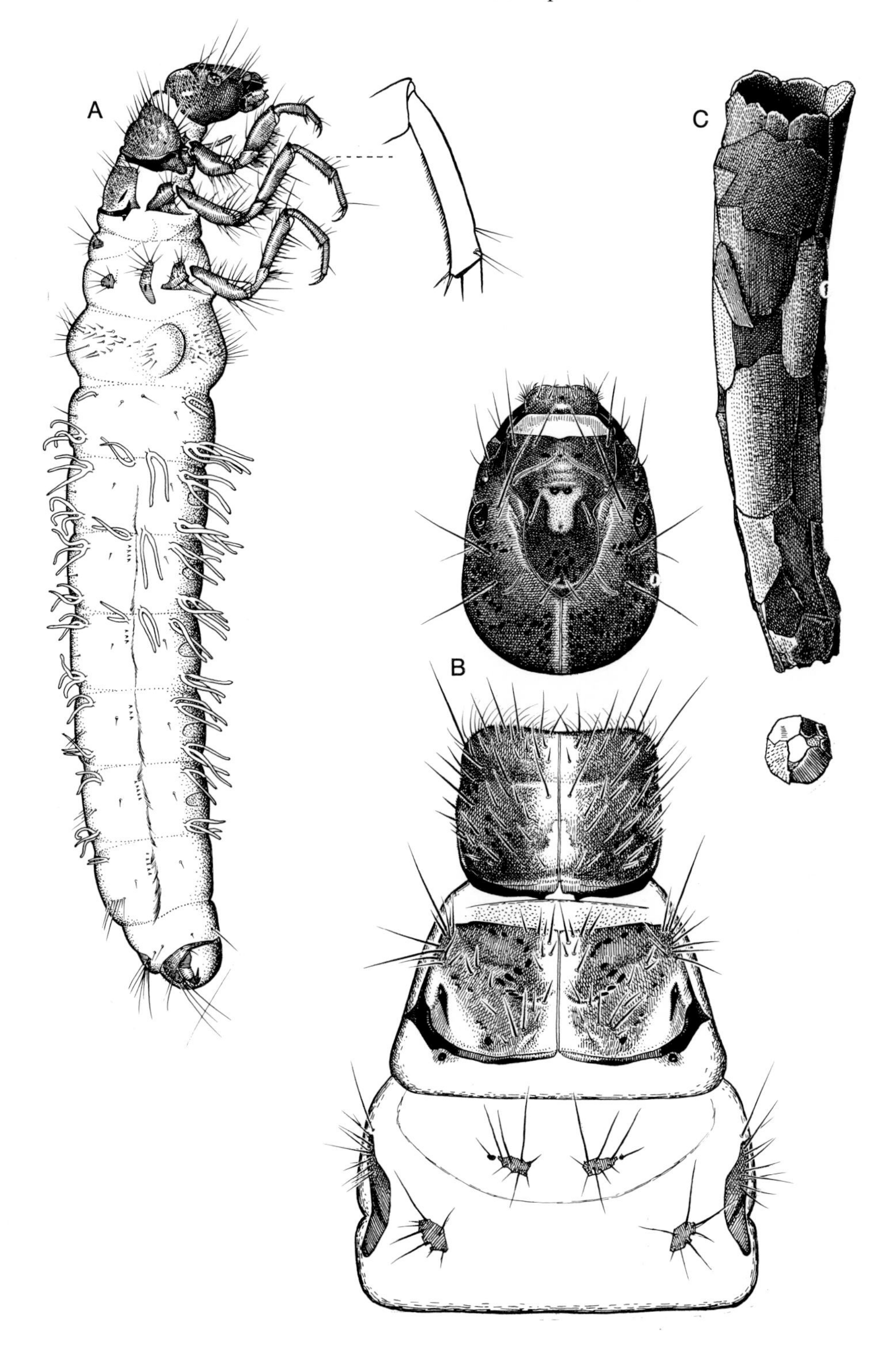
A
B
C

10.40 Genus **Pedomoecus**

DISTRIBUTION AND SPECIES *Pedomoecus* is another of the unusual Nearctic genera in the Limnephilidae. The single species, *P. sierra* Ross, has been recorded from California and Washington; we collected adult specimens from Oregon, British Columbia, and Alberta as well.

Although we have larvae from British Columbia and California, their association with adults of *Pedomoecus* has not been proved. Identity is founded on larvae determined by H.H. Ross and described by Flint (1960), although the basis for that determination was not stated.

MORPHOLOGY *Pedomoecus* larvae have a number of unusual characters. Many of the primary setae on the anterior part of the head are exceptionally stout and their bases swollen; the labrum bears three dorsal tubercles (B), and the blade of each mandible appears to be extended as a scraper but is notched at the apex (D). The pronotum is broad anteriorly in dorsal aspect, a transverse internal ridge incomplete dorsally (A,B); the prosternum bears an ovoid sclerite and the prosternal horn is well developed (E). Each mesonotal plate is subdivided longitudinally into two sclerites. Metanotal *sa*1 sclerites are lacking, although setae are abundant on the dorsum. The fore and middle femora bear a ventral row of short, stout setae; the hind femur is heavily setate ventrally and the hind tibia and tarsus bear long, stout setae on the dorsal edge; the basal seta of each tarsal claw (A), although extending nearly to the tip of the claw, differs from the long basal seta in such genera as *Neophylax*. Gills are single, and the first three abdominal segments unusually hairy. The anal claw bears an accessory hook (F). Length of larva up to 8.5 mm.

CASE The larval case is constructed of small rock fragments and tapers strongly posteriorly; the posterior opening is reduced to a small central hole (C). Length of larval case up to 7 mm.

BIOLOGY We collected larvae in cool, rapid streams; they were local and never numerous. The unusual morphological features suggest that the larvae have some specialized way of life, not yet understood.

REMARKS Diagnostic characters for the male of *P. sierra* were given by Ross (1947).

Pedomoecus sierra (California, Shasta Co., 17 Sept. 1946, ROM)
A, larva, lateral x19, tarsal claw enlarged; B, head and thorax, dorsal; C, case, lateral x15; D, mandible, ventral; E, prothorax, ventral; F, segment IX and anal prolegs, dorsal

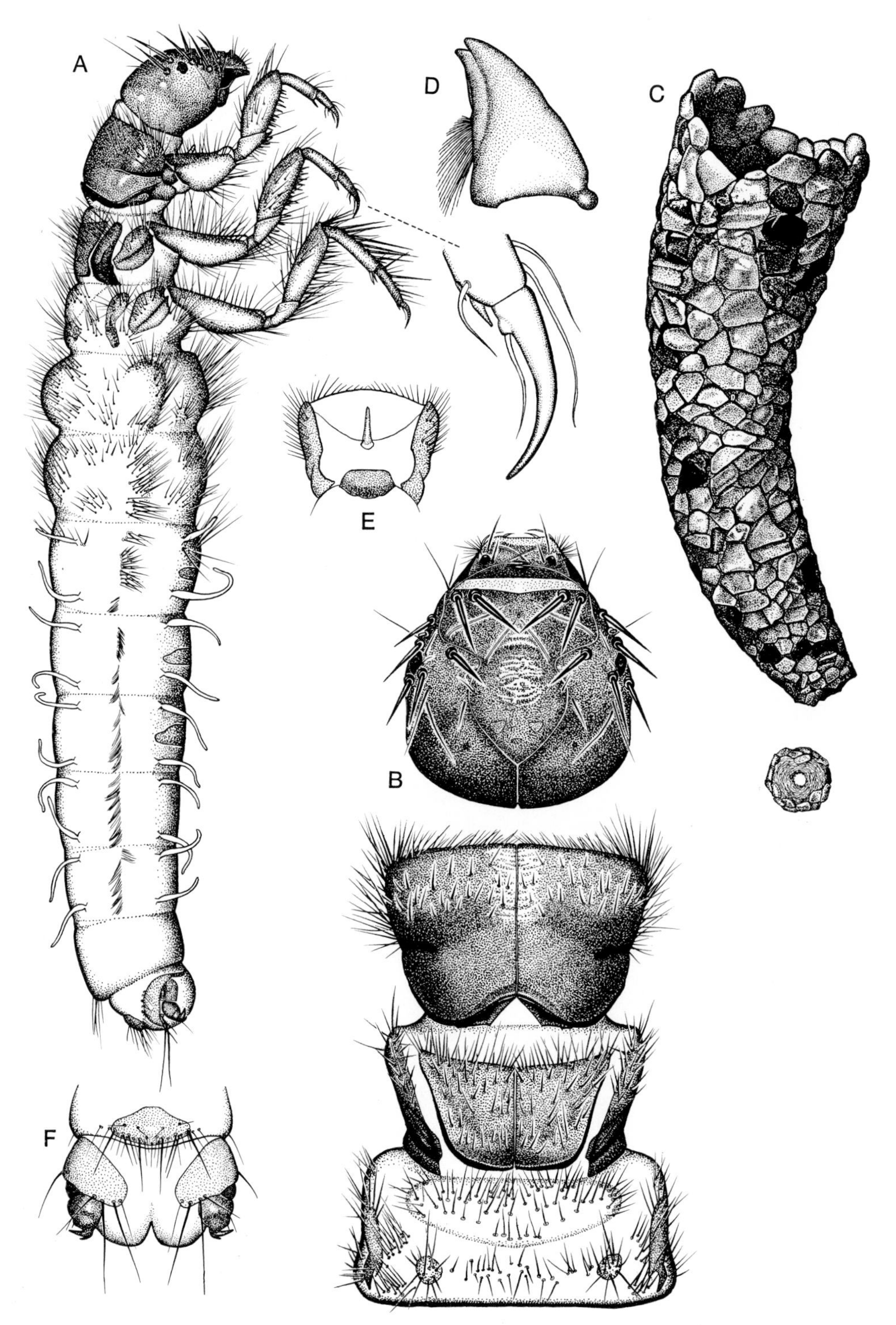
A
D
C
E
B
F

10.41 Genus **Philarctus**

DISTRIBUTION AND SPECIES This is a northern genus of the Holarctic region, with only one North American species, *P. quaeris* (Milne), which is known from Manitoba to British Columbia and south to Colorado.

Larvae we associated in Manitoba have been described elsewhere (Wiggins 1963).

MORPHOLOGY Because of the close morphological similarity between larvae in *Philarctus* and *Limnephilus*, and of the large number of *Limnephilus* spp. for which larvae are not known, separation between the two genera is illusory at this stage in our knowledge. The three light-coloured areas on the narrowed portion of the frontoclypeal apotome occur in *Asynarchus, Clistoronia*, and *Limnephilus* as well as in *Philarctus*. The accessory hook on the anal claw is single and lacks basal spines. Length of larva up to 18.5 mm.

CASE Larvae of *P. quaeris* construct cylindrical cases of little taper or curvature. Usually cases are constructed of small rock pieces or sedge seeds but frequently shells of snails and sphaeriid clams are used; entire cases of snail shells (C) are not uncommon in certain habitats. Length of larval case up to 24 mm.

BIOLOGY We have found large populations of *P. quaeris* in the small ponds and slow streams of the aspen parkland in southern Manitoba and Saskatchewan (Wiggins 1963). Pupation occurred in late June and early July, the larvae tending to congregate in the shallows before fastening their cases to some firm substrate.

REMARKS Diagnostic characters for the male of *P. quaeris* were given by Ross and Merkley (1952).

Philarctus quaeris (Manitoba, Erickson, 10 June 1962, ROM)
A, larva, lateral x8; B, head and thorax, dorsal; C, case, x5

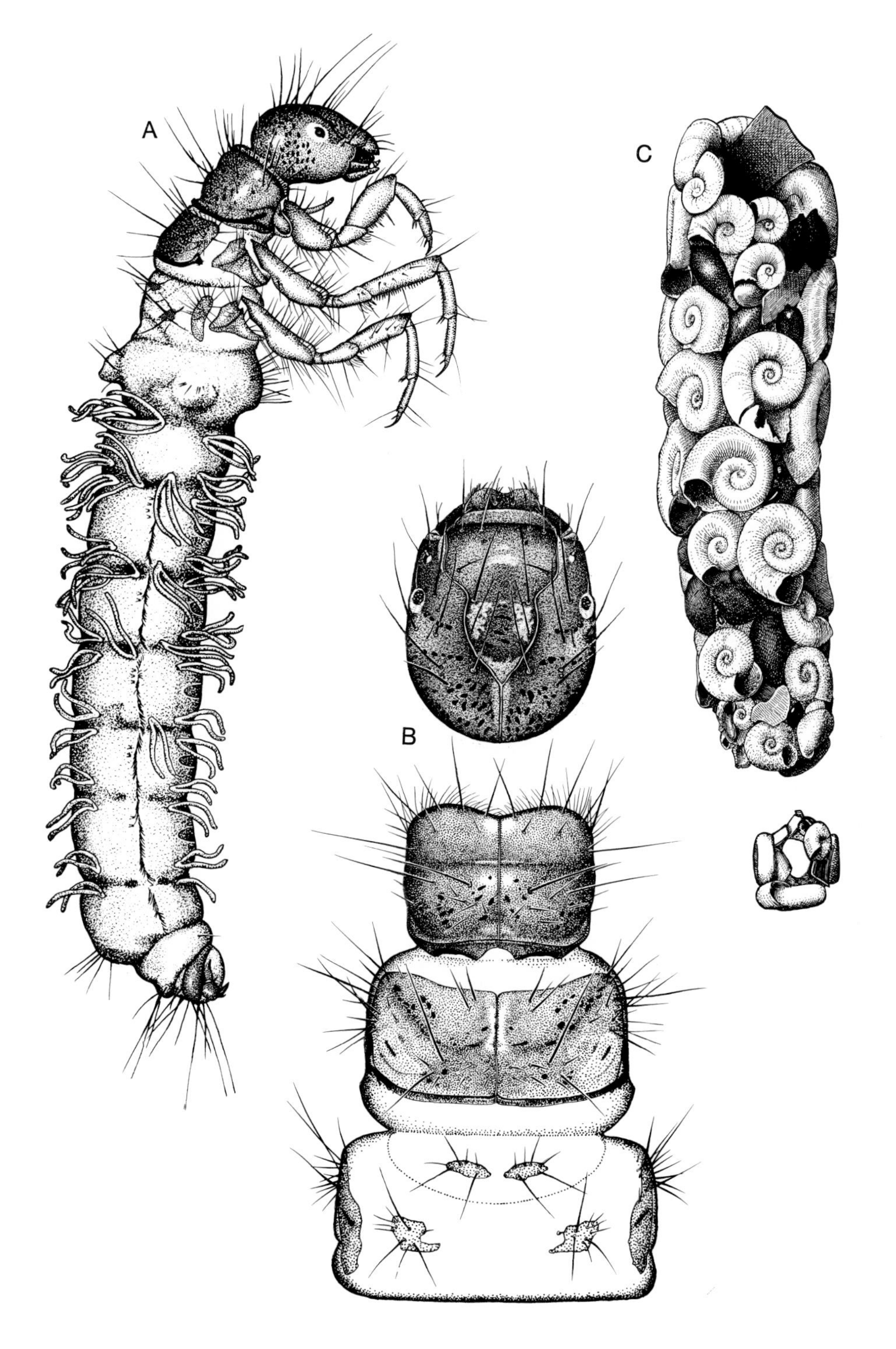
A
B
C

10.42 Genus **Philocasca**

DISTRIBUTION AND SPECIES Six species are now known in this rather obscure Nearctic genus; records extend from Alberta through Montana, Idaho, Washington, Oregon, and California.

Larvae have been associated for two species, *P. demita* Ross and *P. rivularis* Wiggins (Wiggins and Anderson 1968); we have other larvae, almost certainly of this genus, but as yet not identified.

MORPHOLOGY Larvae of *Philocasca* are characterized by enlarged and flattened scale-hairs at least along the anterior edge of the pronotum (B). The dorsum of the head is flattened, often very much so and with a prominent carina as illustrated (A), and usually bears some stout, flattened setae, too. The pronotum and usually some part of the head have a roughened texture; sclerotized parts are reddish brown in colour. Abdominal gills are single and arranged differently in at least some of the species. Length of larva up to 18 mm.

CASE All larval cases known are constructed of small rock fragments forming a cylinder of slight curvature; in the final instar, at least, cases have little or no taper, and the posterior opening is closed for all but a small central hole. Length of larval case up to 23 mm.

BIOLOGY Larvae now known indicate that *Philocasca* is largely characteristic of small, mountain spring streams; usually they are found in the stream in gravel beneath larger rocks. But larvae of *P. demita* have been collected only in soil and leaf litter up to 6 m from a spring stream in Oregon (Anderson 1967a; Wiggins and Anderson 1968). Larger instars of *P. demita* were captured in pitfall traps sunk in the forest floor, smaller ones extracted with Berlese funnels from samples of leaf litter; no larvae of *P. demita* were taken in the near-by stream. Gut contents of the stream-dwelling *P. rivularis* (3) that we examined were largely vascular plant pieces.

REMARKS A key to adults was given by Wiggins and Anderson (1968); one additional species was described from Alberta by Nimmo (1971). *P. antennata* (Banks), now assigned to this genus, was originally treated under *Stenophylax* (see Ross 1944: 299).

Philocasca rivularis (Oregon, Benton Co., 24 Sept. 1964, ROM)

A, larva, lateral x11; B, head and thorax, anterolateral; C, case x6; D, anterior edge of pronotum

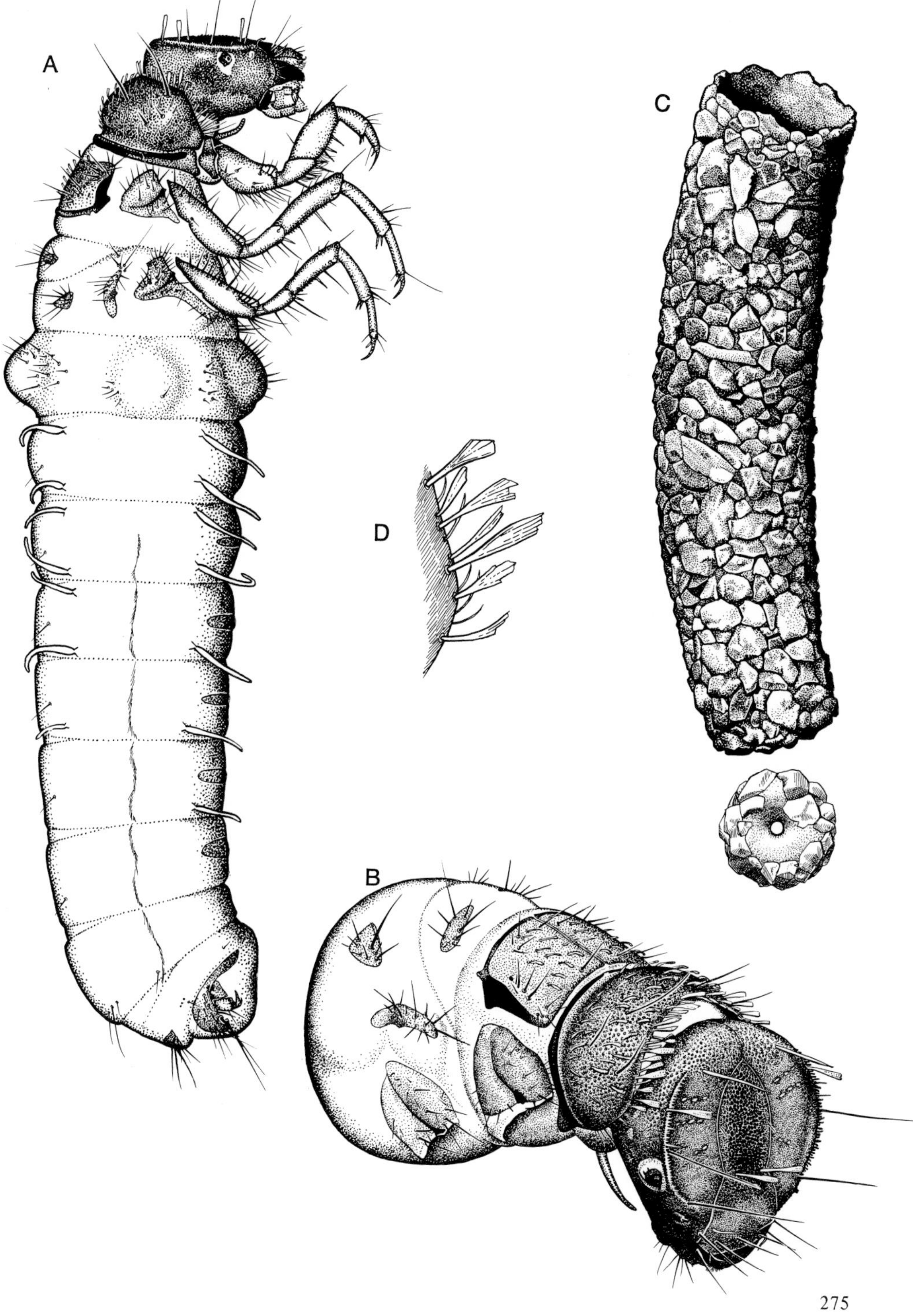
A
B
C
D

10.43 Genus **Platycentropus**

DISTRIBUTION AND SPECIES *Platycentropus* is a Nearctic genus of three species, all confined to the eastern half of the continent: *P. amicus* (Hagen) (syn. *P. plectrus* Ross); *P. indistinctus* (Walker) (syn. *P. fraternus* (Banks)); and *P. radiatus* Say (syn. *P. maculipennis* Kol.). *Platycentropus* spp. are known within the area from Manitoba to Newfoundland, and south to Louisiana.

Larvae have been described for *P. radiatus* (Flint 1960; Ross 1944; Lloyd 1921); we have associated material for all three species.

MORPHOLOGY There seems still to be no better diagnostic character for larvae of this genus than the unusually long prosternal horn cited in most keys, which in *Platycentropus* larvae extends ventrally to the mentum and the distal edges of the stipital sclerites (D). Species differ in the markings on the dorsum of the head: *P. indistinctus* and *P. radiatus* have small, sparse muscle spots, as illustrated; muscle scars in *P. amicus* are more numerous, coalescing into a pattern similar to that illustrated for *Anabolia*. Most abdominal gills of the dorsal and ventral series are three-branched. Length of larva up to 22 mm.

CASE Cases for all larvae we have collected are much the same as the one illustrated. Plant materials such as grasses and sedges are arranged transversely to produce a straight, fuzzy cylinder (C). Pupal cases of sand grains were attributed to *P. indistinctus* (Denning 1937); all *Platycentropus* pupal cases known to me have been of the same construction as the larval cases. Length of case up to 25 mm.

BIOLOGY Habitats of *Platycentropus* larvae range from cool streams to warm ponds, and the two extremes are attributed to the same species, *P. radiatus* (Lloyd 1921; Flint 1960). Species of this genus are among the most tolerant of all limnephilids to warm, quiet waters at the edges of slow streams, marshes, and lake margins where there are dense growths of aquatic plants. No *Platycentropus* species are known in temporary vernal pools, although cases of *Limnephilus indivisus*, which is abundant in these habitats, are often somewhat similar to *Platycentropus* cases.

Platycentropus radiatus (New Hampshire, Grafton Co., 18 May 1957, USNM)
A, larva, lateral x7; B, head and thorax, dorsal; C, case x4; D, head and prosternal horn, lateral

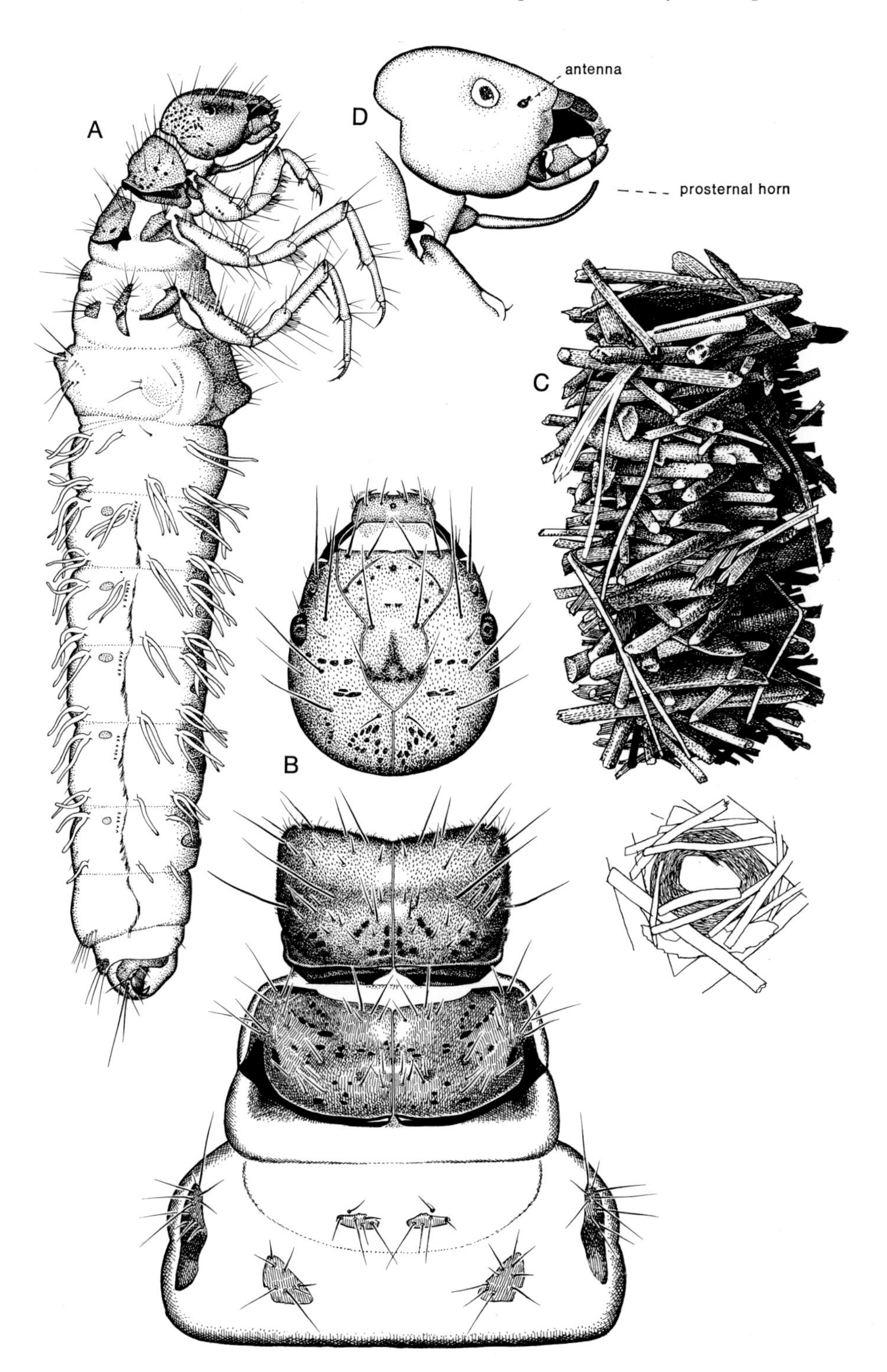
A
D
antenna
prosternal horn
C
B

10.44 Genus **Pseudostenophylax**

DISTRIBUTION AND SPECIES Although primarily a genus of the Oriental region, *Pseudostenophylax* is sparingly represented in the Asian Palaearctic and in North America. In the eastern half of the continent there are two species, *P. uniformis* (Betten) and *P. sparsus* (Banks), within the area from Minnesota to Maine and south to Tennessee. In the west, only *P. edwardsi* (Banks) is known, from British Columbia to California.

Larvae of *P. uniformis* have been described (Flint 1957, 1960) and of another species that is probably *P. sparsus* (Flint 1960, *Pseudostenophylax* species I); we reared larvae for both. Larvae of *P. edwardsi* have also been described (Wiggins and Anderson 1968), and this is the species referred to Limnephilid Genus A by Ross (1959) and Flint (1960).

MORPHOLOGY The flat head of the western *P. edwardsi* (D) compared with the normal head of the eastern larvae (A,B) introduces discordance in some characters of the genus, but in all species mesonotal setal areas are confluent through *sa*1 to *sa*2 to *sa*3, and there is a transverse band of setae between the metanotal *sa*2 sclerites. Sclerotized parts of the head and thorax are uniform reddish or yellowish brown, broken only by muscle scars. Abdominal gills are single. Length of larva up to 16 mm.

CASE Larval cases are constructed mainly of small rock fragments, the exterior uniform in outline. Length of larval case up to 19 mm.

BIOLOGY Observations on the biology of *P. uniformis* have been made by Flint (1957) and Mackay (1969); the biology of *P. edwardsi* was studied by Anderson (1974a). Generalizing, larvae live in cool spring runs or small streams of intermittent flow; they burrow into sand and gravel substrate as water recedes and presumably are protected there until the return of surface water. Larval development proceeds through autumn; winter is passed as final-instar larvae.

REMARKS Species of this genus were treated for some time in the literature under *Drusinus*. Diagnostic characters of males of the two eastern species were given by Ross (1944), *P. sparsus* treated under *virginicus* (see Flint 1966); characters for the male of *P. edwardsi* were given by Denning (1956).

Pseudostenophylax sparsus (Ontario, Algonquin Prov. Park, 12 June 1963, ROM)
A, larva, lateral x8; B, head and thorax, dorsal; C, case, lateral x7
D, *P. edwardsi* (Oregon, Benton County, 16 April 1964, ROM), head and thorax, anterolateral

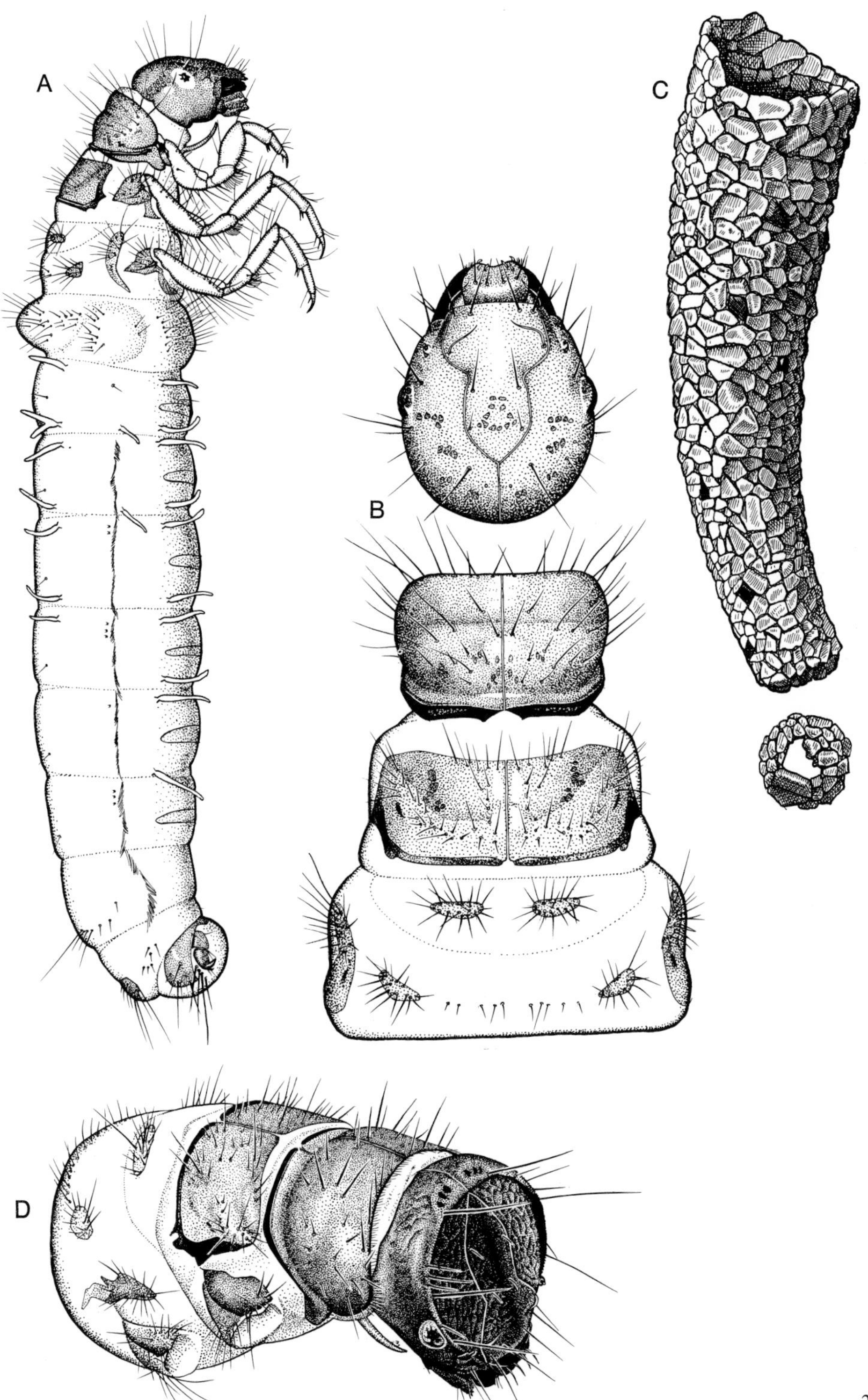
A
B
C
D

10.45 Genus Psychoglypha

DISTRIBUTION AND SPECIES *Psychoglypha* is an exclusively Nearctic genus in which 15 species are now recognized. One species, *P. subborealis* (Banks), cited in the literature as *P. alaskensis* (Schmid 1952c) until the genus was reviewed by Denning (1970), is transcontinental in a broad band extending roughly from Maine through southern Ontario and Michigan to Alaska, and south to California. All other species are known only in western montane areas where as a generic group they are common and widespread.

The larva of only *P. subborealis* has been described (Flint 1960); we have associated material for 10 species.

MORPHOLOGY Larvae of *Psychoglypha* spp. are sufficiently divergent morphologically that it is difficult to find characters diagnostic for the genus. The two sclerites lying close to the dorsal edge of the lateral hump of abdominal segment I (A) provide the only consistent character discovered thus far for all the larvae we have collected. Pigmentation in these sclerites is often little darker than the surrounding membrane, but they can be distinguished by their hard, shiny texture. Among larvae now known, legs with alternate dark and light bands occur only in the transcontinental *P. subborealis* (A); in a few species stout spines and setae occur on the pronotum (B) and on the sclerite of segment IX and the lateral sclerites of the anal prolegs (D). Head markings usually consist of dark brown spots and blotches on a yellowish brown background, except in *P. bella* (Banks) where the head and legs are uniform dark brown. Abdominal gills are single. Length of larva up to 26 mm.

CASE Larval cases are usually constructed of small rock fragments and pieces of wood combined into a straight tube of little taper (C). Length of larval case up to 43 mm.

BIOLOGY *Psychoglypha* larvae occur in a wide range of cool-water habitats, ranging from spring runs to larger streams and their marginal pools. In the autumn, as time for pupation approaches, larvae burrow into bottom gravel for metamorphosis. Observations on the biology of *P. subborealis* in Oregon were contributed by Anderson (1967b); this species is believed to overwinter in the egg stage, or at least confined to the egg-matrix. Gut contents of larvae of *P. subborealis* were fine particulate detritus and animal fragments (Winterbourn 1971a).

Psychoglypha subborealis (Ontario, Durham Co., 12 Sept. 1962, ROM)
A, larva, lateral x10, segment I enlarged; B, head and thorax, dorsal; C, case x5; D, segment IX and anal prolegs, dorsal; E, anal proleg, mesial

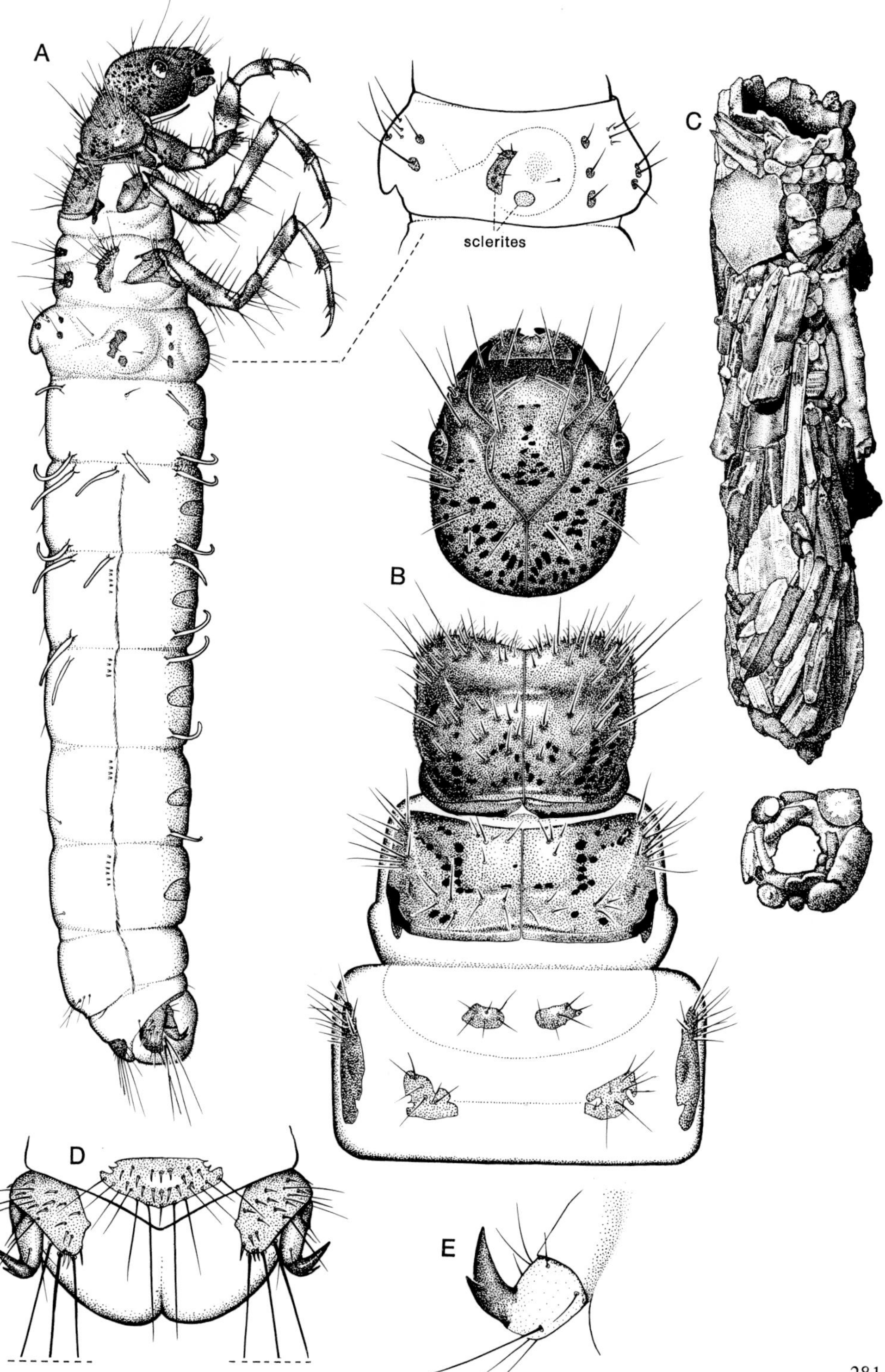
A
sclerites
B
C
D
E

10.46 Genus **Psychoronia**

DISTRIBUTION AND SPECIES *Psychoronia* is an obscure Nearctic genus with two species recorded from New Mexico and Colorado. The taxonomic status of this genus is uncertain and its merger with *Limnephilus* has been proposed; but what is now known of larval morphology in both groups supports retention of *Psychoronia* as a distinct genus (Wiggins 1975).

Diagnostic characters for the genus were given by Ross (1959), based on an associated series of adults and larvae of *P. costalis* (Banks), here illustrated, from Colorado.

MORPHOLOGY The larva of *P. costalis* is one of the few among North American limnephilids in which most gills of the dorsal and ventral series have four or more branches (A); in the lateral series, gills are of single filaments and occur only on segment II and sometimes III. Except for a few muscle scars, the head is uniform brown. Neither of these characters occurs in any *Limnephilus* now known. Mesonotal setal areas tend to be separate (B), but only the separation between *sa*1 and *sa*3 is distinct. On the metanotum a few setae arise between the primary sclerites (B). The dorsum and venter of abdominal segment I bear many setae; only ventral chloride epithelia are apparent, and they are somewhat shorter than in other genera (A). Length of larva up to 12 mm.

CASE The case of *P. costalis* is made of rock fragments, for the most part coarse (C). The posterior end of the case illustrated was closed off for pupation, as were all cases in these series. Length of larval case up to at least 12 mm.

BIOLOGY Larvae were collected in streams at high elevation; specimens of adults and larvae have been taken from 8,000 to 12,000 ft (2,400 to 3,600 m).

REMARKS Material for study and illustration from the Illinois Natural History Survey was provided by J.D. Unzicker.

Psychoronia costalis (Colorado, Teller Co., 22 Aug. 1941, INHS)
A, larva, lateral x15; B, head and thorax, dorsal; C, case x11; D, segments VIII and IX, dorsal

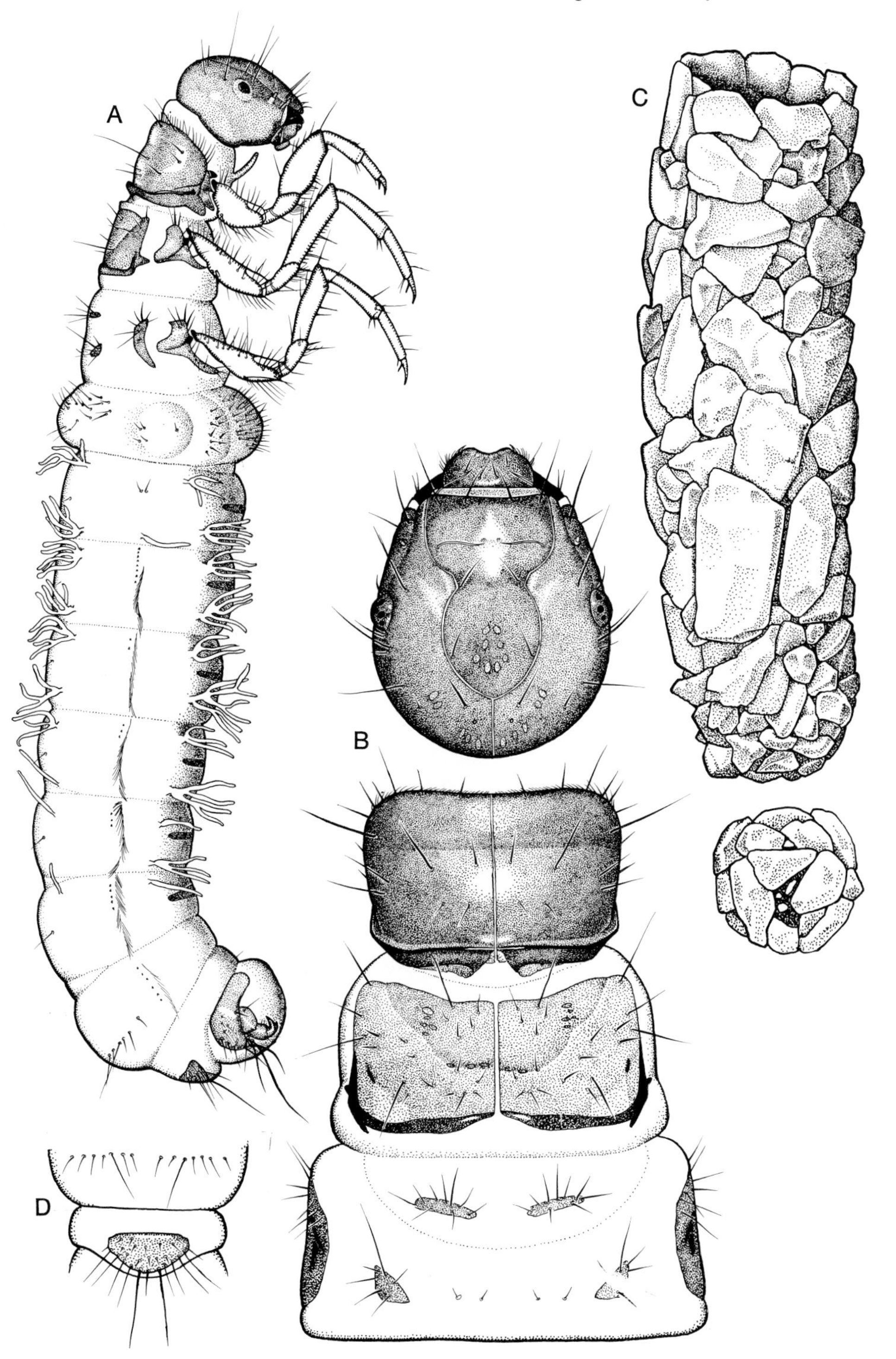
A
B
C
D

10.47 Genus **Pycnopsyche**

DISTRIBUTION AND SPECIES *Pycnopsyche* is confined to North America, and comprises a dominant group of 16 species over much of the eastern half of the continent. Two species, *P. subfasciata* (Say) and *P. guttifer* (Walker), range farther westward than the others to the Rocky Mountains.

Flint (1960) provided a key to larvae of seven species of *Pycnopsyche*.

MORPHOLOGY *Pycnopsyche* larvae have an elongate sclerite adjacent to the posterior edge of the lateral hump (A). Distinction between larvae of this genus and of *Hydatophylax* may be difficult because metanotal *sa*1 sclerites are close together in certain species of *Pycnopsyche*; diagnostic characters are given in the key. Usually the base colour of the head in *Pycnopsyche* is yellowish brown with darker spots and blotches. Sclerites of head and thorax are frequently beset with small spines, which in *P. gentilis* (McL.) are very stout; this species is also unusual in having a flattened head and in lacking a median dorsal hump on segment I. Gills are single. Length of larva up to 29 mm.

CASE Early instars of all species construct cases of plant materials, usually bark and twigs (C) or leaf discs; during the terminal instar the leaf cases are usually converted to cases of bark or rock fragments. Length of larval case up to 59 mm.

BIOLOGY Larvae of *Pycnopsyche* live in cool, woodland streams and small rivers largely in deciduous forest areas. The time of hatching coincides with autumnal leaf fall in September and October, and the larvae are important detrital processors in the autumn-winter stream community. The early instars of all species, with cases either of leaf discs or of bark and twigs, are found in slow-flowing parts of the stream where leaves and organic detritus accumulate. After construction of the final-instar case, larvae pass a prepupal aestivation period of 1-6 months either buried in the stream gravel or with the case attached to the underside of rocks and logs, the time and place characteristic of each species. Pupation and emergence occur between July and October. Pairs of *Pycnopsyche* species commonly occur together in the same stream, one species having a leaf case and the other a twig case (Cummins 1964; Mackay 1972).

REMARKS A review of adults of species of *Pycnopsyche* was given by Betten (1950).

Pycnopsyche guttifer (Ontario, Leeds Co., 26 May 1965, ROM)
A, larva, lateral x6, segments I and IX enlarged; B, head and thorax, dorsal; C, case x3

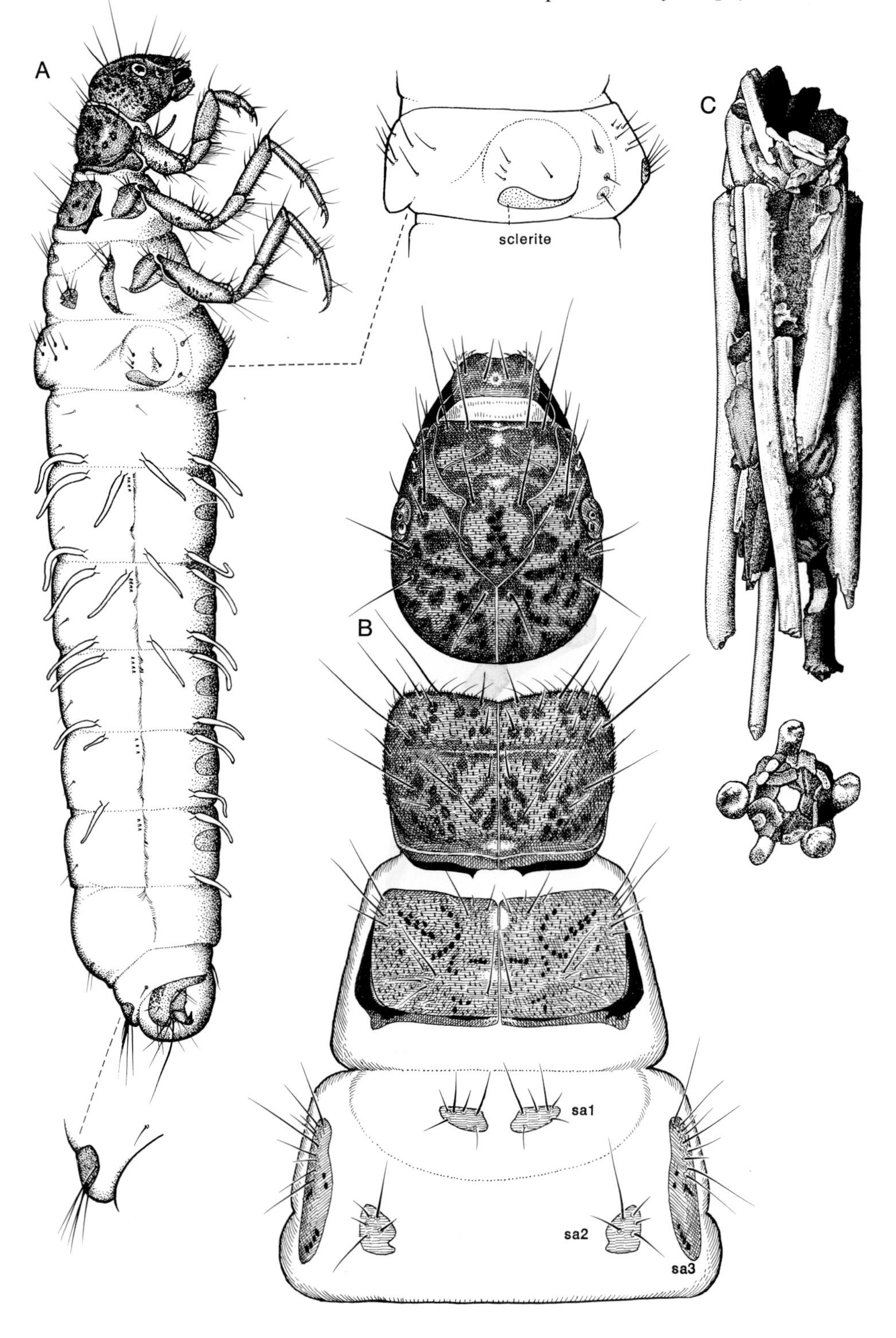

A
sclerite
C
B
sa1
sa2
sa3

10.48 Genus **Rossiana**

DISTRIBUTION AND SPECIES *Rossiana* is another of the unusual Nearctic limnephilid genera; a single species, *R. montana* Denning, is known from Montana, Washington, and British Columbia.

Larvae of *R. montana* have not been previously identified in the literature; we collected an associated series of larvae and mature pupae in Montana. Morphology of adults was reviewed by Schmid (1968b).

MORPHOLOGY Structurally, larvae of *R. montana* are among the most unusual in the Limnephilidae. The posterolateral margins of the head are extended as prominent flanges, and the pronotum is concave on each side (B,E); the head and pronotal sclerites are brownish red and coarsely pebbled. The mandibles have separate tooth-like points (H), and the labrum is sclerotized and unpigmented anteriorly (G); the ventral apotome is uniquely peg-shaped (D). Each mesonotal plate is subdivided by a longitudinal gap in the sclerotization, and the membranous separation on the mid-dorsal line between the sclerites is unusually wide (B); metanotal sclerites are indistinct. The basal seta of the middle and hind tarsal claw (A) is longer than in most genera, but does not reach the end of the claw as in *Apatania.* Abdominal gills are single and long, chloride epithelia are absent, and the anal claw bears an accessory hook (F). Length of larva up to 9 mm.

CASE The larval case (C) is constructed of small rock fragments, somewhat curved, but with little taper. Length of larval case up to 10.5 mm.

BIOLOGY The species is characteristic of small, cold streams in mountainous terrain; it is very local in occurrence. We have taken larvae in gravel under moss in shallow water at the edge of streams, and on vertical rock surfaces in a thin layer of running water. Guts of larvae (3) we examined contained mostly plant fibres, suggesting that woody materials were ingested, and fungi. Pupae and larvae of the last three instars were taken in a single collection in June.

Rossiana montana (Montana, Missoula Co., 24 June 1968, ROM)
A, larva, lateral x12, tarsal claw enlarged; B, head and thorax, anterolateral; C, case x9; D, head, ventral; E, head, frontal; F, segment IX and anal prolegs, dorsal; G, labrum, dorsal; H, mandible, dorsal

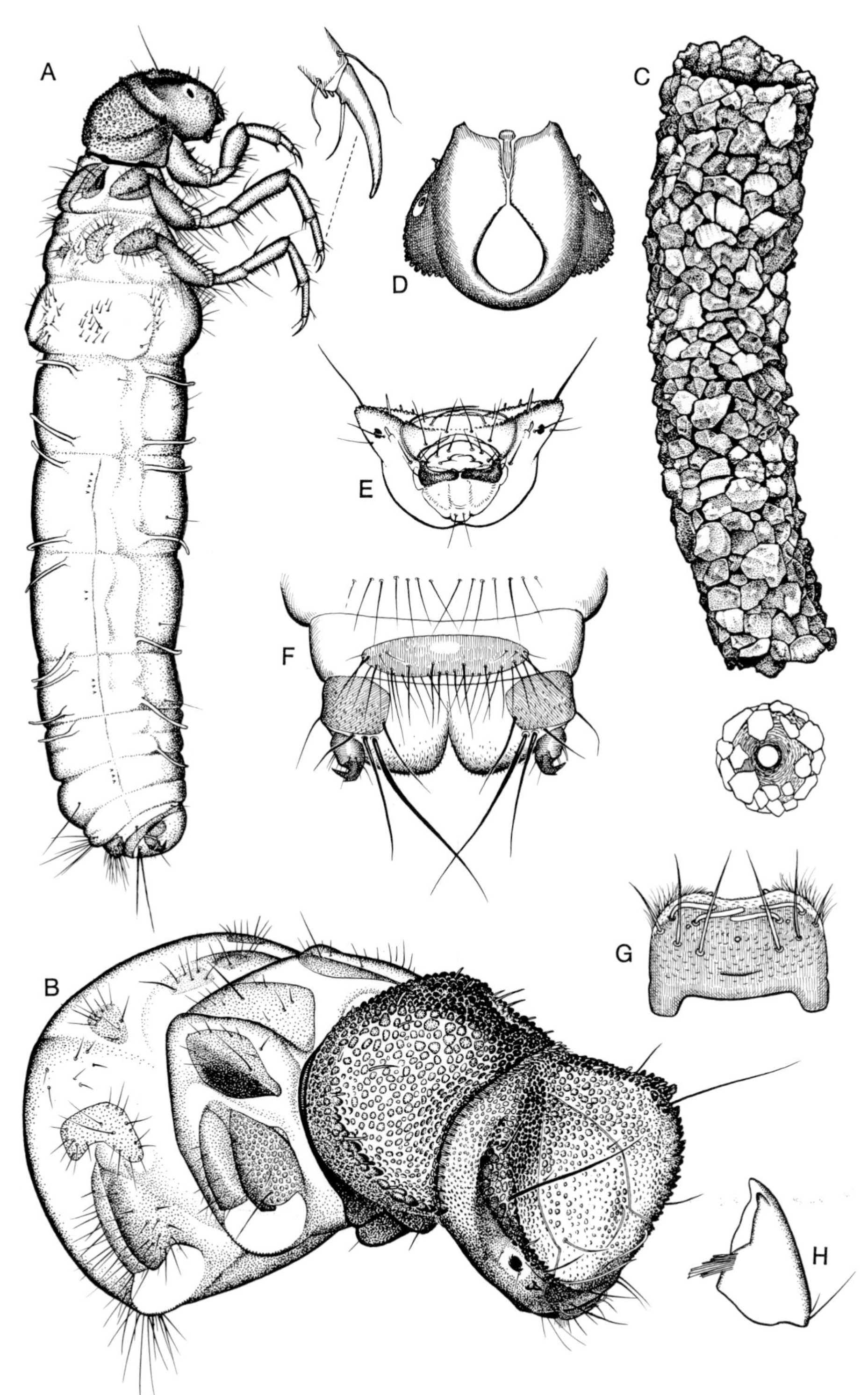
A
B
C
D
E
F
G
H

11
Family Molannidae

This is a small but distinctive family of the Holarctic and Oriental faunal regions. Two genera are represented in the Nearctic fauna, with only seven species.

Molannid larvae can be recognized immediately in the field by their cases, which are largely of rock pieces but flattened and with lateral flanges and a dorsal hood over the anterior opening. Since the cases blend with the surrounding substrate and no part of the insect is visible from above, larvae are usually detected only when they move their case. Somewhat similar flanged cases are made by some species in the leptocerid genus *Ceraclea* but there the posterior opening is reduced with silk to a small hole on the dorsal surface of the central tube; in molannid cases the posterior opening is ragged, ill-defined, and not reduced with silk, and is largely dorsal (Fig. 11.1C).

Structurally, molannid larvae are distinguished by aberrant hind tarsal claws of two types, as described under the generic headings; the function of these unusual claws is unknown. The hind legs are substantially longer than the others, and the tibiae secondarily subdivided near the middle; the fore tibia is extended distoventrally as a long process with a single spur at the apex. The mesonotum bears a lightly sclerotized plate, divided along the median line, in some species completely but in others only through the anterior half to intersect with a transverse suture (Fig. 11.2B). Primary setal areas are represented on the metanotum but sclerites are absent. Humps on abdominal segment I are prominent; abdominal gills are present. The lateral fringe is well developed, as are lateral tubercles on segment VIII. Chloride epithelia are present on the abdomen but difficult to distinguish; in *Molanna blenda* they occur as anterolateral oval areas immediately dorsal to the lateral fringe on segments III to VII.

Most larvae live in lakes, ponds, and areas of slow current in rivers, but *Molanna blenda* occurs in cold spring runs. The insects live on sand and mud substrates, where they feed on diatoms, vascular plant tissue, and small invertebrates (Balduf 1939; Lepneva 1964); generally omnivorous feeding emerges from all food studies.

Molannid larvae bury themselves in the substrate before pupation. As in the Leptoceridae, the larval exuviae are ejected through the posterior opening of the case shortly after larval-pupal ecdysis.

11 Family **Molannidae**

Key to Genera

1 Tarsal claw of hind leg modified into setose lobe, much shorter than tarsus (Fig. 11.1A). Widely distributed through the east and westward to Colorado and British Columbia **11.1 Molanna**

Tarsal claw of hind leg modified into slender filament of about same length as tarsus (Fig. 11.2A). Alaska and the Yukon **11.2 Molannodes**

11.1 Genus **Molanna**

DISTRIBUTION AND SPECIES This rather small group, confined to the Holarctic and Oriental regions, comprises six species in North America. All occur within the eastern half of the continent, fewer species to the south than the north, but one, *M. flavicornis* Banks, extends west to Colorado and British Columbia.

Diagnoses were given for three species by Sherberger and Wallace (1971): *M. tryphena* Betten, *M. blenda* Sibley, and *M. uniophila* Vorhies. Other data were contributed for *M. blenda* by Sibley (1926) and Ross (1944); for *M. flavicornis* Banks by Neave (1933) and Denning (1937); for *M. tryphena* by Betten (1902, as *M. cinerea*); and for *M. uniophila* by Vorhies (1909) and Ross (1944).

MORPHOLOGY *Molanna* larvae are readily distinguished by the short, stout claw of the hind leg (A); the basal seta of the claw originates from about its centre, and the upper surface of the claw bears short setae. Larvae of all the Nearctic species have contrasting dark bands on the dorsal ecdysial lines of the head. In some species, although not in the one illustrated, the mesonotal sclerite is subdivided by a transverse suture (Ross 1944, fig. 709). Most abdominal gills have 2-4 filaments. Length of larva up to 19 mm.

CASE *Molanna* cases (C) are made of rock fragments, sometimes with organic materials, and have a prominent lateral flange and anterior hood which conceals the larva completely. Larvae of some species incorporate a marginal row of relatively larger pieces into their cases (Sherberger and Wallace 1971). Length of larval case up to 27 mm.

BIOLOGY Larvae live in lakes or the slower currents of rivers and streams, and inhabit the sand and mud substrates of these sites. One species, *M. blenda*, is confined to small, cold springs. *M. flavicornis* was found to 20 m in the deepest parts of Lake Winnipeg (Neave 1933). Diatoms, filamentous algae, vascular plant tissue, and small invertebrates are reported as the major items ingested (Neave 1933; Sibley 1926; Solem 1970).

REMARKS Diagnostic characters of *Molanna flavicornis*, illustrated here, have not previously been available in the literature; it is distinguished from all other North American species by a long, stout spur at the base of each anal proleg (E).

Molanna flavicornis (Manitoba, Minnedosa, 10 June 1962, ROM)
A, larva, lateral x8, with portions of leg segments enlarged; B, head and thorax, dorsal; C, case, ventral x6, posterior opening in dorsal view; D, head, ventral; E, anal prolegs, dorsal

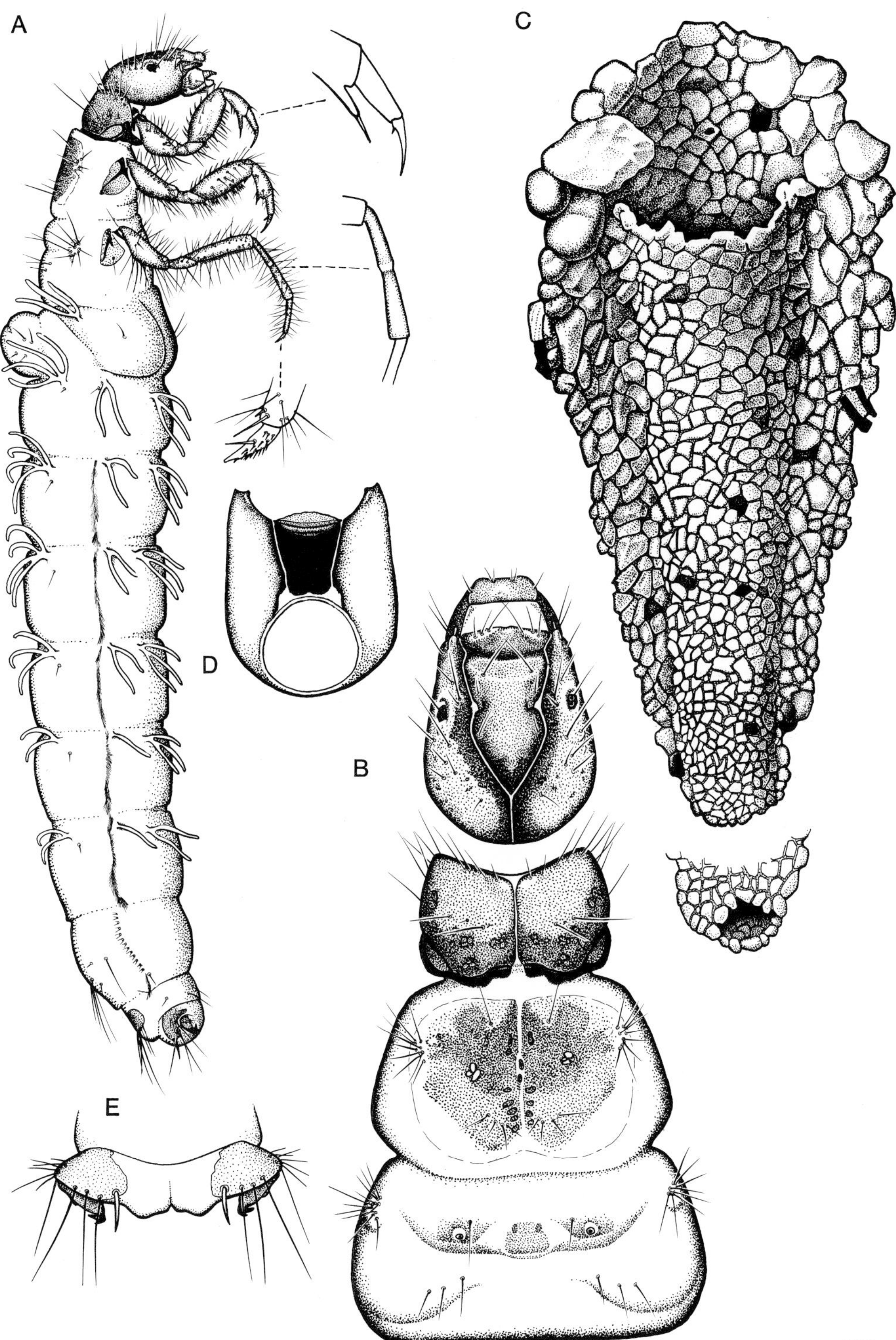
A
C
D
B
E

11.2 Genus **Molannodes**

DISTRIBUTION AND SPECIES The single species *M. tinctus* Zett. assigned to this genus occurs across northern Europe and Asia into Alaska and the Yukon. The North American populations, originally described as *M. bergi* Ross (1952), were found to be conspecific with the Eurasian populations (Wiggins 1968).

Molannodes larvae have not been identified in North America, but are well known in European literature (Lepneva 1966).

MORPHOLOGY In general features *Molannodes* larvae are similar to those of *Molanna* but are clearly distinguished by the slender filament-like tarsal claw of the hind leg (A). As in *Molanna* the basal seta of the claw arises at about its mid-point, but is very long. The dorsum of the head is uniformly dark, and the mesonotal plate is subdivided by a transverse suture as in some species of *Molanna*. Abdominal gills are mostly single. Length of larva up to 12.5 mm.

CASE *Molannodes* larvae incorporate more pieces of plant detritus into their cases (C) than do *Molanna*, and sometimes build a case entirely of detritus (Lepneva 1966). The lateral flanges and anterior hood are smaller in *Molannodes* cases. Length of larval case up to 20 mm.

BIOLOGY Larvae live in sites similar to those frequented by *Molanna*. Food of this species comprised insects and vascular plant tissue (Siltala 1907b). Distributional records to date suggest that in North America the range of *Molanna* does not extend to the extreme northwestern part of the continent where *Molannodes* occurs.

REMARKS The illustration was prepared from larvae obtained from the Zoological Institute, Academy of Sciences, USSR, loaned through the cooperation of L. Zhiltzova.

Molannodes tinctus (European USSR, River Belaja, 20 Oct. 1949, Acad. Sci. USSR)
A, larva, lateral, hind tarsus enlarged; B, head and thorax, dorsal; C, case, ventral

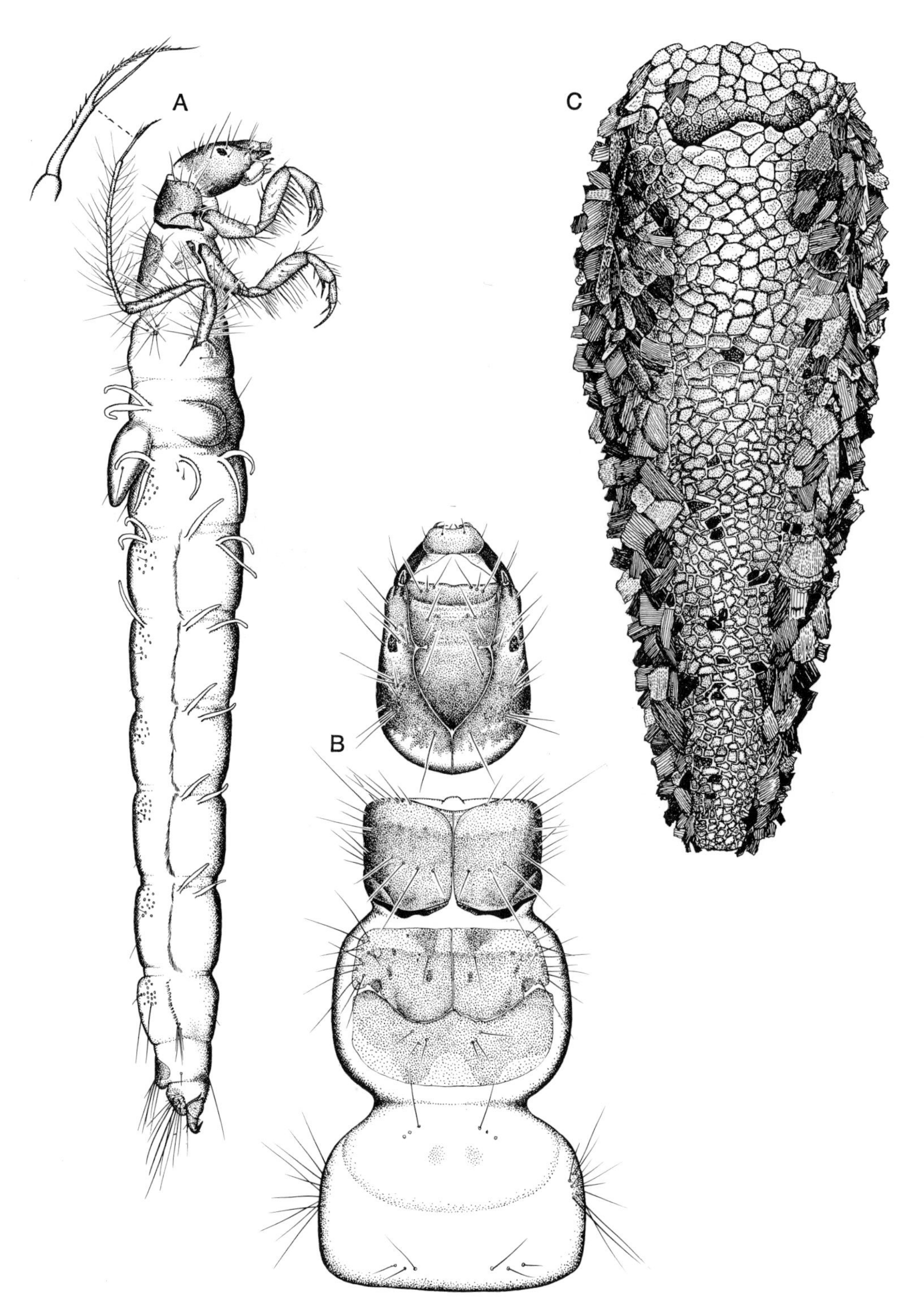
A
B
C

12
Family Odontoceridae

This is a small family represented in all faunal regions except the Ethiopian, and usually by a small number of species and genera in each. Twelve species are known north of Mexico representing the six Nearctic genera, three of which are monotypic; apart from *Psilotreta*, the North American species tend to be local and obscure. These can be interpreted as characteristics of a senescent group, once widespread but with the passing of time increasingly confined in distribution and habitat.

Structurally, there is such marked diversity among the North American genera that it is difficult to find diagnostic characters at the family level. The mesonotal plates are heavily sclerotized, separated in most genera only by a median ecdysial line but in *Marilia* each plate is subdivided into three sclerites. Metanotal sclerites range from the condition in *Namamyia* and *Nerophilus* where all are small and separate with *sa*2 in two parts, through larger, separate sclerites in *Pseudogoera* and *Marilia*, to the condition in *Parthina* and *Psilotreta* where the *sa*1 and *sa*2 pairs are fused into transverse plates. The ventral apotome of the head is also diverse, ranging from a convex sclerite completely separating the genae in *Marilia* and *Parthina*, through a long, narrow sclerite in *Namamyia*, to a short sclerite partially separating the genae in *Nerophilus* and *Psilotreta*, to the minute ventral apotome in *Pseudogoera.* Large prosternal sclerites are present in all genera except *Parthina*; somewhat weaker sclerotization of the mesosternum is also evident. Tarsi of the middle and hind legs have a small distal cluster of spines (Fig. 12.4A). Abdominal gill filaments are usually short, thin, and always in tufts. The anal prolegs usually lack curved accessory hooks, but tend to have straight spines on both the claw and the lateral sclerite; and the lateral sclerite is often extended ventrally. The lateral fringe is variously developed, and lateral tubercles usually are confined to segment VIII.

For the most part North American odontocerid larvae are burrowers in gravel, sand, and silt deposits of running waters, and are not usually encountered unless these deposits are searched. The eastern genus *Psilotreta* might appear to be an exception because the larvae pupate gregariously in layers on the underside of rocks, but observations indicate that all instars burrow. Larvae of the European *Odontocerum albicorne* (Scop.) found under stones during the day crawled about at night, scavenging plant and animal materials (Elliott 1970).

Case-making behaviour in the Odontoceridae is different from that in other families in a way that is probably related to the burrowing habit. Examination reveals that cases in most trichopteran families are held together by means of some silk around the edges of component pieces as a kind of mortar joint, and by a continuous lining sheet of silk around the whole interior of the case. Odontocerid cases appear to be constructed with more conspicuous mortar joints but with no internal lining of silk; instead, most have interior silken brace bands between adjacent rock pieces (Fig. 12.6c). This method of construction is evident in cases of *Psilotreta, Namamyia*, and *Pseudogoera* where larger rock fragments are combined with small pieces; on the inside of these cases it can also be seen that small rock pieces are incorporated into the silken mortar joints almost as a mosaic between larger pieces (Fig. 12.6c). The remarkable resistance of *Psilotreta* cases to crushing between one's fingers suggests that these modifications in case construction are effective safeguards against case breakage in shifting gravel deposits of running waters. Sieve-like closures of the posterior openings of their cases with rock fragments by larvae of *Psilotreta* and *Namamyia* seem also to contribute to the solidity of the case. Larvae making cases of smaller and more uniform rock fragments, such as *Nerophilus* and *Parthina* in North America, are more characteristic of fine-particle deposits in small streams; these cases do not seem to be unusually resistant to breakage. Larvae of *Parthina* cover the external surface of their cases with a thin layer of silk. Thus, among the North American odontocerids, and perhaps throughout the family, there are examples of somewhat different case-making behaviour; and this appears to have furthered diversification of habitats to a burrowing mode of life. Perhaps also related to a burrowing habit is the tendency in odontocerid larvae for sclerotized sternal plates on the first and second thoracic segments.

Absence of a complete silk lining in odontocerid cases is possibly correlated with the broadly pointed anal claws of the larvae and with their tendency to bear straight accessory spines. Larvae in other families constructing portable cases generally have acutely pointed anal claws with curved accessory hooks. The implication is that the odontocerid structures anchor the larva more effectively in its case when the interior is not lined with silk. Transverse discontinuities in the arrangement of rock pieces in some odontocerid cases (Fig. 12.1c) are also an unusual feature of this family.

Gut contents from larvae of North American genera indicate that the family is largely omnivorous; insect parts are found in the guts in some genera with some consistency, but it is not clear whether this reflects predacious or scavenging feeding. The wide range of instars represented in single collections of larvae throughout our material suggests that life cycles of more than one year are the general condition.

Two subfamilies are recognized. The Pseudogoerinae comprise only the unusual genus *Pseudogoera*, and are so characterized (Wallace and Ross 1971). All other genera remain in the Odontocerinae, probably only because no extensive review of the group has been made. Larval morphology has been little used in assessing relationships within the Odontoceridae, but the diversity among larvae of the North American genera indicates that it has much to offer.

Key to Genera

1 Fore tibia broad, approximately four times as long as tarsus, single apical tibial spur expanded and blade-like (Fig. 12.5A). Southeastern (subfamily Pseudogoerinae) **12.5 Pseudogoera**

Fore tibia normal, approximately same length as tarsus, apical spurs slender and not expanded (Fig. 12.1A) (subfamily Odontocerinae) 2

2 (1) Anterolateral corner of pronotum produced into sharp point (Fig. 12.6A,B) 3

Anterolateral corner of pronotum rounded and not pointed (Fig. 12.2A,B) 4

3 (2) Ventral apotome of head long, separating genae completely (Fig. 12.4D); claw of anal proleg stout, claw and lateral sclerite armed with straight spines (Fig. 12.4A). Western montane areas **12.4 Parthina**

Ventral apotome of head short, genae largely contiguous along median ventral ecdysial line (Fig. 12.6D); claw of anal proleg slenderer, claw and lateral sclerite without spines. Eastern half of the continent **12.6 Psilotreta**

4 (2) Each mesonotal plate subdivided into three sclerites, metanotal *sa*1 sclerites large and rectangular, closely appressed along mid-dorsal line (Fig. 12.1A,B). Widespread **12.1 Marilia**

Each mesonotal plate undivided, metanotal *sa*1 sclerites small, ovate, and widely separated across the mid-dorsal line (Fig. 12.2A,B) 5

5 (4) Venter segment I with two clusters of gill filaments and two pairs of setae (Fig. 12.3E). Western **12.3 Nerophilus**

Venter segment I without gills but with many setae (Fig. 12.2E). Western **12.2 Namamyia**

12.1 Genus **Marilia**

DISTRIBUTION AND SPECIES Apart from two species known to occur north of Mexico, *Marilia* is primarily a Neotropical genus in the New World; species assigned to *Marilia* are also recorded from the Oriental and Australian regions and from China. *Marilia nobsca* Milne is known from Texas; and *M. flexuosa* Ulmer is recorded from California, Arizona, Texas, and Arkansas. We have associated larvae for both and also for a population from southeastern Ontario (Hastings Co.) similar to, and perhaps identical with, *M. flexuosa*.

Diagnostic larval characters for the genus were given by Ross (1959).

MORPHOLOGY *Marilia* larvae are the only North American odontocerids in which each mesonotal plate is subdivided into three sclerites (A,B). Metanotal *sa*1 sclerites are separated mesially and posterior to each is a transverse line of setae in the *sa*2 position. The head bears one or two ridges along each side, and the genae are entirely separated by the ventral apotome (D). The abdomen is without a lateral fringe and the lateral sclerite of the anal proleg is edged mesially with stout spines (A). Length of larva up to 9.5 mm for *M. flexuosa*; up to 17.5 mm for *M. nobsca.*

CASE Larval cases (C) are made of sand grains, smoothly contoured, curved and tapered posteriorly. The silken membrane reducing the size of the posterior opening has an oval hole at the dorsal edge. On the case illustrated, the arrow indicates an apparent transverse discontinuity in the addition of new pieces, a condition often found in odontocerid cases. Length of larval case up to 10.5 mm for *M. flexuosa*; up to 24 mm for *M. nobsca.*

BIOLOGY Our collections of *Marilia* larvae have been taken in small streams in arid areas of the southwest and in a shallow river 50 m or so wide in Ontario. Gut contents of both *M. flexuosa* and *nobsca* (3 larvae of each species) were largely arthropods, but there were also filamentous algae and pieces of vascular plants.

REMARKS The Ontario larvae of *Marilia* were taken literally in the same sample of bottom gravel as specimens *Psilotreta.* Diagnostic characters for the male of *M. flexuosa* were given by Denning (1956).

Marilia flexuosa (Arizona, Cochise Co., 25 June 1966, ROM)
A, larva, lateral x17, anal proleg enlarged; B, head and thorax, dorsal; C, case, lateral x12, arrow indicates discontinuity; D, head, ventral

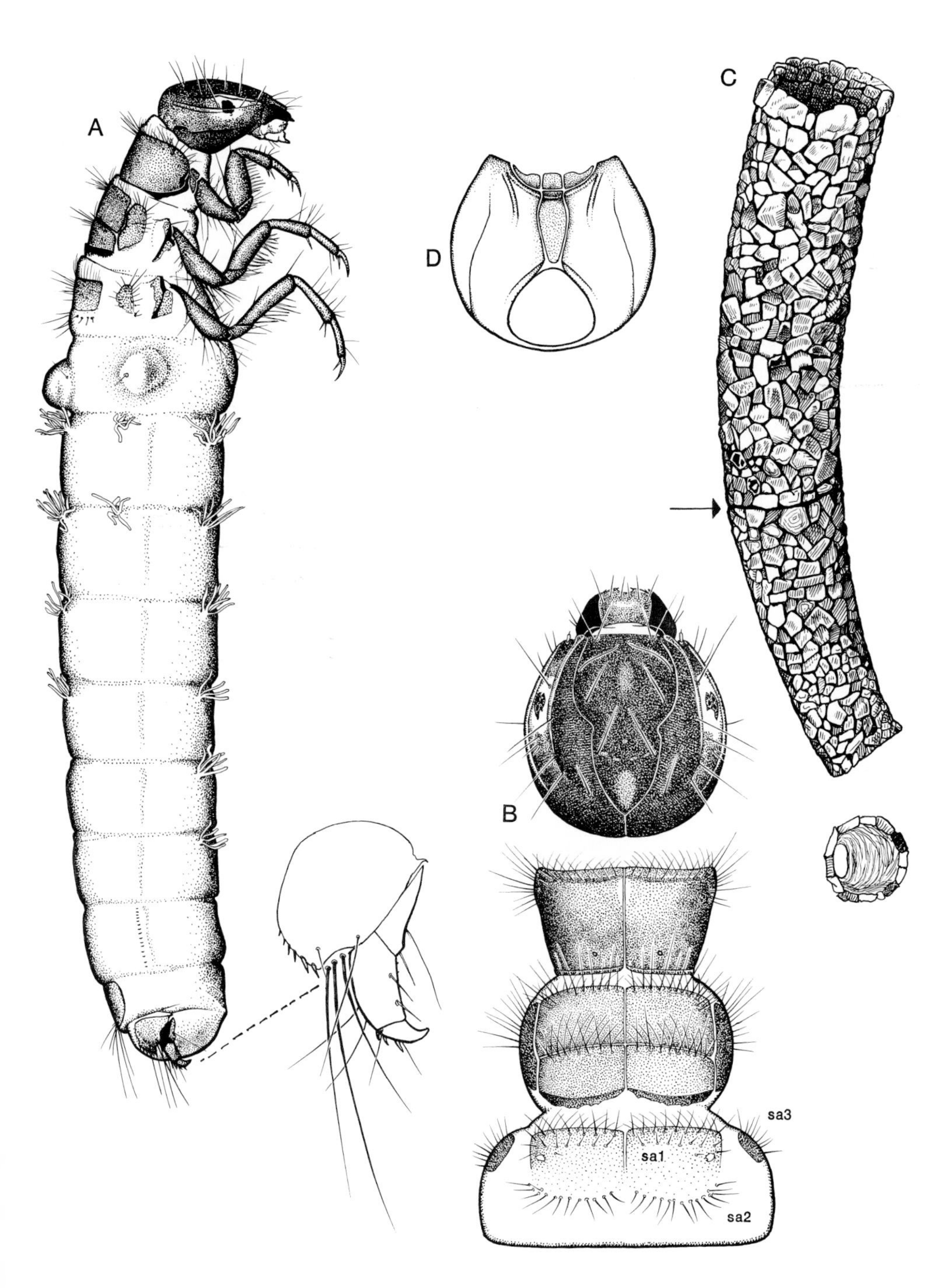
A
B
C
D
sa1
sa2
sa3

12.2 Genus Namamyia

DISTRIBUTION AND SPECIES *Namamyia* occurs only in western North America where a single species, *N. plutonis* Banks, is recorded from Oregon and California.

No positive larval association has been established for this genus. But with generic identity of all other Nearctic odontocerid larvae now known, the one type still outstanding, here illustrated, is almost certain to be *Namamyia*; this is probably the larva referred to Odontocerid Genus A by Ross (1959). We have collected several series of larvae in California and Oregon and have adults from the Oregon locality (in Benton Co.) as well.

MORPHOLOGY Larvae of *Namamyia* as well as *Nerophilus* are distinctive among North American odontocerids in having small, separate metanotal sclerites in the *sa*1 position and a second group of setae in *sa*2 (B). But in *Namamyia* abdominal segment I is heavily setate dorsally and ventrally, and lacks ventral gills (A,E). The head in *Namamyia* is pebbled in texture dorsally, bears a ridge along each side (B), and has a slender ventral apotome separating the genae entirely (F). The prosternum bears a heavily sclerotized plate (D), shorter than in *Pseudogoera* but subdivided mesially as in *Nerophilus*. Length of larva up to 20 mm.

CASE The larval case (C) is made of small rock fragments and is somewhat coarser textured than that of *Nerophilus*. The case is curved, and in some examples more tapered than illustrated. The sieve-like closure of the posterior end gives the impression of a pupal case, but cases of subterminal instars in our collections are similar in this respect; case-building behaviour is therefore different from that in *Nerophilus*. Length of larval case up to 30 mm.

BIOLOGY Larvae we have collected were living in gravel substrates of small, cool streams. Gut contents of larvae (3) examined were largely pieces of vascular plants with some fine particulate organic material.

REMARKS Diagnostic characters for adults of *N. plutonis* were summarized by Schmid (1968b).

Namamyia prob. *plutonis* (California, Shasta Co., 29 July 1950, ROM)
A, larva, lateral x7, anal claw enlarged; B, head and thorax, dorsal; C, case, lateral x5; D, prothorax, ventral; E, segment I, ventral; F, head, ventral

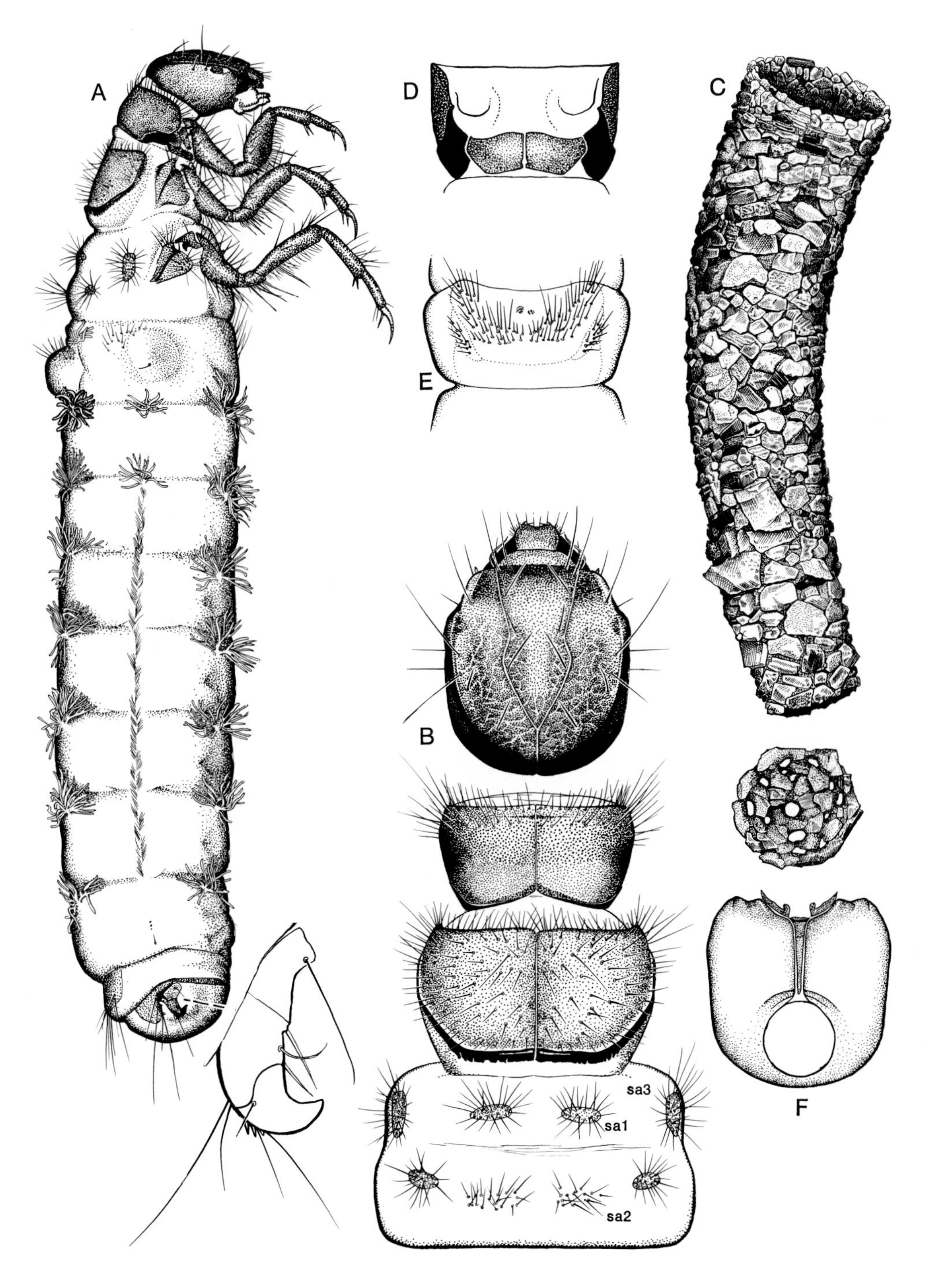
A
D
C
E
B
sa3
sa1
sa2
F

12.3 Genus **Nerophilus**

DISTRIBUTION AND SPECIES This is another genus confined to mountainous parts of western North America; there is a single species, *N. californicus* (Hagen), recorded from California and Oregon.

Larvae have not been identified in the literature previously; the description here is based on material we reared in Oregon.

MORPHOLOGY *Nerophilus* larvae are generally similar to those of *Namamyia*, but can be distinguished by the presence of two gill tufts on the venter of abdominal segment I, and by the much sparser setation on both dorsum and venter of segment I (A,E). Another character distinctive for *Nerophilus* is the fringe of short setae along the posterior margin of the ventral extension of the anal proleg's lateral sclerite (F); the anal claw bears stout, straight spines dorsally (A). The head and pronotum have a dark median band. The ventral apotome of the head separates the genae only partially (D). A heavily sclerotized prosternal plate is present, divided along the mid-ventral line as in *Namamyia*. Length of larva up to 16.5 mm.

CASE Larval cases in *Nerophilus* (C) are built of sand grains, smoothly textured, curved and tapered. The posterior end of the case is evidently left entirely open until some time during the final instar when it is reduced to the curved central opening illustrated. Length of larval case up to 25 mm.

BIOLOGY Larvae we collected in several Oregon localities were all living in sand and silt deposits of small, cool streams. Guts of larvae (3) we examined contained vascular plant pieces and fine organic particles, but each one also contained arthropod remains.

REMARKS Diagnostic characters for the male of *N. californicus* were given by Denning (1956).

Nerophilus californicus (Oregon, Lincoln Co., 15 April 1964, ROM)
A, larva, lateral x10, anal claw enlarged; B, head and thorax, dorsal; C, case, lateral x8; D, head, ventral; E, segment I, ventral; F, segment IX and anal proleg, lateral

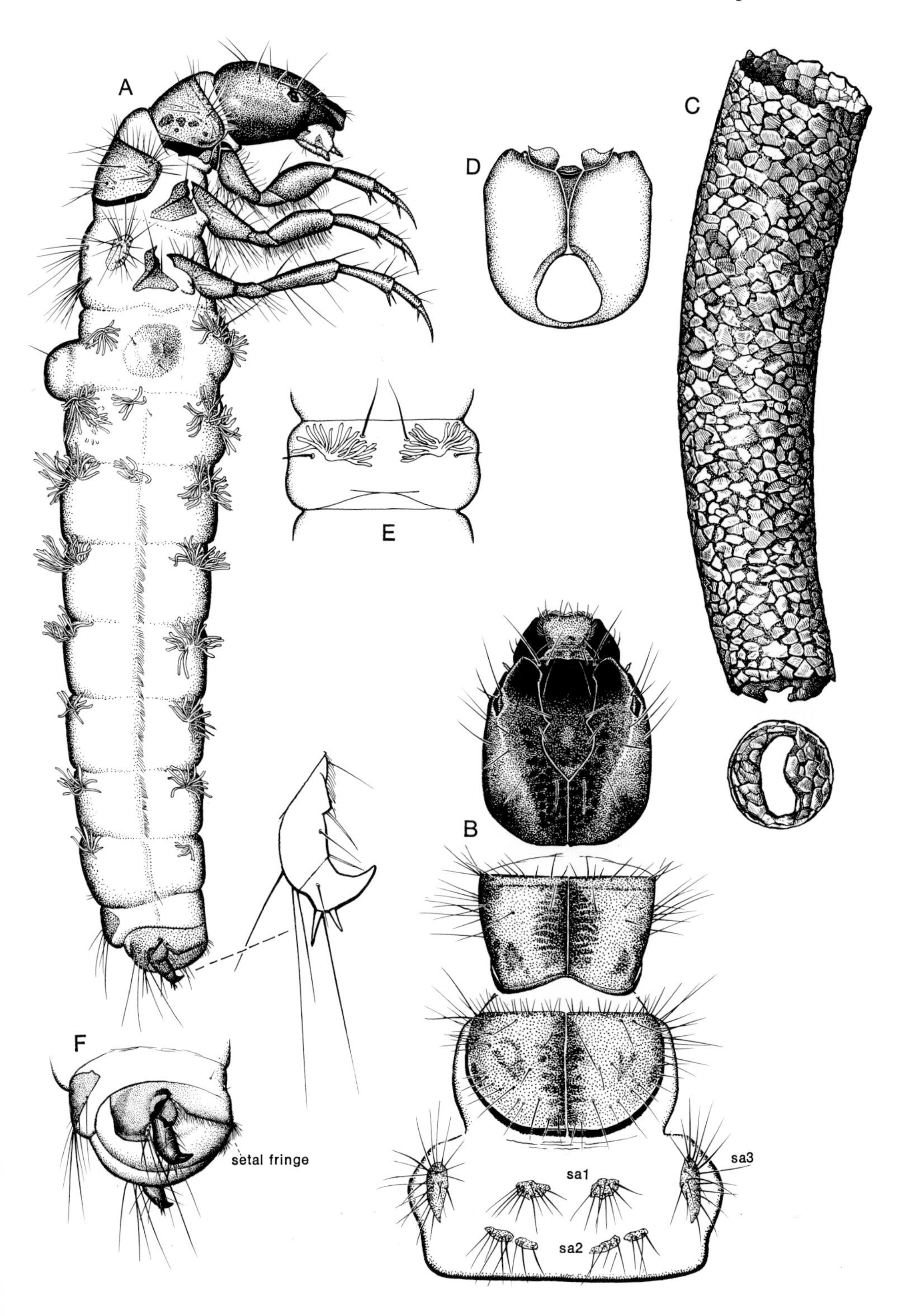
A
B
C
D
E
F
setal fringe
sa1
sa2
sa3

12.4 Genus **Parthina**

DISTRIBUTION AND SPECIES *Parthina* is known only from western North America: *P. linea* Denning from California, Arizona, and Oregon; and *P. vierra* Denning from California.

No larva of *Parthina* has heretofore been identified in the literature; we have collected associated series of both species.

MORPHOLOGY Larvae of *Parthina* are similar to those of the eastern genus *Psilotreta*, but can be distinguished by the broad anal claw, both claw and lateral sclerite beset with stout spines (A). On the head the genae are separated almost entirely by the ventral apotome (D), much as in *Marilia*. The prothorax is without a transverse ventral sclerite, but the notum is extended into a prominent anterolateral point at each side (A). Metanotal *sa*1 and *sa*2 sclerites are each fused with their opposite members into single transverse sclerites (B). The abdomen is without a lateral fringe. Length of larva up to 8.5 mm.

The larva of *P. vierra* differs from that of *P. linea* (illustrated) in having a crenate anterior pronotal margin and prominent extensions on each side of the head.

CASE Larval cases in *Parthina* (C) are made of sand grains but are unique among North American odontocerids in having a layer of silk applied to the exterior, giving a varnished aspect to the case. Evidently the posterior opening of the case is not restricted until the final instar when a small rock fragment is fastened to a silken tab at one side (C); a closure membrane is ultimately spun across the entire opening. Length of larval case up to 10 mm.

BIOLOGY Larvae of *Parthina* occur in small, cold springs in substrate materials or on rocks or moss. Contents examined from guts of larvae (3) were largely vascular plant pieces and fine organic particles, with some filamentous algae.

REMARKS Diagnostic characters for adults were given by Denning (1956, 1973).

Parthina linea (California, Nevada Co., 19 July 1966, ROM)
A, larva, lateral x13, middle tarsus, claw and anal proleg enlarged; B, head and thorax, dorsal; C, case, lateral x10; D, head, ventral

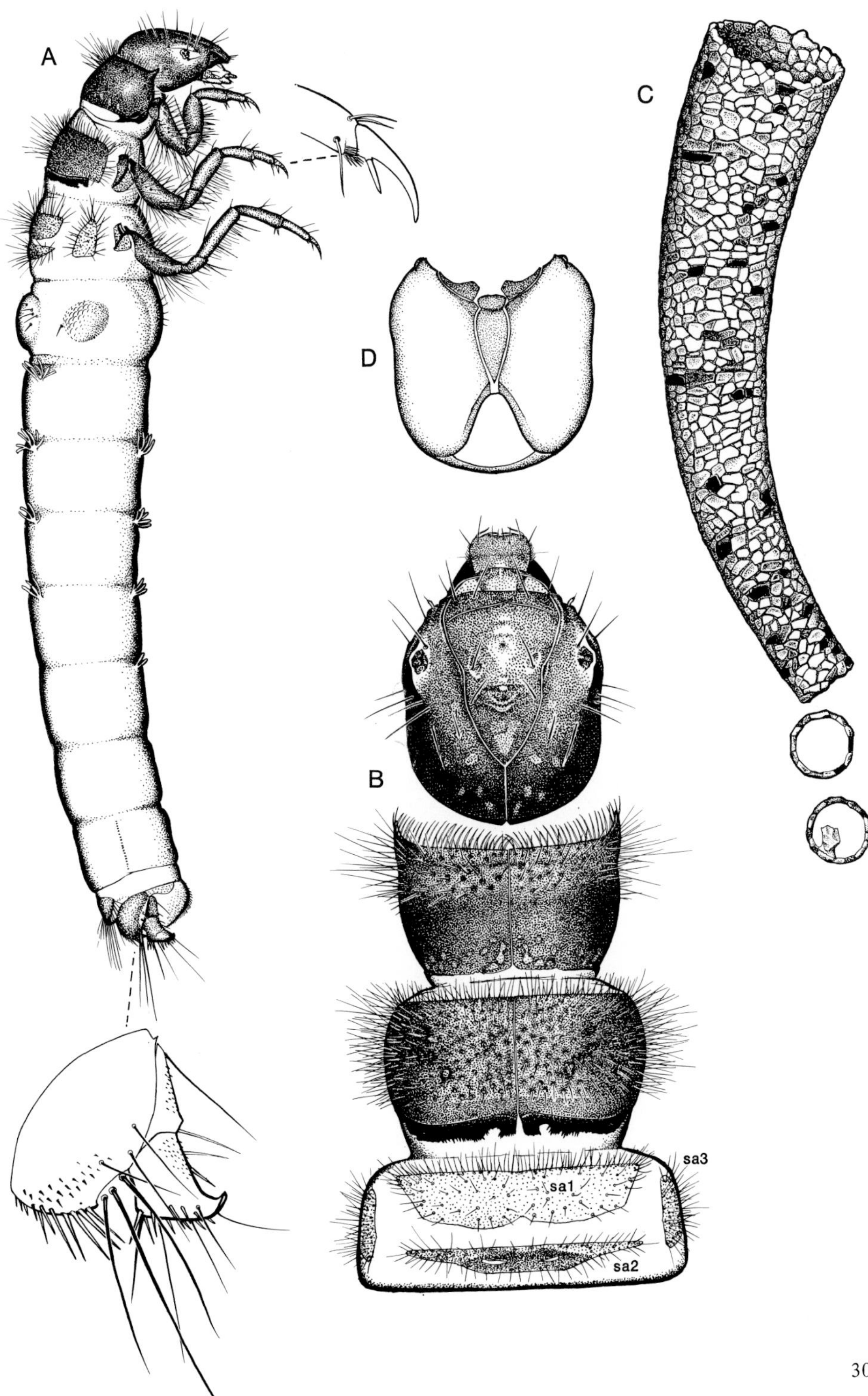
A
B
C
D
sa1
sa2
sa3

12.5 Genus **Pseudogoera**

DISTRIBUTION AND SPECIES The rather unusual species *Pseudogoera singularis* Carpenter, the sole member of this genus, has been found only in the southeastern United States (Georgia, North and South Carolina). Originally assigned to the Goeridae on the basis of the male, and long an enigma (Flint 1966), *P. singularis* was recently shown to be an aberrant member of the Odontoceridae when identity of its larval and pupal stages was established (Wallace and Ross 1971). We collected larvae in North Carolina.

MORPHOLOGY The larva can be distinguished by the enlarged tibia of the fore leg (A), which is about four times as long as the tarsus and unusually wide; the movable apical tibial spur is expanded into a blade-like structure with a terminal point; the legs are unusually long. The head is elongate, the anterior ventral apotome very small (E). The prothorax bears a sclerotized sternal plate (D) which is not subdivided along the mid-ventral line as in *Namamyia* and *Nerophilus*, and the fore trochantin is reduced; the mesosternum also bears a sclerotized plate. Metanotal setal areas are located on large discrete sclerites (B). Abdominal segment I bears a pair of large setal patches ventrally (A), and the lateral fringe of abdominal setae is lacking; tiny lateral tubercles occur on segment VIII and forward at least to IV. The lateral sclerite of each anal proleg is heavily spined (F). Length of larva up to 8.5 mm.

CASE The larval case of *P. singularis* (C) is constructed of rock fragments, curved and tapered. The posterior opening is reduced with silk to a rosette shape. Length of larval case up to 10 mm.

BIOLOGY Larvae live in small, cool streams where they are most commonly found associated with small waterfalls (Wallace and Ross 1971); generally cases with larvae are buried in moss growing on rocks, often under ledges of the waterfalls. Gut contents of larvae (2) we examined were fine organic particles and arthropod parts.

Pseudogoera singularis (North Carolina, Great Smoky Mountains Nat. Park, 21 May 1970, ROM 700366)

A, larva, lateral x18, fore leg enlarged; B, head and thorax, dorsal; C, case, lateral x9; D, pro- and meso-thorax, ventral; E, head, ventral; F, segment IX and anal prolegs, dorsal

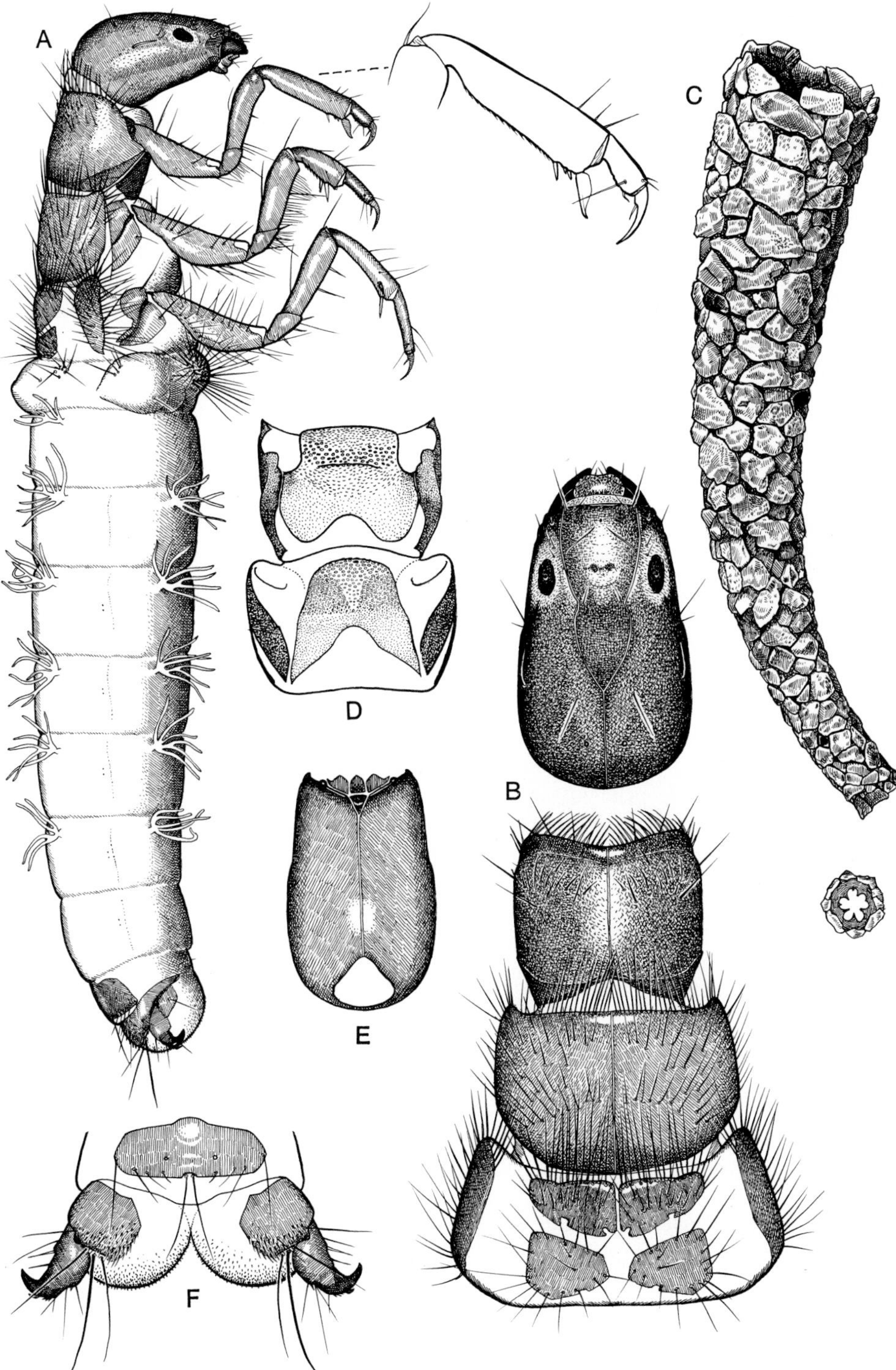
A
C
D
B
E
F

12.6 Genus **Psilotreta**

DISTRIBUTION AND SPECIES *Psilotreta* is known from the Nearctic and Oriental regions. Seven species are recorded from North America, collectively from the area bounded by Tennessee and North Carolina to New Hampshire and Quebec, through Ontario to Wisconsin.

Larval stages have been characterized in the literature only for *P. frontalis* Banks (Lloyd 1921), although frequently at the generic level. We have larval associations for four species.

MORPHOLOGY Larvae of *Psilotreta* are generally similar to those of the western *Parthina*; but the slenderer claw of the anal proleg and the absence of stout spines will distinguish *Psilotreta* (A), as will the small ventral apotome separating the genae for only a short distance (D). The anterolateral pronotal margins are sharply pointed, and the prosternum bears a short transverse sclerite; much of the mesosternum is also sclerotized; metanotal *sa*1 and *sa*2 sclerites are fused to form two transverse plates (B). The abdomen bears a lateral fringe. Length of larva up to 15 mm.

CASE *Psilotreta* cases (C) are curved, tapered, and constructed of rather coarse fragments of rock; they are exceedingly sturdy and are more resistant to crushing than cases of any other North American caddisfly that we have found. As discussed previously in the introduction to the Odontoceridae, sturdiness appears to be gained by reinforcing the depressions between larger pieces in the interior wall with small rock fragments, and by strengthening interior connections between pieces with bands of silk (C). The posterior opening is reduced with rock and silk to only a small eccentric hole. Length of larval case up to 20 mm.

BIOLOGY Larvae live in riffle areas of streams both large and small, where they crawl over bottom substrates or burrow in them; similar observations were reported by Mackay (1969). Food is largely plant materials: algae were the main items ingested by *P. indecisa* (Coffman et al. 1971); gut contents from larvae (3) we examined were mainly pieces of vascular plants and fine organic particles. Lloyd (1921) described the marked gregarious behaviour that causes *Psilotreta* larvae to fasten their cases for pupation in layers on the underside of rocks; no comparable behaviour is known in any other North American odontocerid.

Psilotreta sp. (Virginia, Shenandoah Nat. Park, 18 Sept. 1958, ROM)
A, larva, lateral x13, anal proleg enlarged; B, head and thorax, dorsal; C, case, lateral x7, portion of case showing construction of interior wall and two rock fragments with silken cross braces enlarged; D, head, ventral

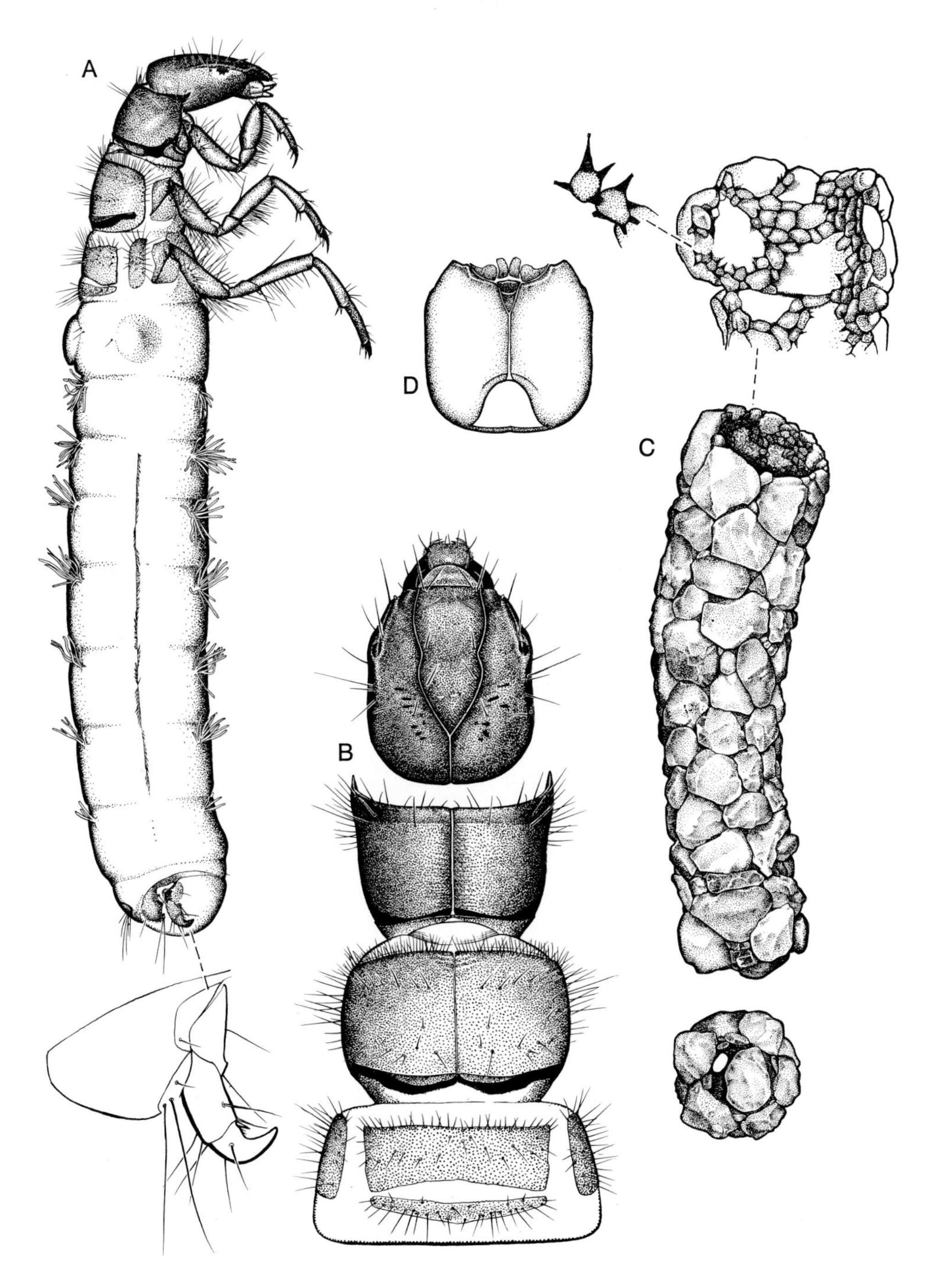
A
D
C
B

13
Family Philopotamidae

The Philopotamidae occur in running waters in all faunal regions of the world; the three Nearctic genera are common and widespread over much of the continent, with approximately 40 species known north of Mexico. Larvae spin elongate, sack-like nets of silken mesh to filter fine particulate organic matter from currents, and among all the filter-feeding Trichoptera, philopotamids evidently utilize the finest particles.

Larvae usually live on the undersides of rocks, the capture nets collapsing as amorphous silt-covered masses when the rock is removed from water. Normally positioned, however, the nets would be seen fastened by the anterior edge to the rock in such a way that they are distended as the current flows through them and the downstream ends move freely (Fig. 13.2F). Beneath the silty coating are the elegantly woven rectangular meshes of the net, with openings smaller than those for any other net-spinning family in North America (Wallace and Malas 1976). These authors calculated that in a single net of a final-instar larva of *Dolophilodes distinctus* there are over 100 million mesh openings and over 1 km of silk strands; series of grooves in the opening of the silk gland are thought to enable the larva to simultaneously spin about 70 of the finest silk strands. A small opening at the downstream end of each net is large enough to permit exit of the larva, and ensures a current of water through the net for respiration and removal of faeces even when the fine meshes are largely obstructed with particulate matter. Nets commonly occur in groups, but one larva lives within each net and feeds by cleaning the fine particles from the inside with its highly specialized membranous labrum.

The membranous labrum (Fig. 13.1D) is diagnostic for the family, although in preserved specimens it is often retracted under the anterior edge of the frontoclypeal apotome and is not immediately apparent. The head and pronotum are brownish orange in colour, the pronotum bounded posteriorly by a pronounced black line; meso- and metanota are not sclerotized. Basal setae of the tarsal claws arise from a slender extension of each claw (Fig. 13.3A). The abdomen is whitish; gills and lateral fringe are absent, and anal papillae are present. Segment IX lacks a dorsal sclerite.

Taxonomy, phyletic relationships, and zoogeographic dispersal of the Philopotamidae, based largely on evidence from adults, were reviewed by Ross (1956).

13 Family **Philopotamidae**

Key to Genera

1 Anterior margin of frontoclypeal apotome with prominent notch (Fig. 13.1B); coxa of fore leg with long, slender process arising near distal end, bearing seta (Fig. 13.1C); venter of head with seta no. 18 located at level of posterior edge of ventral apotome (Fig. 13.1F). Widespread **13.1 Chimarra**

Anterior margin of frontoclypeal apotome lacking prominent notch, although there may be some asymmetry (Fig. 13.2B); coxa of fore leg lacking long slender process; venter of head with seta no. 18 located approximately half-way between posterior edge of ventral apotome and occipital foramen (Fig. 13.2E) 2

2 (1) Fore trochantin projecting freely anteriorly to form elongate, finger-like process (Fig. 13.2C,D); venter of head with seta no. 18 approximately same thickness as stoutest seta on dorsum of head (Fig. 13.2E). Widespread **13.2 Dolophilodes**

Fore trochantin projecting freely only a short distance, thus forming very short process (Fig. 13.3D,E); venter of head with seta no. 18 stouter than any seta on dorsum of head (Fig. 13.3C). Widespread **13.3 Wormaldia**

13.1 Genus **Chimarra**

DISTRIBUTION AND SPECIES *Chimarra* is a large genus represented in all major faunal regions, but for the most part in tropical and warm temperate latitudes. The genus is richly represented in Mexico and Central America, and about 17 species are now known to the north. Species diversity decreases markedly with higher latitude, but species of *Chimarra* are known from Maine, Manitoba, and Montana.

A key to larvae of four eastern species was given by Ross (1944); the larva of *C. betteni* Denning was described by Edwards and Arnold (1961).

MORPHOLOGY Larvae are readily identified to this genus by the prominent, often asymmetrical notch along the anterior margin of the frontoclypeal apotome (B), and by the elongate process of the fore coxa (C). Evidently also distinctive among the three North American genera is the location of ventral head seta no. 18 close to the level of the posterior tip of the ventral apotome (F). Length of larva up to 12.5 mm.

RETREAT Some of the nets associated with *Chimarra* larvae have rectangular meshes similar to those in *Dolophilodes*, although of somewhat less uniform dimensions; but in others the strands are irregularly arranged throughout and show no regular mesh structure (Wallace and Malas 1976). Capture nets of *Chimarra* were described by Noyes (1914, pl.38, fig. 1); the nets were rarely found singly, but generally in a row of five or six, with average size of one net about 25 mm long by 3 mm in diameter.

BIOLOGY The biology of *C. atterrima* Hagen was studied in Ontario by N.E. Williams and Hynes (1973). The species was univoltine and larvae subsisted largely on detritus filtered by their capture nets throughout the year, growing rapidly during the summer. Although living on the undersides of rocks in riffle areas most of the time, larvae migrated to deeper water in winter and to submerged roots of streamside grasses in warm weather. In another study, Coffman et al. (1971) found that algae comprised about two-thirds of the material ingested by this species, detritus about one-third, with a small animal component as well.

Chimarra sp. (Ontario, Peel Co., 14 Oct. 1951, ROM)
A, larva, lateral x17; B, head and thorax, dorsal; C, fore coxa, anterior; D, labrum dorsal; E, segment IX and anal prolegs, showing anal papillae, dorsal; F, head, ventral

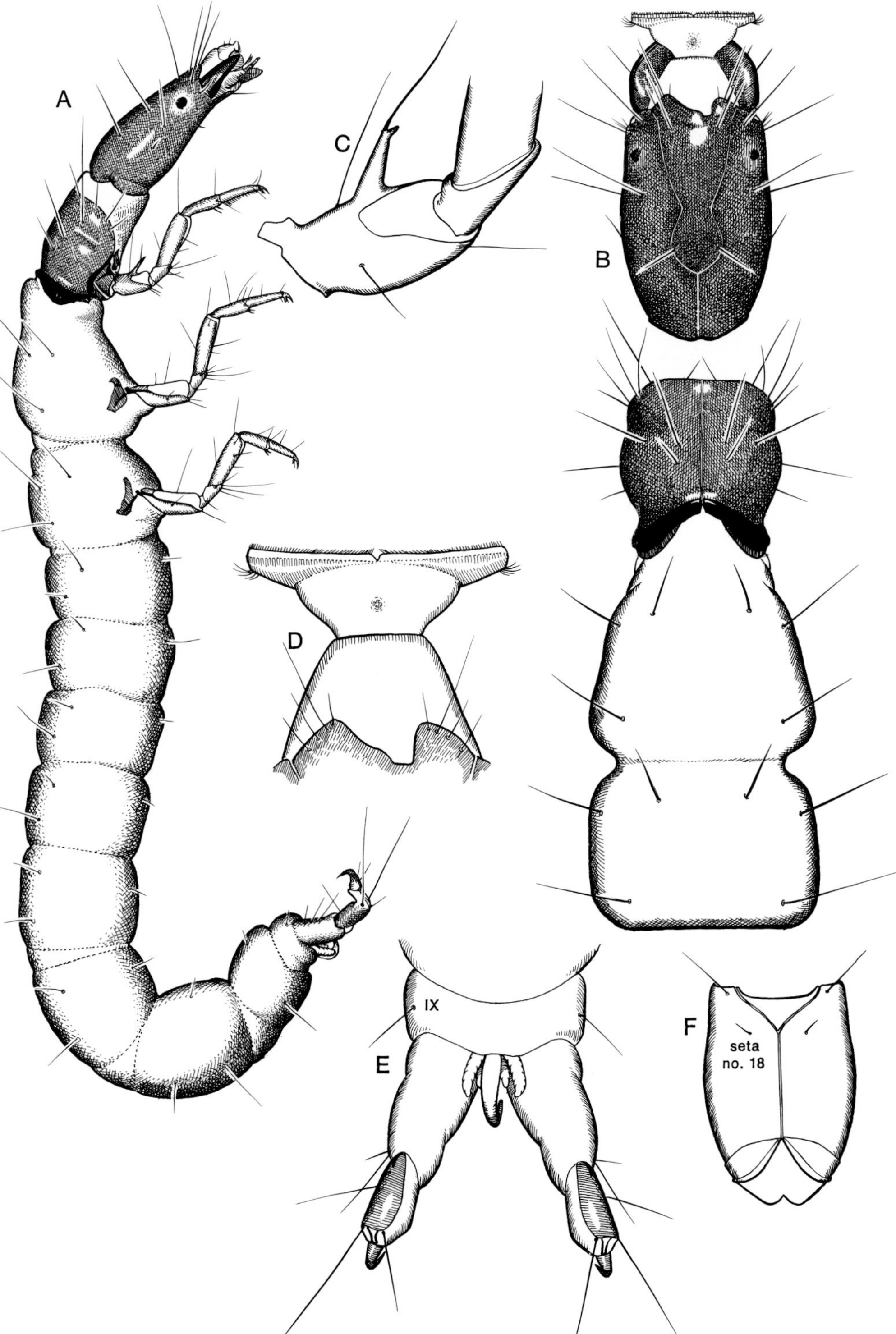
A
B
C
D
E
IX
F
seta
no. 18

13.2 Genus **Dolophilodes**

DISTRIBUTION AND SPECIES The genus *Dolophilodes*, here taken to be the equivalent of *Sortosa* as defined by Ross (1956), is represented in all major faunal regions. Eight species are known in North America north of Mexico, largely restricted to eastern and western montane areas; one species *D. distinctus* (Walker) (syn. *Trentonius distinctus*), is common and widespread in the east from Newfoundland and Minnesota to Georgia.

Some structures of *D. distinctus* larvae were illustrated by Ross (1944) and Betten (1934). We have series of *Dolophilodes* larvae from many localities throughout the continent.

MORPHOLOGY The anterior margin of the frontoclypeal apotome in *Dolophilodes* larvae is somewhat asymmetrical (B), but is not as prominently notched as in *Chimarra.* The fore trochantin bears a finger-like process freely projecting anteriorly (C,D), which is distinctive among the North American genera. In contrast to *Wormaldia*, ventral head seta no. 18 is relatively small (E). The outer edges of the mandibles are transversely serrate in at least some species. Length of larva up to 16.5 mm.

RETREAT Capture nets constructed by larvae of *Dolophilodes distinctus* are elongate sacks with extremely fine, narrow meshes, approximately 0.5 x 5.5 μm in final instars; usually the net consists only of a single layer (Wallace and Malas 1976). The fine meshes are formed by closely spaced slender strands fastened transversely across stouter double support strands that run parallel with the long axis of the net (F). Silk strands comprising the one-third or so of the net around the upstream opening are arranged randomly, presumably to strengthen the mooring to the rock. Both nets and meshes are increased in size as the larva grows, the nets of final instars ranging from 25 to 60 mm in length and 2.5 to 4.5 mm in diameter (Wallace and Malas 1976).

BIOLOGY Contents of the foreguts of *D. distinctus* in a Georgia river were largely fine particulate organic matter and diatoms (Wallace and Malas 1976). Similar findings were reported by Mecom (1972b) for another species of the genus in Colorado, although vascular plant pieces were also ingested in the autumn.

REMARKS The eastern species *D. distinctus* is notable for the wingless females that emerge during winter months (Ross 1944, fig. 171).

Dolophilodes distinctus (Ontario, Peel Co., 1 Oct. 1970, ROM)
A, larva, lateral x16; B, head and thorax, dorsal; C, fore trochantin, lateral; D, fore trochantin, ventral; E, head, ventral; F, two capture nets, with section of silken mesh enlarged approx. x 1,000, arrow indicating direction of current

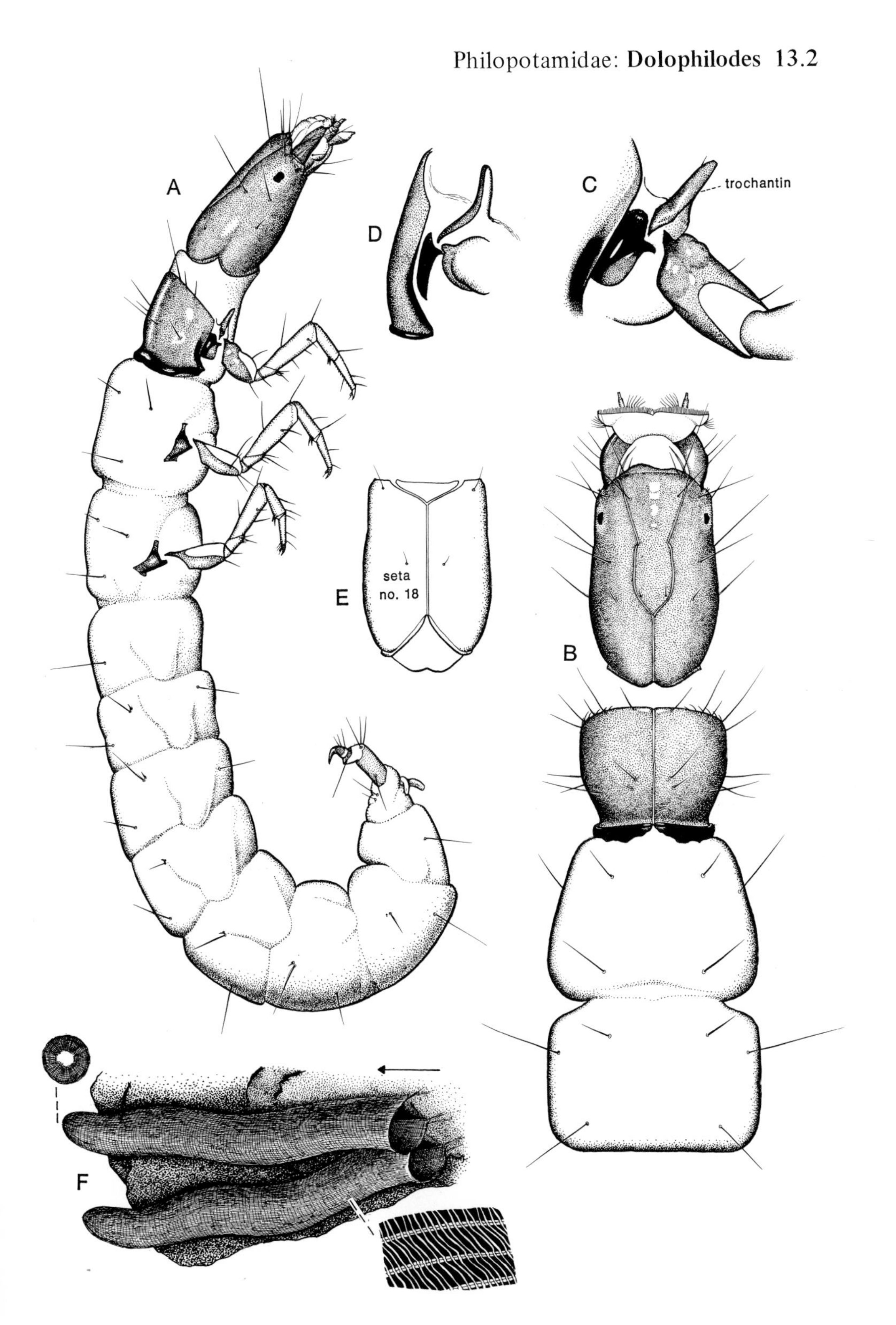
A
B
C
trochantin
D
E
seta
no. 18
F

13.3 Genus Wormaldia

DISTRIBUTION AND SPECIES *Wormaldia* as defined by Ross (1956) includes species treated under *Dolophilus* (*sensu* Ross 1944). The genus is represented in all faunal regions except the Australian; approximately 13 species are known north of Mexico, most in the west.

Diagnostic characters were given by Ross (1944) for larvae of two widespread eastern species, *W. moesta* (Banks) and *W. shawnee* (Ross). We have associated material for one western species, *W. occidea* (Ross).

MORPHOLOGY The anterior margin of the frontoclypeal apotome is symmetrical, but may be either convex (B) or concave. The fore trochantin projects freely as a small, knob-like process (D,E); coxal setae are very stout in some species. On the ventral surface of the head, seta no. 18 is located approximately half-way between the ventral apotome and the occipital foramen as in *Dolophilodes*, but is stouter than any of the dorsal head setae (C). Length of larva up to 14.5 mm.

RETREAT Nets of final instars of *Wormaldia* larvae evidently comprise several layers of mesh; meshes of each layer are narrow and rectangular as in *Dolophilodes*, although slightly smaller - 0.4 x 3.7 μm (Wallace and Malas 1976). Separate layers of net are superimposed in such a way that the fine meshes overlay each other at an angle, thereby reducing even further the effective mesh openings. Nets described for European species of *Wormaldia* (Nielsen 1942; Philipson 1953b, fig. 4) were about 35 mm long by 5 mm in diameter.

BIOLOGY Larvae move about freely within their silken tube, and have been observed cleaning fine organic particles from the inner surface of the net by short downward strokes of the head, the labrum extended (Philipson 1953b). A western species, *W. anilla*, was found to have an emergence period from March to December (Anderson and Wold 1972).

Wormaldia sp. (Arizona, Cochise Co., 24 June 1966, ROM)
A, larva, lateral x23, tarsal claw enlarged; B, head and thorax dorsal, detail of antenna; C, head, ventral; D, fore trochantin, lateral; E, fore trochantin, ventral

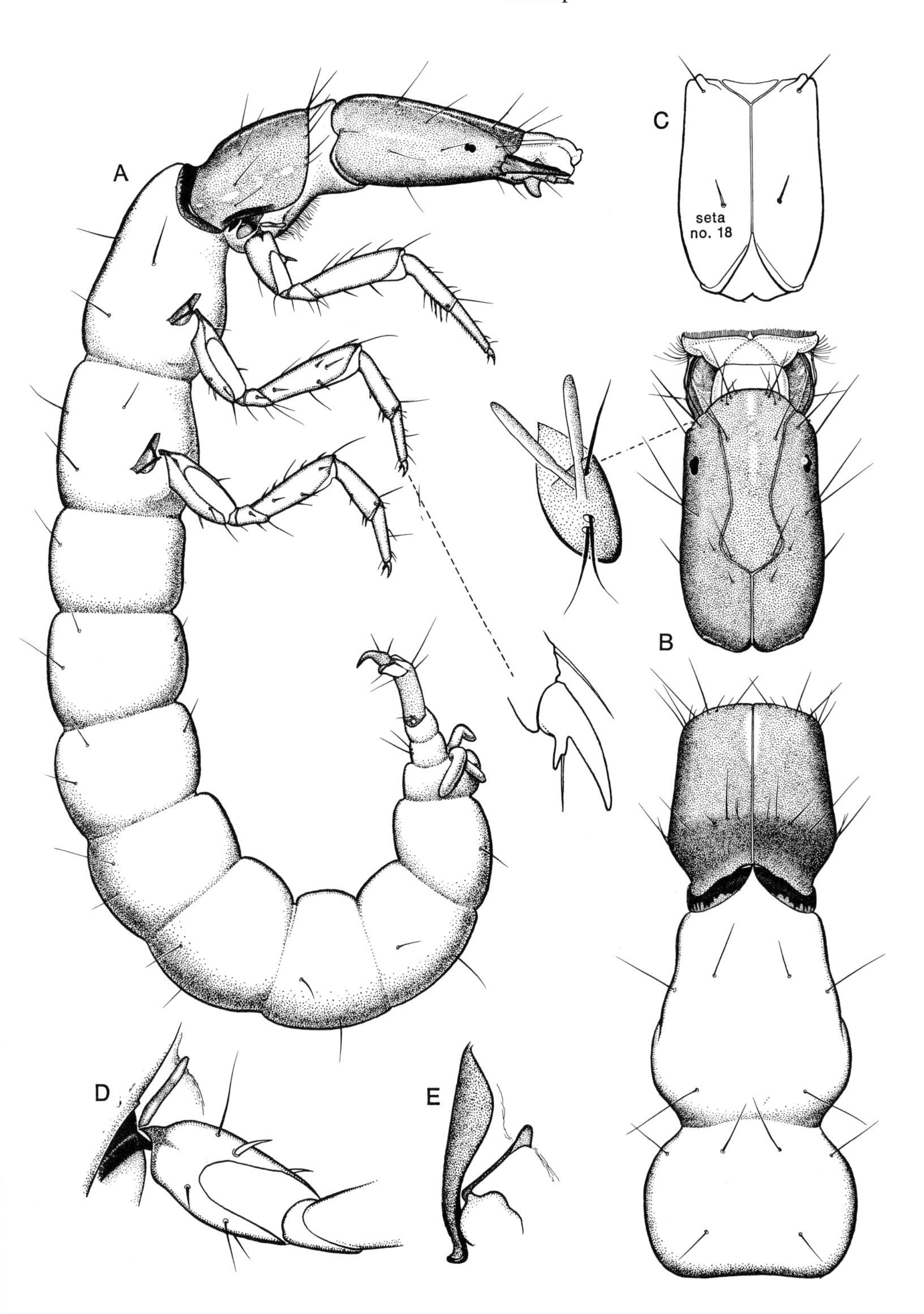
A
B
C
D
E
seta
no. 18

14
Family Phryganeidae

The Phryganeidae are a family of the Holarctic region, numbering nearly 70 species in 15 genera. For the most part, they are insects of northerly latitudes; in North America there are 10 genera with 27 species. Larvae of different genera live in a wide range of waters: spring streams, marshes, lakes even to depths of 100 m, and temporary vernal pools. Available evidence suggests that over all the larvae are omnivorous, although some are largely predacious for at least part of their life cycle.

Larvae are large, often over 40 mm long, but remarkably homogeneous in structure. Intersegmental constrictions are more prominent, the larvae more slender and agile than in other families of the Limnephiloidea, and the head more prognathous than in other tube-case families; these are characteristics conveyed in the term *suberuciform* generally applied to the Phryganeidae. The head and pronotum in almost all species are conspicuously marked with dark bands on a light yellowish brown background, although species in the same genus often differ in these markings. A prominent prosternal horn is present in all, and there is a median sternellum (Fig. 14.1F) on the prosternum in all Nearctic genera except *Phryganea*. In contrast to most other limnephiloid families, the mesonotum is largely unsclerotized, although small *sa*1 and *sa*2 sclerites occur in a few genera; both meso- and meta-nota have small, usually rounded, *sa*3 sclerites bearing several setae. The characteristic arrangement of light markings on both meso- and meta-nota, well expressed in Figure 14.1B although also represented on abdominal segments, appears to coincide with attachments of muscles to the cuticle. Comb-like spines, the coxal combs, appear to be present on the basal leg segments in all genera, but are seen in most only as raised points under moderate magnification of approximately 50x (Fig. 14.8E); the coxal combs are larger and of taxonomic use in two genera, *Agrypnia* (Fig. 14.1D,E) and *Phryganea* (Fig. 14.7E,F). Mouthparts of the Phryganeidae (Fig. 14.4D) have a slender, sclerotized galea similar to that of several groups in the Hydropsychoidea, and rather different from the short, rounded galea of the many other families in the Limnephiloidea. Humps of segment I are prominent and gills are single. Segment I always bears two ventral gills on each side; segments II–V usually have a complete set of six single gills on each side, but gills are frequently absent from succeeding segments. Arrangement of gills often provides characters of taxonomic value, but is subject to variation within species. The lateral abdominal fringe is well developed, but lateral tubercles are absent; segment IX bears a small dorsal sclerite with at least one pair of major setae (Fig. 14.1G), sometimes two.

There are two subfamilies: the Yphriinae, with a single genus and species *Yphria californica* (Banks), are markedly different in all stages from the others (Wiggins 1962), and are probably the most primitive group in the family; the Phryganeinae comprise the remaining genera.

Larvae of the Phryganeinae usually use pieces of leaves and bark to construct cases of two types; in spiral construction (Fig. 14.7C,D) the pieces are fastened in a continuous spirally wound ribbon; and in ring construction (Fig. 14.8C,D) the pieces are fastened to form discrete sections joined end to end. As in spiral cases of the leptocerid genus *Triaenodes*, cases of *Phryganea* can be either dextral or sinistral within the same species; this is probably a variable feature of behaviour in all phryganeids constructing spiral cases. Phryganeid larvae removed from their cases have greater ability for re-entry than do those of most other families (Merrill 1969), thus compensating for the speed with which they voluntarily abandon their cases when disturbed. All other larvae that construct tubular cases show marked reluctance to abandon them. Larvae leave the posterior end of their cases completely open until just prior to pupating; at that time the larva conceals its case by burrowing into bottom sediments or into rotting logs or plant stalks. The posterior end is sealed with a perforate sieve membrane of silk, as is the anterior end of the case in several genera; this is the normal behaviour for Trichoptera. But in some phryganeid genera the anterior membrane is not spun, and only a loose plug of debris is fastened with silk over the opening (Wiggins 1960a); correlated with this specialized behaviour are the reduced pupal mandibles in such genera as *Banksiola, Ptilostomis, Oligostomis,* and *Hagenella*, some of the very few examples of adecticous or non-functional mandibles in endopterygote insect pupae (Hinton 1971). It is not clear why larvae of these genera can dispense with part of the protective enclosure that seems essential for most other Trichoptera during pupation.

Observing living larvae, one sees that the suberuciform phryganeids are more active and agile than larvae of other tube-case families, and that they are less dependent on their case than the others. On evidence derived from all stages, the Phryganeidae are regarded as among the most primitive among the tube-case makers (Ross 1967); and perhaps these unusual features should be interpreted as representing some of the transitional stages leading to eruciform larvae and their greater dependence on cases. In this connection it is interesting to note that arrangement of the tracheal system of phryganeid larvae is unique and distinct from that of other tube-case-making families (Novák 1952). Phylogenetic relationships of phryganeid genera have been studied and summarized by Wiggins (1972).

Key to Genera

1 Ventral surface of head with genae almost completely separated by ventral apotome (Fig. 14.2D) (subfamily Phryganeinae) **2**

Ventral surface of head with genae mostly contiguous, separated only near anterior border by small ventral apotome (Fig. 14.9D); case of both plant and rock pieces (Fig. 14.9C). Western (subfamily Yphriinae) **14.9 Yphria**

2 (1) Each mesonotal *sa*1 seta arising near anterior edge of rounded sclerite of diameter several times larger than *sa*3 sclerite (Fig. 14.5B); case of ring construction (Fig. 14.5C). Eastern **14.5 Oligostomis**

Mesonotal *sa*1 seta lacking rounded sclerite, or seta arising from centre of small sclerite of diameter smaller than *sa*3 sclerite (Fig. 14.4B); case usually of ring or spiral construction **3**

3 (2) Head and pronotum uniform light brown, without contrasting light and dark markings except for muscle scars on head (Fig. 14.4B); case of ring construction (Fig. 14.4C); northeastern **14.4 Hagenella**?

Head and pronotum with dark markings contrasting prominently with light ground colour (Fig. 14.1B) **4**

4 (3) Coxal combs of fore legs well developed, their structure (Fig. 14.1D,E) apparent at relatively low magnification of approximately 50x **5**

Coxal combs of fore legs small (Fig. 14.8E), appearing as tiny raised points at a magnification of approximately 50x **6**

5 (4) Small, pigmented sclerite (sternellum) usually present between prothoracic coxae (Fig. 14.1F); coxal combs on prothoracic legs with axis of base transverse to long axis of coxa (Fig. 14.1D), coxal combs of mesothoracic legs with basal axes both transverse and parallel to long axis of coxa (Fig. 14.1E); case usually of spiral construction. Widespread **14.1 Agrypnia**

Sternellum between prothoracic coxae lacking; coxal combs on both pro- and meso-thoracic legs with basal axis transverse to long axis of coxa (Fig. 14.7E,F); case of spiral construction (Fig. 14.7D). Widespread **14.7 Phryganea**

6 (4) Meso- and meta-nota with pair of longitudinal, irregular, dark bands (Fig. 14.2B); case of spiral construction (Fig. 14.7D) **7**

Meso- and meta-nota mostly uniform in colour without dark bands (Fig. 14.3B) **8**

7 (6) Segments VI and VII with anterodorsal gills present, segment VII with posteroventral gills absent (Fig. 14.2A). Throughout most of North America, but not recorded from western Alaska **14.2 Banksiola**

Segments VI and VII with anterodorsal gills absent, segment VII with posteroventral gills present (Fig. 14.6A). In North America only in western Alaska **14.6 Oligotricha**

8 (6) Pronotum with dark line only along anterior border (Fig. 14.3B); case basically of spiral construction, but trailing ends give it a bushy, irregular appearance (Fig. 14.3C). Central **14.3 Fabria**

Pronotum with transverse dark band on each side, more central in position and not along anterior border (Fig. 14.8B); case of ring construction (Fig. 14.8C,D). Transcontinental **14.8 Ptilostomis**

14.1 Genus Agrypnia

DISTRIBUTION AND SPECIES This is essentially a genus of the northern Holarctic region with nine species known in North America. Most Nearctic species are transcontinental and northern in distribution, but some follow montane areas southward to Georgia and California; two species are Holarctic.

We have associated material for six species of *Agrypnia*; larval descriptions and a key were given for *A. improba* (Hagen), *A. pagetana* Curtis, *A. straminea* Hagen, and *A. vestita* (Walker) by Wiggins (1960b). The larva of *A. vestita* was described by Ross (1944), but the one described by Lloyd (1921, as *Phryganea vestita*) and supposed to be that species is probably *Banksiola crotchi.*

MORPHOLOGY The coxal combs of North American *Agrypnia* larvae differ widely in size, those of *A. vestita* being the smallest; but as a group the combs of *Agrypnia* species are larger than those of the other genera except *Phryganea*, and are confined to a smaller area (D,E; cf. *Ptilostomis* Fig. 14.8E). Markings of the head and pronotum are not consistent for even the larvae known (Wiggins 1960b, figs. 7-10); in some species the median dark band of the frontoclypeal apotome is lacking and in others the pronotum bears dark markings on the anterior margin but lacks the oblique bands of the one illustrated. Most, but not all, *Agrypnia* larvae have characteristic meso- and meta-notal markings of dark blotches broken up by light areas along the lines of muscle attachments as in B. Length of larva up to 30 mm.

CASE Larval cases of *Agrypnia* species are usually constructed of pieces of leaves and bark arranged spirally (C), individual pieces closely fitted and without trailing ends. *A. pagetana* occasionally uses a piece of hollow stem as a case. Length of larval case up to 50 mm.

BIOLOGY These are species of lakes, marshes, and slow-flowing rivers; larvae were found as deep as 45 m in Lake Superior (Selgeby 1974). Guts of larvae (3) we examined all contained arthropod parts.

Agrypnia improba (Ontario, Peel Co., 29 May 1955, ROM)
A, larva, lateral x5; B, head and thorax, dorsal; C, case x3; D, fore coxa, ventral; E, middle coxa, ventral; F, prothorax, ventral; G, sclerite of segment IX, dorsal

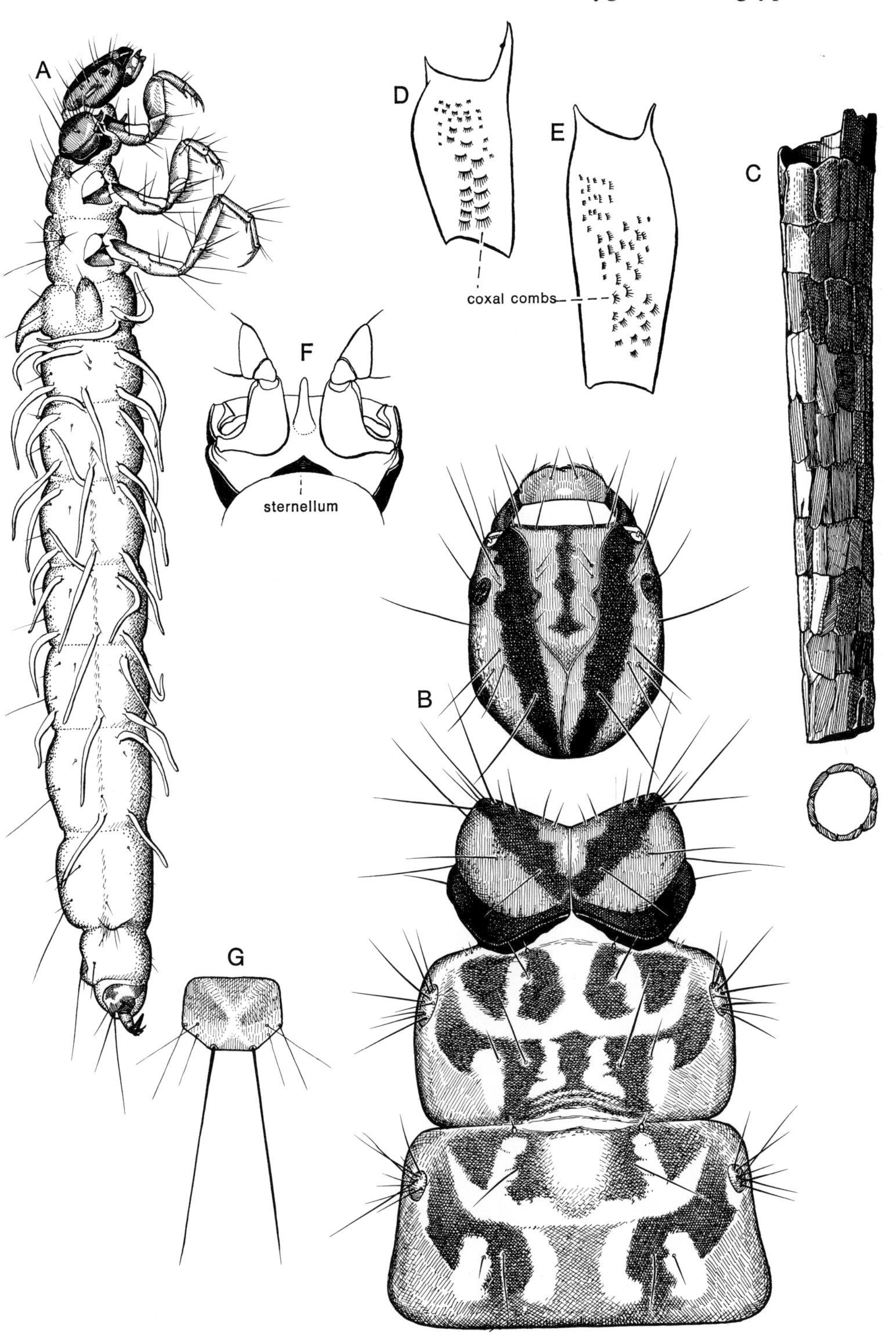
A
D
E
C
coxal combs
F
sternellum
B
G

14.2 Genus Banksiola

DISTRIBUTION AND SPECIES *Banksiola* is exclusively a Nearctic genus. Five species are known; four are restricted to eastern parts of the continent from Nova Scotia to Minnesota and Florida, and the most common species, *B. crotchi* Banks (syn. *B. selina* Betten), is transcontinental.

Larvae of three species have been described by Wiggins (1960b): *B. crotchi, B. dossuaria* (Say), and *B. smithi* (Banks). The larva assigned to Phryganeid Genus A by Ross (1944, 1959) is probably *B. crotchi*, but that assigned to *Banksiola selina* is probably *Agrypnia straminea.*

MORPHOLOGY Larvae of some species have a median dark band on the frontoclypeal apotome (Wiggins 1960b, figs. 11, 12), but in *B. crotchi* the band is absent (B); the dark pronotal bands are parallel in *B. dossuaria.* The dark meso- and meta-notal bands (B) are characteristic of all *Banksiola* now known, and provide a ready character for field identification of even early instars, especially since *Oligotricha*, the only other Nearctic genus characterized by the bands, is confined to western Alaska. Tibiae of the fore and middle legs are somewhat more expanded distally in *Banksiola* and *Oligotricha* (A) than in some other phryganeid genera. Gill characters given in the key for separating these two genera in North America may not hold for subterminal instars or for the two eastern species of *Banksiola* not yet known as larvae. Length of larva up to 22 mm.

CASE Larval cases are constructed of pieces of plant materials arranged spirally (C) and often with some ends trailing. Length of larval case up to 45 mm.

BIOLOGY *Banksiola* larvae occur in lakes, marshes, and sluggish streams. Larvae of *B. crotchi* in a British Columbian lake hatched from eggs in August and September, and most passed through the first three or four instars before November, feeding largely on filamentous algae; fifth instars were almost exclusively predacious (Winterbourn 1971b). Larvae burrowed in bottom sediment where they pupated, emergence beginning in July.

REMARKS Taxonomy of *Banksiola* adults was reviewed by Wiggins (1956).

Banksiola crotchi (Oregon, Jefferson Co., 1 June 1968, ROM)
A, larva, lateral x7, tibiae and tarsi enlarged; B, head and thorax, dorsal; C, case x3; D, head, ventral

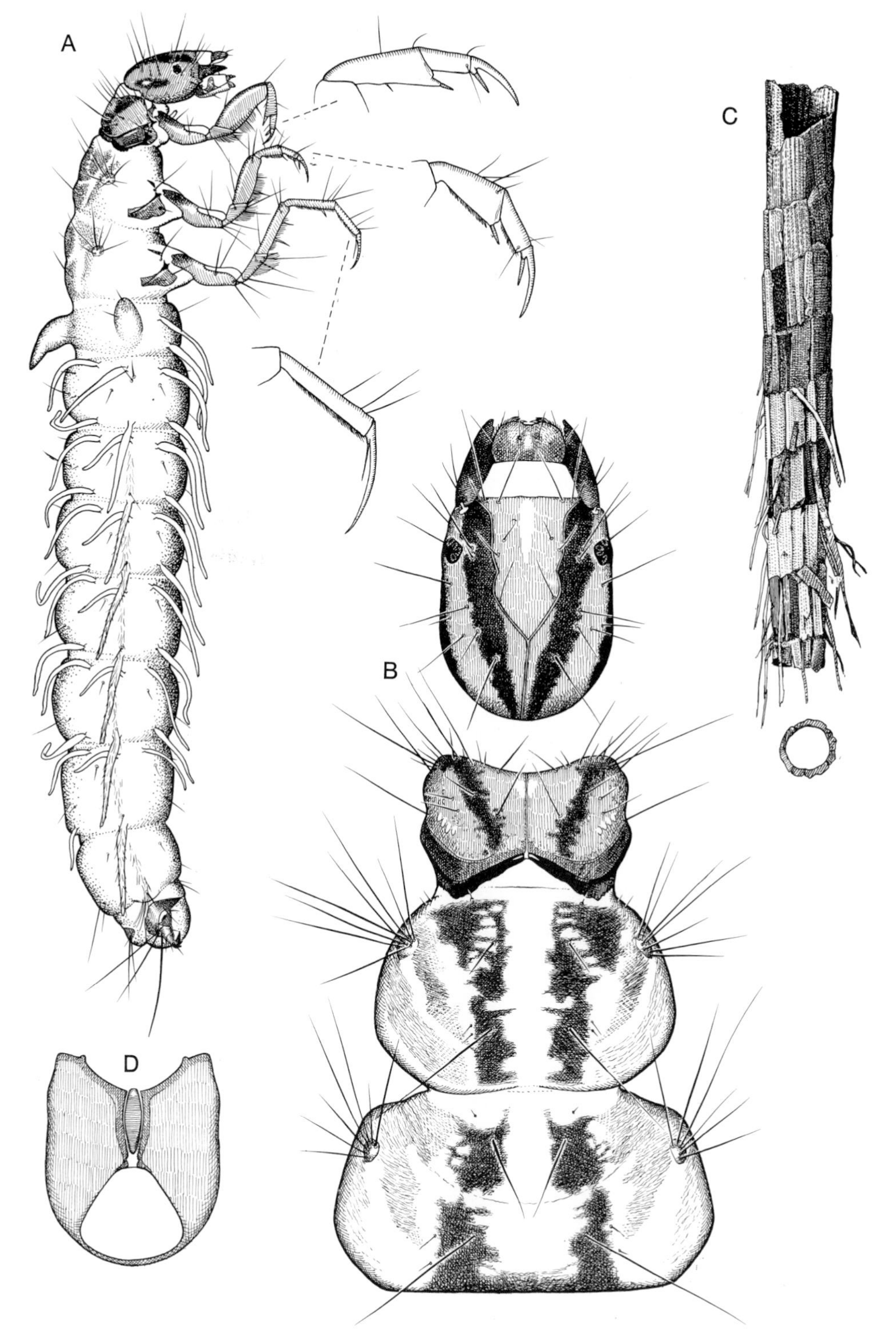
A
B
C
D

14.3 Genus **Fabria**

DISTRIBUTION AND SPECIES Two species have been assigned to the genus *Fabria*, and both are confined to North America. The type species of the genus, *F. inornata* (Banks), is known from Illinois, Manitoba, Minnesota, Ontario, and Quebec; it is local in occurrence and rather rare in collections. The second species, *F. complicata* (Banks), although assigned to this genus for many years (Milne 1934; Ross 1944), is not congeneric with *F. inornata* (Wiggins 1961); it is more similar to *Phryganea* and probably represents a new genus (Wiggins 1958). *F. complicata* is a rare species, with only a few specimens known from Ontario and Newfoundland (Wiggins 1961), and more recently Alberta and Wisconsin.

The larva of *F. inornata*, not identified in the literature previously, is characterized here from specimens we reared in Ontario. The larva of *F. complicata* remains unknown; because of the structural features of the adults, I suspect that the case of this species will prove to be a spiral, but otherwise there is no way of predicting where this larva will fall in the generic key.

MORPHOLOGY The larva of *F. inornata* (A,B) is basically similar to all others in the Phryganeinae. Markings of head and pronotum are similar to those of *Agrypnia vestita* (Wiggins 1960b, fig. 10), but the meso- and meta-nota in *F. inornata* are more uniformly pigmented and lack the dark blotches of that species; in comparison with other North American phryganeine larvae now known, this seems the most useful diagnostic character. Gills are all present on abdominal segments I–VII; segment VIII bears the three anterior gills and a prominent posterolateral lobe on each side (A). Other characters are generally typical for the subfamily. Length of larva up to 32 mm.

CASE The case of *F. inornata* (C) is a characteristic feature and is readily distinguished from that of any other North American caddisfly. The basic spiral arrangement of many of the phryganeine cases is evident, but trailing ends of component pieces give this case a unique bushy aspect. Length of larval case up to 55 mm.

BIOLOGY Our collections indicate that larvae of *F. inornata* live in dense beds of submerged aquatic plants in standing or slowly moving waters; shallow areas where *Ceratophyllum* and *Potamogeton* flourish are typical of these sites. Larvae tend to use fresh green pieces of these same plants in construction, and the resultant cases are difficult to distinguish from the plants; the trailing ends of plant pieces probably also give additional buoyancy to the case, helping the larva to maintain its position near the surface. Gut contents of larvae (3) examined were vascular plant pieces and fine organic particles.

Fabria inornata (Ontario, Leeds Co., 27 Aug. 1966, ROM)
A, larva, lateral x7; B, head and thorax, dorsal; C, case x3

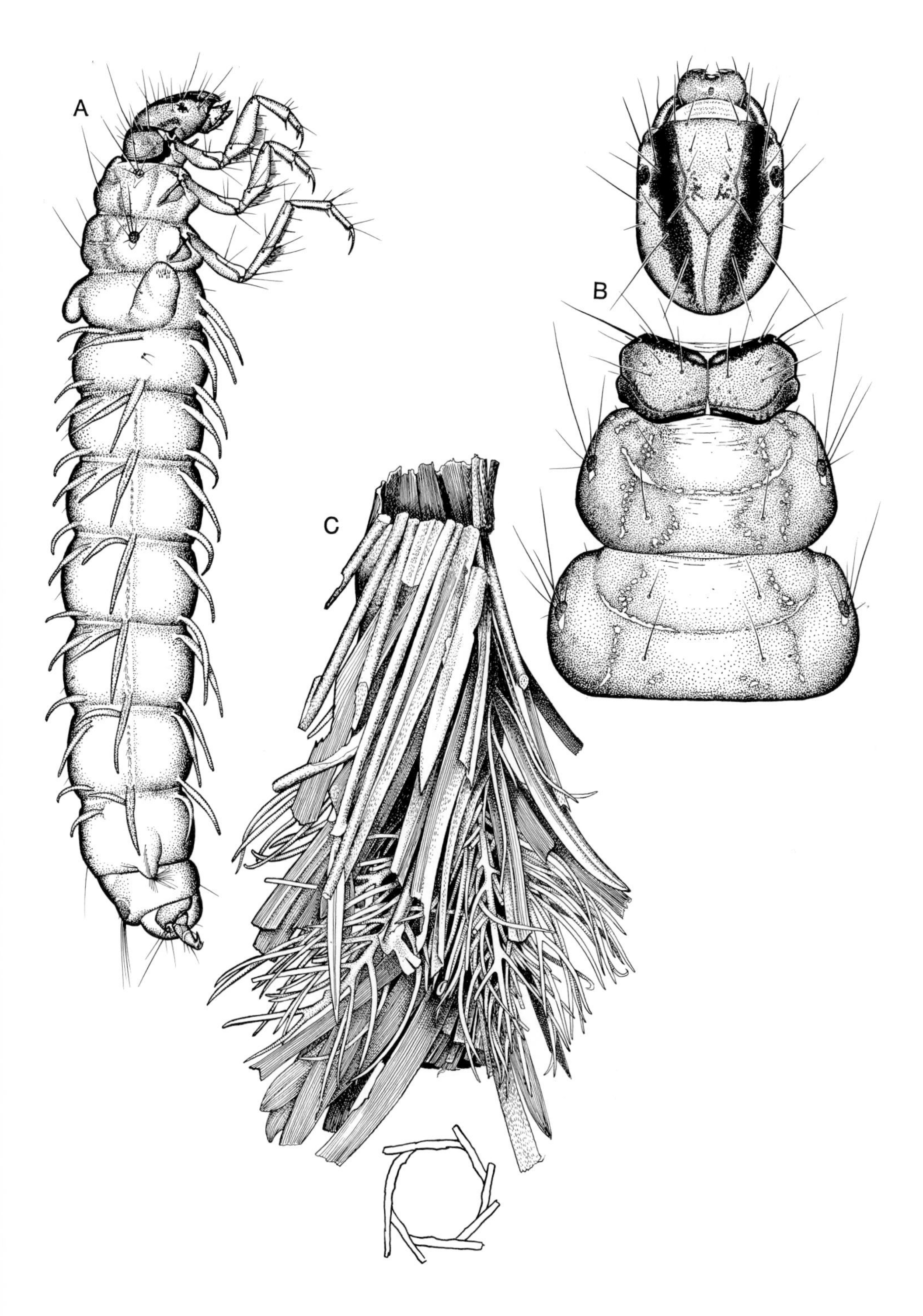
A
B
C

14.4 Genus **Hagenella**

DISTRIBUTION AND SPECIES *Hagenella* is a Holarctic genus with one species in Europe, three in Asia, and one in North America. The Nearctic species, *H. canadensis* (Banks), is known from New Brunswick and New Hampshire through Quebec, Ontario, and Michigan to Wisconsin and Minnesota.

No larva has yet been associated with adults of *H. canadensis*. The larva illustrated here and provisionally assigned to this species is one of two that we collected in Ontario. Since these are considerably different from those of any other North American phryganeids now known, and since their cases are of ring construction as are those of the European *H. clathrata* (Kol.) (Silfvenius 1904), there is a strong possibility that they belong to *Hagenella canadensis*.

MORPHOLOGY This larva is unique among those of Nearctic phryganeids now known in having the head and pronotum uniform brown and without contrastingly coloured markings (A,B); rounded muscle scars are evident on the head. Meso- and meta-notal setae in positions *sa*1 and *sa*2 have a small sclerotized area around the base; although there is some tendency for this in other genera, the sclerotized bases in this larva are considerably larger and darker than in the others, apart from *Oligostomis*. The ventral apotome of the head (D) is narrower than in other phryganeid genera; segment VIII is without gills. In other characters the larva is typical of the Phryganeinae. Length of larva illustrated 21 mm.

CASE The larval case (C) is constructed of leaf pieces arranged to form ring-like sections joined end to end. Length of case illustrated approximately 20 mm.

BIOLOGY Our single collection was taken in a small pond, apparently of floodwater origin from a nearby stream.

Hagenella canadensis? (Ontario, Nipissing Dist., Samuel de Champlain Prov. Park, 15 May 1972, ROM 720139)
A, larva, lateral x7; B, head and thorax, dorsal; C, case, lateral x5; D, head, ventral

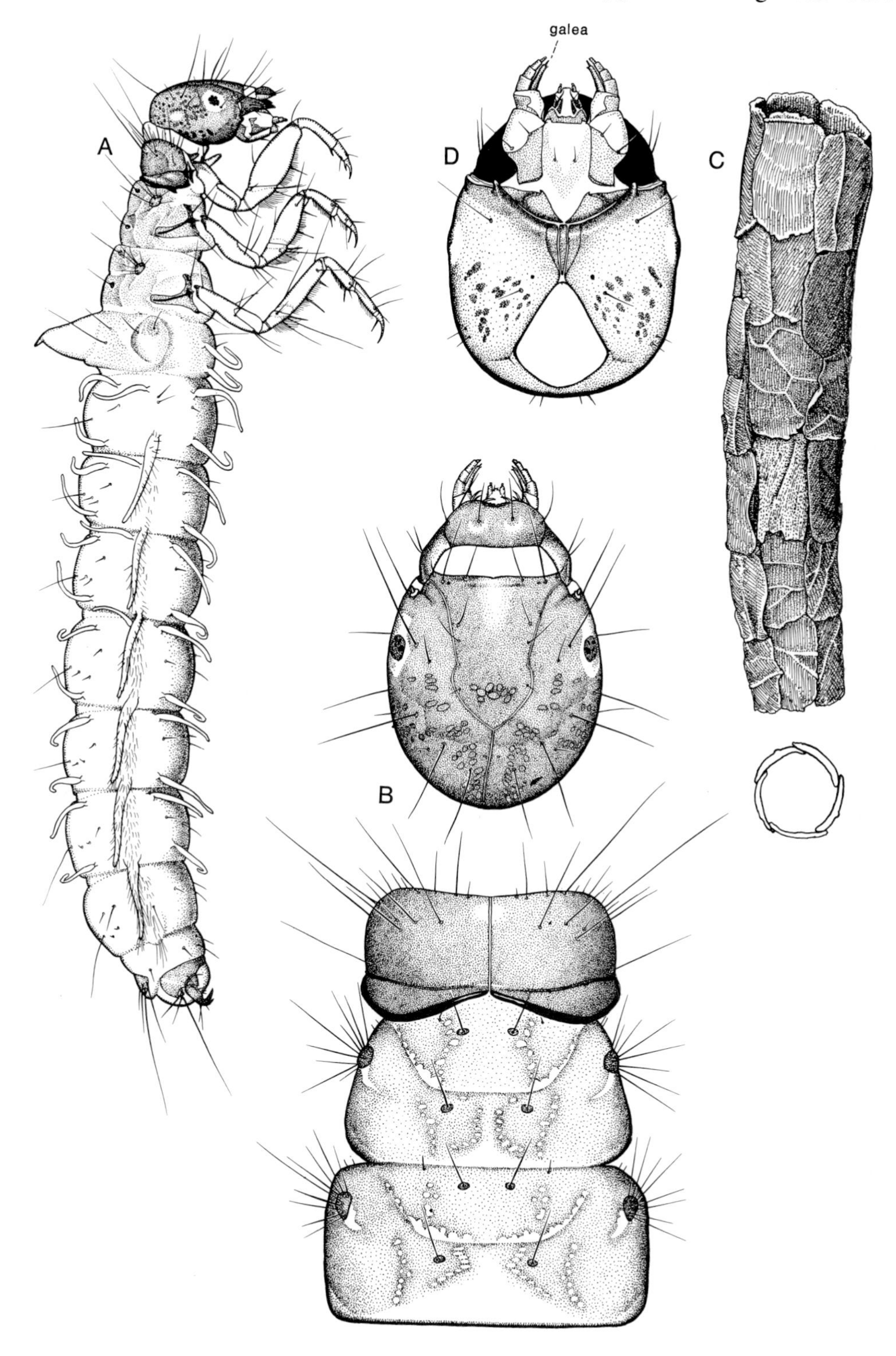
galea
A
D
C
B

14.5 Genus **Oligostomis**

DISTRIBUTION AND SPECIES This is a Holarctic genus, with the North American representatives numbering two species: *O. ocelligera* (Walker) from Wisconsin to Newfoundland, south to New Jersey; *O. pardalis* (Walker) from Ontario to Nova Scotia and Massachusetts.

Larvae of both species were described by Lloyd (1921, under *Neuronia, stygipes= ocelligera*). We have associated material of both species (Wiggins 1960b).

MORPHOLOGY Larvae of *Oligostomis* are distinctive from other North American phryganeids in having a rounded sclerite at the base of each mesonotal *sa*1 seta (B), the diameter of the sclerite approximately three times that of the *sa*3 sclerite. The pronotum has an oblique angular dark mark on each side, but the anterior margin is without dark markings. The largest seta at mesonotal *sa*2 and metanotal *sa*1 and *sa*2 frequently has a very small sclerotized area around the base. Abdominal gills are complete only on the first four segments, anterodorsal gills are absent from segments V to VII inclusive, and segment VIII lacks gills entirely. Length of larva up to 24 mm.

Larvae of the two species appear to differ only in size, but rather than the total length used previously to distinguish them (Ross 1944; Wiggins 1960b), a more effective index would probably be the width of the anterior border of the frontoclypeal apotome. We do not have sufficient material of proven identity to establish a range, but this measurement from exuviae of a reared female of each species is: *O. pardalis* 1.60 mm; *O. ocelligera* 1.13 mm.

CASE Larval cases in *Oligostomis* (C) are composed of discrete, ring-like sections of leaf and bark pieces joined together as in *Ptilostomis* (Fig. 14.8C). Length of larval case up to 50 mm.

BIOLOGY *Oligostomis* larvae live on bottoms of cool, forest streams, generally in sections of slow current where leaves accumulate. Larvae ingest detritus (Lloyd 1921), and guts of *O. pardalis* (3) that we examined contained animal remains and filamentous algae as well as some vascular plant tissue. Food of the European *O. reticulata* (L.) was found to be largely insects (Smirnov 1962).

REMARKS Although in the past frequently assigned to *Eubasilissa, O. pardalis* ought not to be placed in that genus which is confined to Asia (Wiggins 1960a).

Oligostomis pardalis (Ontario, York Co., 20 Oct. 1951, ROM)
A, larva, lateral x6; B, head and thorax, dorsal; C, case x3

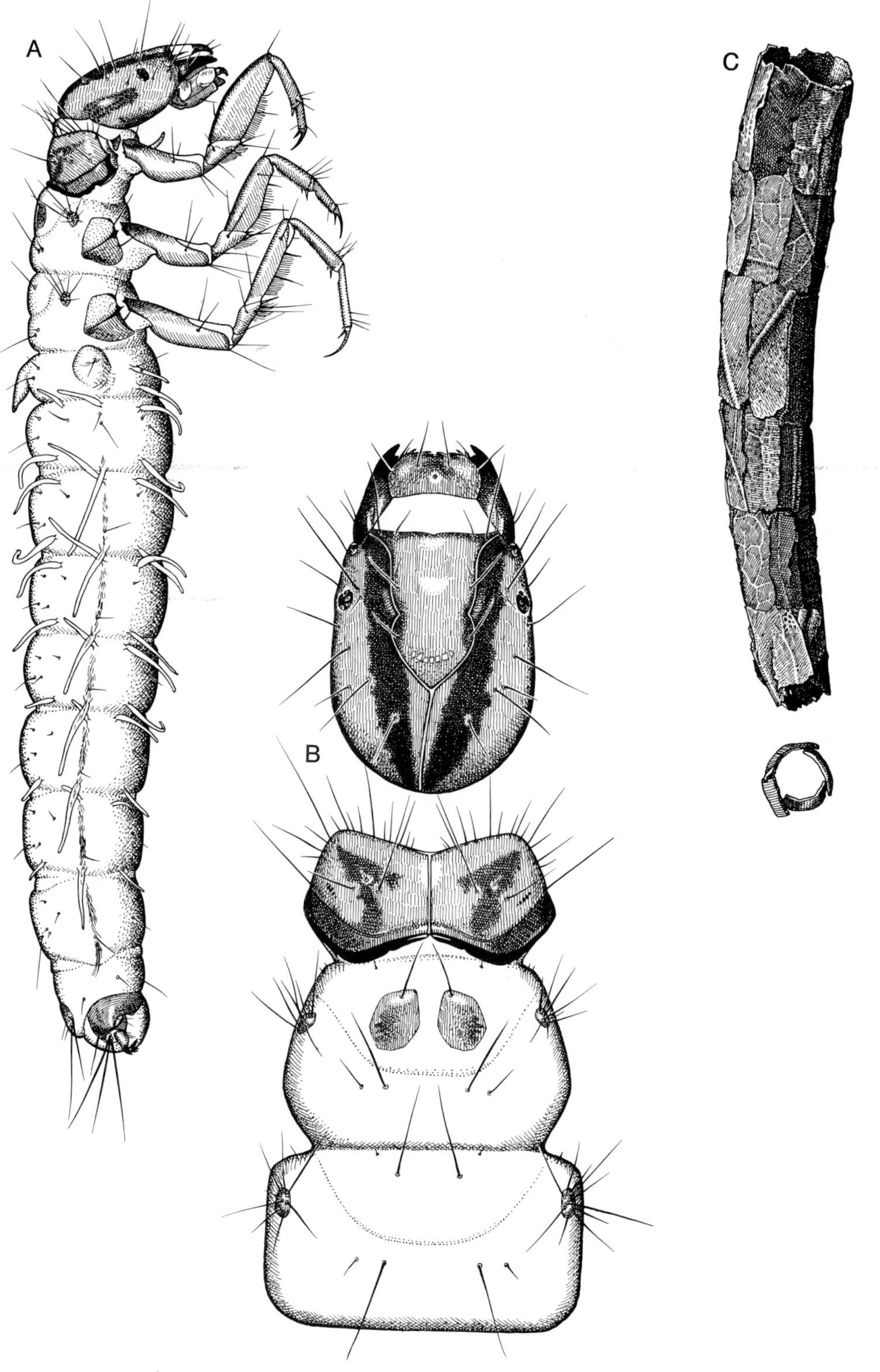
A
B
C

14.6 Genus **Oligotricha**

DISTRIBUTION AND SPECIES *Oligotricha* is entirely Palaearctic except for an extension of the boreal *O. lapponica* (Hagen) into western Alaska. Five species are known, three of them in Japan (Wiggins and Kuwayama 1971).

The larva of *O. lapponica* was described by Silfvenius (1905) and Lepneva (1966), among others.

MORPHOLOGY The larva of *O. lapponica* is similar to those of *Banksiola* species, but in so far as the latter are known, it can be distinguished by the presence of posteroventral gills on segment VII and absence of anterodorsal gills on segments VI and VII (A). Larvae of both *Oligotricha* and *Banksiola* are distinctive in the pair of parallel dark bands on the meso- and meta-nota. Since *Banksiola crotchi* (Fig. 14.2B), the only species of that genus in western North America, lacks a median dark band on the frontoclypeus, *O. lapponica* would be more distinctive among western species because this median dark band is present. Length of larva illustrated 18 mm.

CASE Larval cases of *Oligotricha* species are made of short lengths of plant pieces arranged spirally (C). In contrast to cases of *Banksiola*, no ends are left trailing. Length of larval case illustrated 28 mm.

BIOLOGY The habitat of *O. lapponica* and *O. striata* (L.) (syn. *O. ruficrus* (Scop.)) in Europe and Asia is slowly flowing or standing waters, often the brown water drainage ditches of peat deposits (Lepneva 1966). Food of the European *O. striata* was reported as largely insect larvae (Siltala 1907b).

REMARKS To my knowledge no North American larvae of this species have yet been identified. I am indebted to J.O. Solem, Museum, University of Trondheim, for providing the specimen illustrated here.

Oligotricha lapponica (Norway, 24 Aug. 1968, DKNVS Museet 3106)
A, larva, lateral x8; B, head and thorax, dorsal; C, case x3

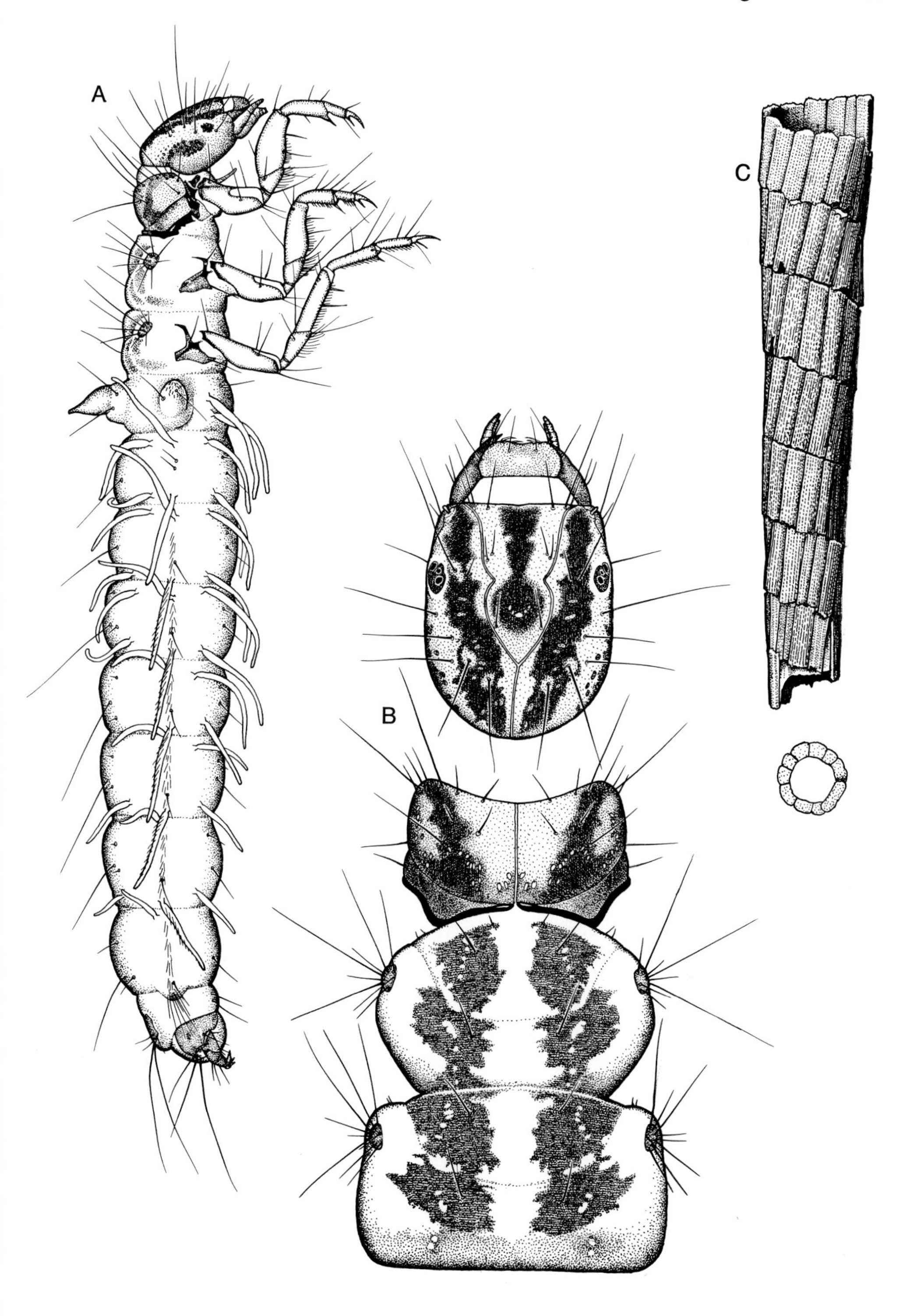
A
B
C

14.7 Genus **Phryganea**

DISTRIBUTION AND SPECIES *Phryganea* is widely distributed through the Holarctic region. Two species occur in North America: *P. cinerea* Walker extends from Newfoundland and Massachusetts to Alaska and California; *P. sayi* Milne (syn. *P. interrupta* Say) is an eastern species occurring from Maine through southern Ontario to Wisconsin, south through Missouri to North Carolina. The two species overlap in a broad belt from New Hampshire through the southern Great Lakes area.

We have associated larvae of both species (Wiggins 1960b), but I am still unable to distinguish between them structurally. The larva of *P. cinerea* was described by Neave (1933), who noted variation in markings of head and thorax, and the larva of *P. sayi* was described by Vorhies (1909); Lloyd's (1921) description of *P. cinerea* is not of that species, but probably pertains to *Banksiola* (Wiggins 1960b).

MORPHOLOGY Phryganeid coxal combs reach maximum development in *Phryganea* (E,F), their basal axes consistently transverse to the long axis of the coxa. This is the only phryganeid genus in North America in which the prosternal sternellum (Fig. 14.1F) is lacking. Length of larva up to 43 mm.

CASE Larval cases in this genus are normally constructed of short lengths of plant materials arranged to form a continuous spiral band (C,D). Larvae taken far from shore in the deepest parts of Lake Winnipeg where plant detritus is rare used animal parts for building cases (Neave 1933): ostracods, cladoceran ephippia, fish scales, sclerotized parts of insects, snails and their opercula. Many of the ostracods were affixed while alive. Length of larval case up to 56 mm.

BIOLOGY *Phryganea* larvae are most often found in marshes and lake margins, but *P. cinerea* has been recorded at depths of 20 m in Lake Winnipeg (Neave 1933), and down to 100 m in Lake Superior (Selgeby 1974). Larvae ingest dead and living plant, as well as animal, materials; the offshore larvae studied by Neave (1933) consumed oligochaetes, copepods, ostracods, cladocerans, and algae.

Phryganea cinerea (Ontario, Peel Co., 18 Nov. 1955, ROM)
A, larva, lateral x4; B, head and thorax, dorsal; C, case x2; D, diagram showing spiral structure of case; E, fore coxa, ventral; F, middle coxa, ventral, coxal comb enlarged

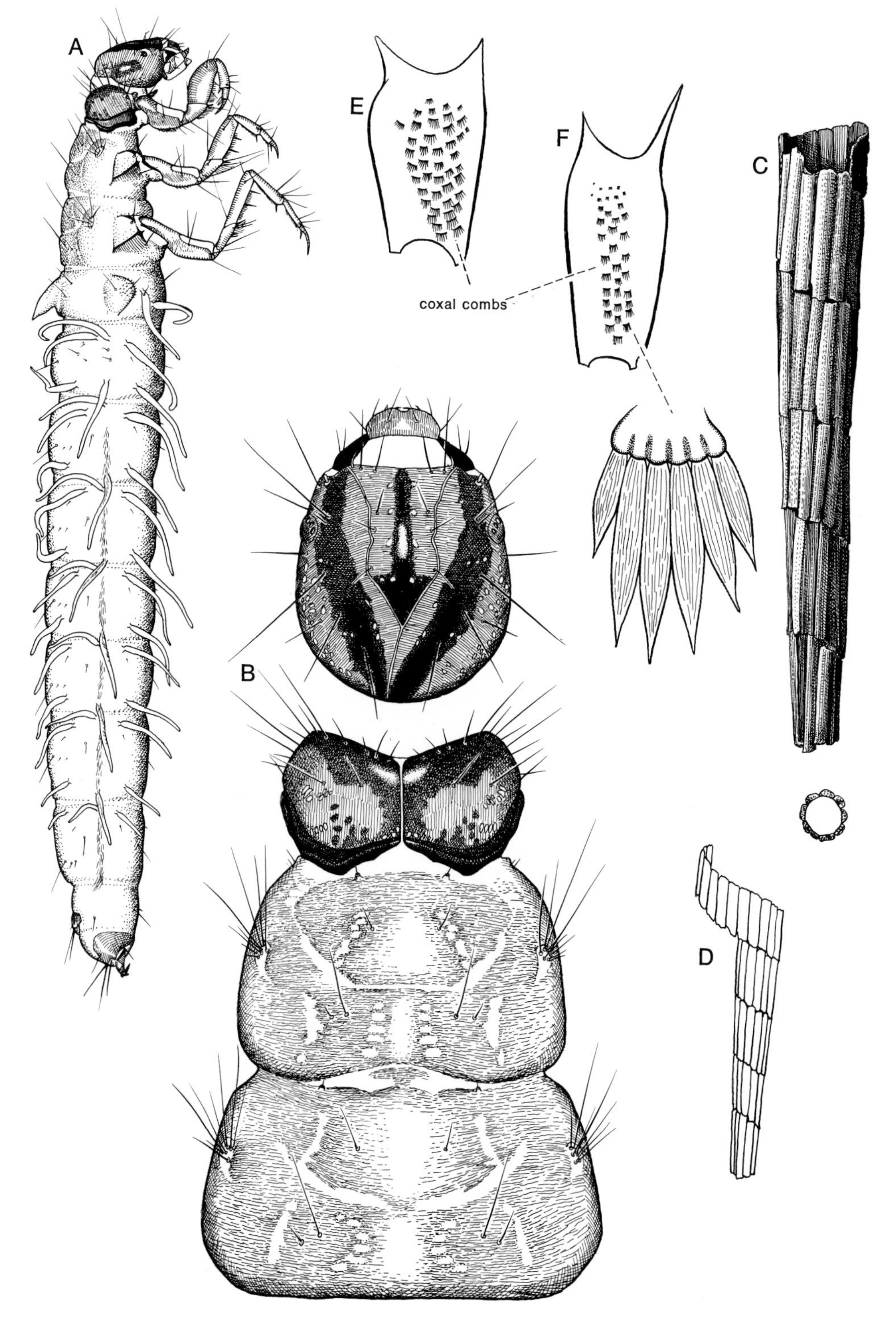
A
E
F
C
coxal combs
B
D

14.8 Genus **Ptilostomis**

DISTRIBUTION AND SPECIES *Ptilostomis* is a Nearctic genus of four species: *P. ocellifera* (Walker) and *P. semifasciata* (Say) are transcontinental; *P. postica* (Walker) and *P. angustipennis* (Hagen) are known only from the eastern half of the continent.

The larva of *P. ocellifera* was described by Vorhies (1909, as *Neuronia postica*) and Ross (1944). We have associated larvae for *P. ocellifera* and *semifasciata* (Wiggins 1960b) and for *P. postica* but diagnostic characters have not been found for separation of any of these species.

MORPHOLOGY *Ptilostomis* larvae (B) have a dark transverse band on each side of the pronotum, separated from the anterior border by an area of light ground colour of approximately the same length. Tibiae of the fore and middle legs are broad distally, much as in *Banksiola*; at magnifications of approximately 50x the coxal combs in *Ptilostomis* are seen only as tiny points spread over much of the ventral surface of the coxa (E). Abdominal gills of the first six segments are complete, on segment VII the anterolateral gill is variable, and on VIII all three anterior gills are present although the anterodorsal gill is variable (A). Other structures are typical for the Phryganeinae. Length of larva up to 35 mm.

CASE Larval cases in *Ptilostomis* are of ring-like sections of leaf pieces joined end to end (C,D). Length of larval case up to 60 mm.

BIOLOGY We have taken *Ptilostomis* larvae in both lentic and lotic waters; *P. ocellifera* has a particularly wide range of habitats, ranging from cool streams to lakes and temporary vernal pools. There is a single generation per year; final instars of *P. ocellifera* were present in the littoral region of a British Columbia lake from November to June (Winterbourn 1971a) with late instars largely predacious in feeding.

Ptilostomis ocellifera (Ontario, York Co., 16 May 1958, ROM)
A, larva, lateral x6; B, head and thorax, dorsal; C, case x2; D, diagram showing ring structure of case; E, fore coxa, ventral, showing minute coxal combs

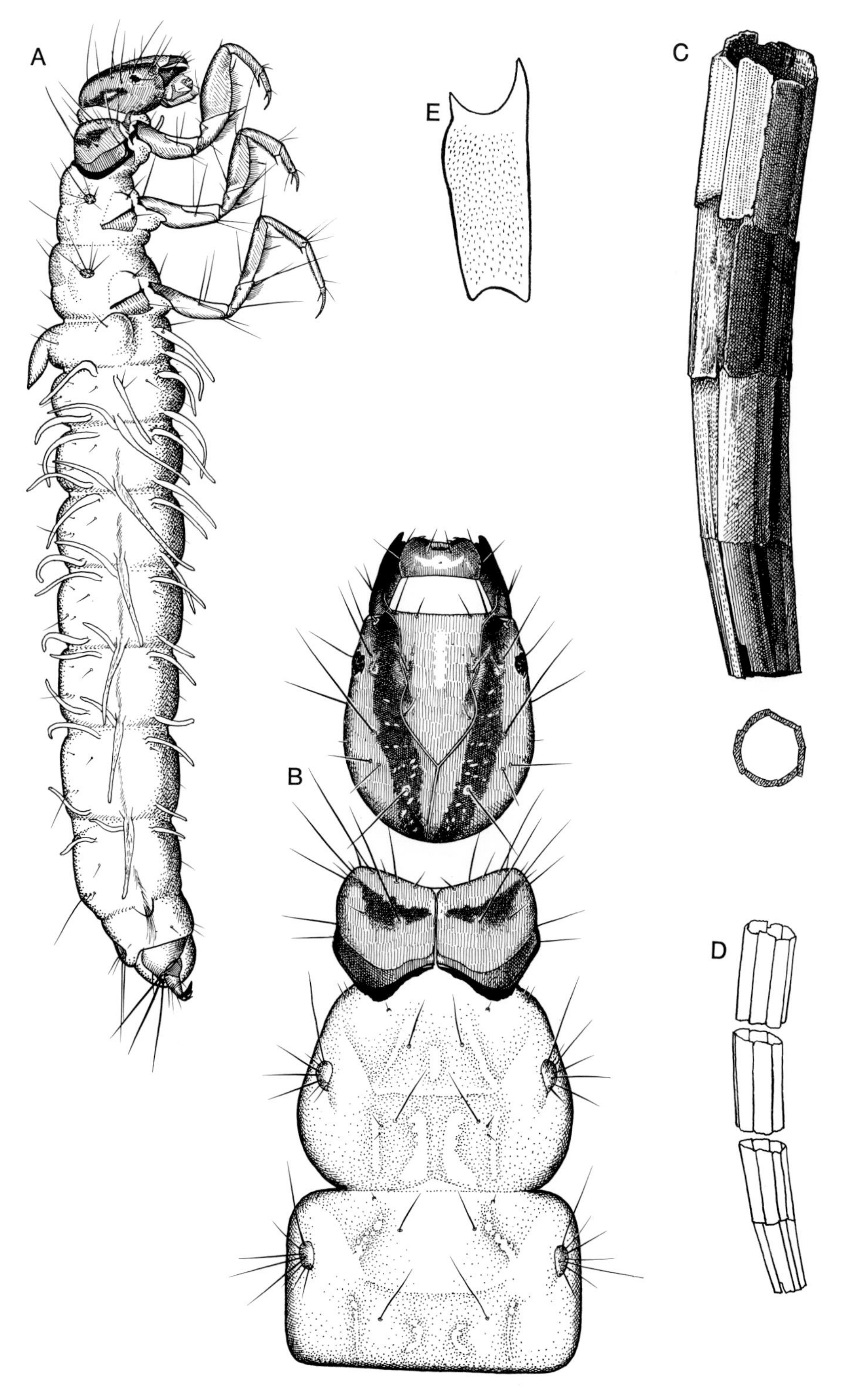
A
B
C
D
E

14.9 Genus **Yphria**

DISTRIBUTION AND SPECIES In this remarkable Nearctic genus there is a single species, *Y. californica* (Banks), recorded from California and Oregon. All stages were described and the systematic relationships of the species assessed by Wiggins (1962); *Y. californica* is the sole member of the subfamily Yphriinae, and the most primitive living phryganeid.

MORPHOLOGY The larva is unique among all of the Phryganeidae in having the ventral apotome reduced to a small anterior sclerite (D). *Yphria* larvae are further distinguished by enlarged mesonotal *sa*1 sclerites contiguous mesially (B); mesonotal *sa*2 sclerites are present but small and separate. Meso- and meta-notal *sa*3 sclerites are more ovoid than circular as they are in the phryganeine genera. Length of larva up to 22 mm.

CASE The larval case is decidedly unlike that of any other North American phryganeid, nor is it likely to be confused with cases in any other genus. Rock fragments, bark, and twigs are used, but the rock pieces are placed anteriorly for the most part, extending posteroventrally (C) in a way that concentrates the weight of the case ventrally in accord with its orientation by the larva. The diameter of the posterior opening of the case is not reduced with silk. Before they pupate, *Yphria* larvae construct an entirely new and more flexible case of mica-like mineral fragments (Wiggins 1962, fig. 7). Length of larval case up to 33 mm.

BIOLOGY We have collected larvae of *Y. californica* from many localities, but always on bottom materials in pools and along the margin of small, cool mountain streams. Larvae move with exceptional agility, and when picked up, their immediate reaction is to leave their cases. The rate of development is variable, but collection of early to late instar larvae, adults, and eggs in June suggests a life cycle of more than one year. Larvae burrow in sandy sediments for pupation. Gut contents of larvae (3) indicate that at least the late instars are largely predacious.

Yphria californica (Oregon, Deschutes Co., 23 April 1964, ROM)
A, larva, lateral x8; B, head and thorax, dorsal; C, case, lateral x4; D, head, ventral, ventral apotome enlarged; E, mandibles, dorsal

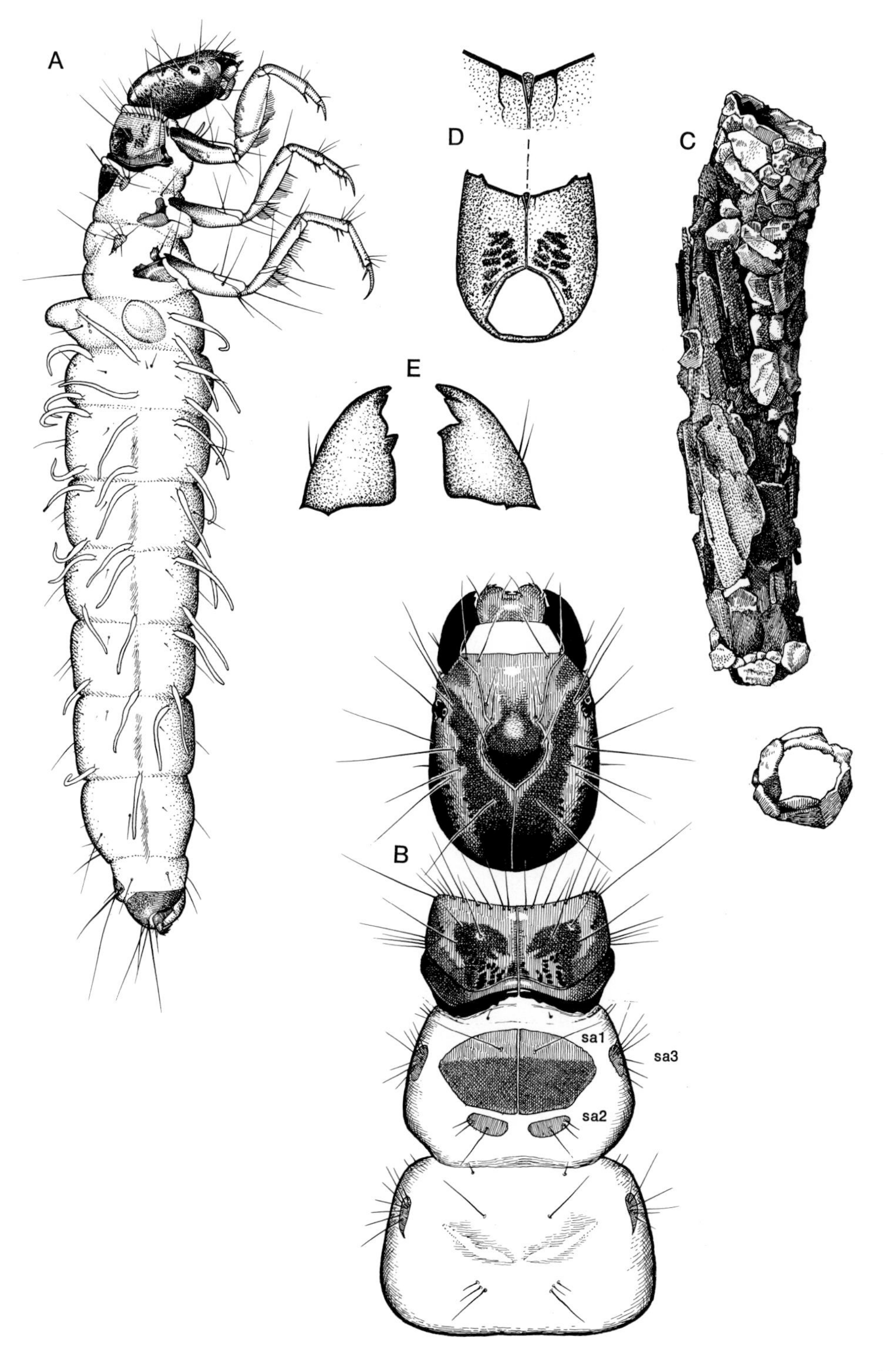
A
D
C
E
B
sa1
sa3
sa2

15
Family Polycentropodidae

The Polycentropodidae are one of the important families of net-spinning Trichoptera in all major faunal regions. The seven Nearctic genera comprise nearly 70 species north of Mexico; the family is widespread over much of the continent.

In general structure this family shows some similarity to the Psychomyiidae, and in fact the Polycentropodidae have often been treated as a subfamily of the Psychomyiidae in North American works (e.g. Ross 1959; Flint 1964a). The Polycentropodidae are most readily distinguished by the pointed fore trochantin, fused without a separating suture to the episternum (Fig. 15.2A). In most Nearctic polycentropodid larvae the tip of the labium is short, extending beyond the anterior margin of the head little if at all, but in *Phylocentropus* the tip of the labium is long; labial palpi are present (Fig. 15.5C) though small, but they are absent in *Phylocentropus*; the galeae are elongate, but flattened in *Phylocentropus*; submental sclerites are fused, and the ventral apotome is short and broad. Only the pronotum is sclerotized, and all three pairs of legs are approximately the same size; tarsal claws are without a basal process. The abdomen usually has a lateral fringe, and larvae ventilate by abdominal undulation; anal papillae are present in at least some genera. The basal segment of the anal proleg is at least as long as the distal, usually longer; the anal claw is well developed, tooth-like points on its concave edge providing useful taxonomic distinction for some genera.

Larval retreats are fixed in position but are much more diverse than in any of the other net-spinning families. In certain genera, such as *Neureclipsis* and some *Polycentropus*, the trumpet-like net is dependent on a gentle current for its effectiveness as a filter; *Phylocentropus* larvae construct long tubes in bottom sediments, and feed on organic particles filtered from the current. Apart from these, larval retreats in the other North American polycentropodid genera are essentially short tubes open equally at each end; these larvae are largely predacious, and in some *Polycentropus* at least, feeding behaviour suggests that prey are detected by their vibrations on outlying silken strands of the retreat, as they are by some spiders. These larvae, with their ability to ventilate the body surface by abdominal undulation, have essentially the same means for creating respiratory currents within their fixed tubes as do other larvae in portable tubes. Probably this is why larvae of some genera in the Polycentropodidae occur regularly in lakes and ponds and, with the possible exception of the psychomyiid *Tinodes*, are the only ones among the North American net-spinning Trichoptera that can exploit lentic habitats in the absence of water circulation induced by wave action.

Five subfamilies are recognized in the Polycentropodidae. The Hyalopsychinae and Pseudoneureclipsinae are not represented in the New World. The aberrant *Phylocentropus* was recently placed in the Dipseudopsinae (Ross and Gibbs 1973), a group otherwise of the Old World tropics and recognized by some authors as a distinct family (Malicky 1973). All other Nearctic genera are placed in the Polycentropodinae. The Ecnominae, although not recorded from the Nearctic region, are represented by a Neotropical genus *Austrotinodes* which extends into southern Mexico (Flint 1973); the larva has been described by Flint (op. cit.). Some authors treat the Ecnominae as a separate family (Lepneva 1964; Malicky 1973).

No larva has yet been positively identified in the Nearctic genus *Cernotina*, although a possible candidate from Puerto Rico was described by Flint (1968b; 1964b, fig. 8A-H). *Cernotina* occurs in the Caribbean islands and Mexico, but to the north is restricted to the eastern half of North America where seven species are now recognized (Ross 1951c).

Key to Genera*

1 Tarsi of all legs broad and flat (Fig. 15.4A); larvae construct branching silken tubes in sand deposits of streams (Fig. 15.4D). Widespread in eastern and central areas (subfamily Dipseudopsinae) **15.4 Phylocentropus**

Tarsi of all legs normal, not broad and flat (Fig. 15.1A); larvae construct silken shelters of various shapes on substrates (Figs. 15.2D, 15.3E) (subfamily Polycentropodinae) **2**

2 (1) Anal claw with several conspicuous pointed teeth arising from ventral, concave margin (Figs. 15.3A, 15.6A) **3**

Anal claw lacking conspicuous pointed teeth on ventral concave margin (Fig. 15.5D), or with row of many tiny spines (Fig. 15.2A) **4**

3 (2) Teeth on anal claw much shorter than apical hook, accessory hook present (Fig. 15.3A); pronotum with short, stout bristle arising near each ventrolateral margin (Fig. 15.3C). Widespread over much of continent, but not in southwest **15.3 Nyctiophylax**

Teeth on anal claw almost as long as apical hook, accessory hook absent (Fig. 15.6A); pronotum without short bristle arising near each ventrolateral margin. Southwestern **15.6 Polyplectropus**

4 (2) Basal segment of anal proleg approximately same length as distal segment and largely without setae, concave margin of anal claw with row of many tiny spines (Fig. 15.2A,C). Widespread in eastern and central areas **15.2 Neureclipsis**

*Larvae are not yet known for the genus *Cernotina*; see above.

Basal segment of anal proleg distinctly longer than distal segment and bearing many setae, concave margin of anal claw without tiny spines (Fig. 15.1C) **5**

5 (4) Dorsal plate between claw and lateral sclerite of anal proleg with two dark bands contiguous mesially (Fig. 15.5E); meso- and meta-notal *sa*1 seta not more than one-third as long as longest *sa*2 seta (Fig. 15.5B). Widespread **15.5 Polycentropus**

Dorsal plate between claw and lateral sclerite of anal proleg with two dark bands not contiguous mesially (Fig. 15.1D); meso- and meta-notal *sa*1 setae approximately same length as longest *sa*2 setae (Fig. 15.1B). Eastern and central **15.1 Cyrnellus**

15.1 Genus Cyrnellus

DISTRIBUTION AND SPECIES *Cyrnellus* is a New World genus, up to now with only a single Nearctic species *C. fraternus* (Banks) (syn. *C. marginalis* (Banks)) which is recorded from Minnesota to Texas and eastward (Ross 1944; Flint 1964a); the species is evidently much more common in southeastern and mid-western areas than in northeastern. *C. fraternus* has also been recorded from the Amazon River in South America (Ross 1944), along with several others of the genus (Flint 1964a).

Identity of *Cyrnellus* larvae was established by Ross (1959), and the larva of *C. fraternus* described by Flint (1964a).

MORPHOLOGY Distinction between larvae of *Cyrnellus* and *Polycentropus* has always been difficult, and all the more so since our material of *Polycentropus* indicates that muscle scars of the head can be lighter than the base colour as well as darker as stipulated in other diagnoses (e.g. Flint 1964a). But up to now in all *Polycentropus* larvae that we have found, the two dark, sclerotized bands of the dorsal plate between the claw and lateral sclerite of the anal proleg touch to form an X-shaped sclerite. In *C. fraternus* these two dark, sclerotized bands do not touch (D); the anal claw lacks teeth on the concave, ventral surface (C), as it does in *Polycentropus*. Length of larva up to 9 mm.

RETREAT Larval retreats of specimens we collected in Tennessee (E) were somewhat similar to those of *Nyctiophylax*. A depression in a rock is covered by a flattened silk roof, roughly circular in outline and covered with fine organic particles; round front and rear exit holes open through the flat roof. The larval chamber beneath has a silk floor and the larva can reverse its position within. Viewed from above, the silk roof of the retreat has a diameter of approximately 20 mm.

BIOLOGY Although *C. fraternus* shows a marked preference for large rivers, it occurs in a wide range of waters from smaller streams to lakes and reservoirs (Ross 1944; Flint 1964a). We collected the larval series illustrated on rocks in water approximately 60 cm deep along the edge of an impoundment (Old Hickory Reservoir, Cumberland R., Tenn.) and larvae were also collected at that locality in the river below the dam at a depth of 15 ft or 4.6 m (Flint 1964a). Larval guts (3) we examined all contained fine organic particles, with arthropod remains in one.

Cyrnellus fraternus (Tennessee, Davidson Co., 13 May 1970, ROM 700335)
A, larva, lateral x20, tibiae and tarsi enlarged; B, head and thorax, dorsal; C, anal proleg, lateral; D, dorsal plate of anal proleg, dorsal; E, diagram of retreat on rock based on field notes, approx. x1.4

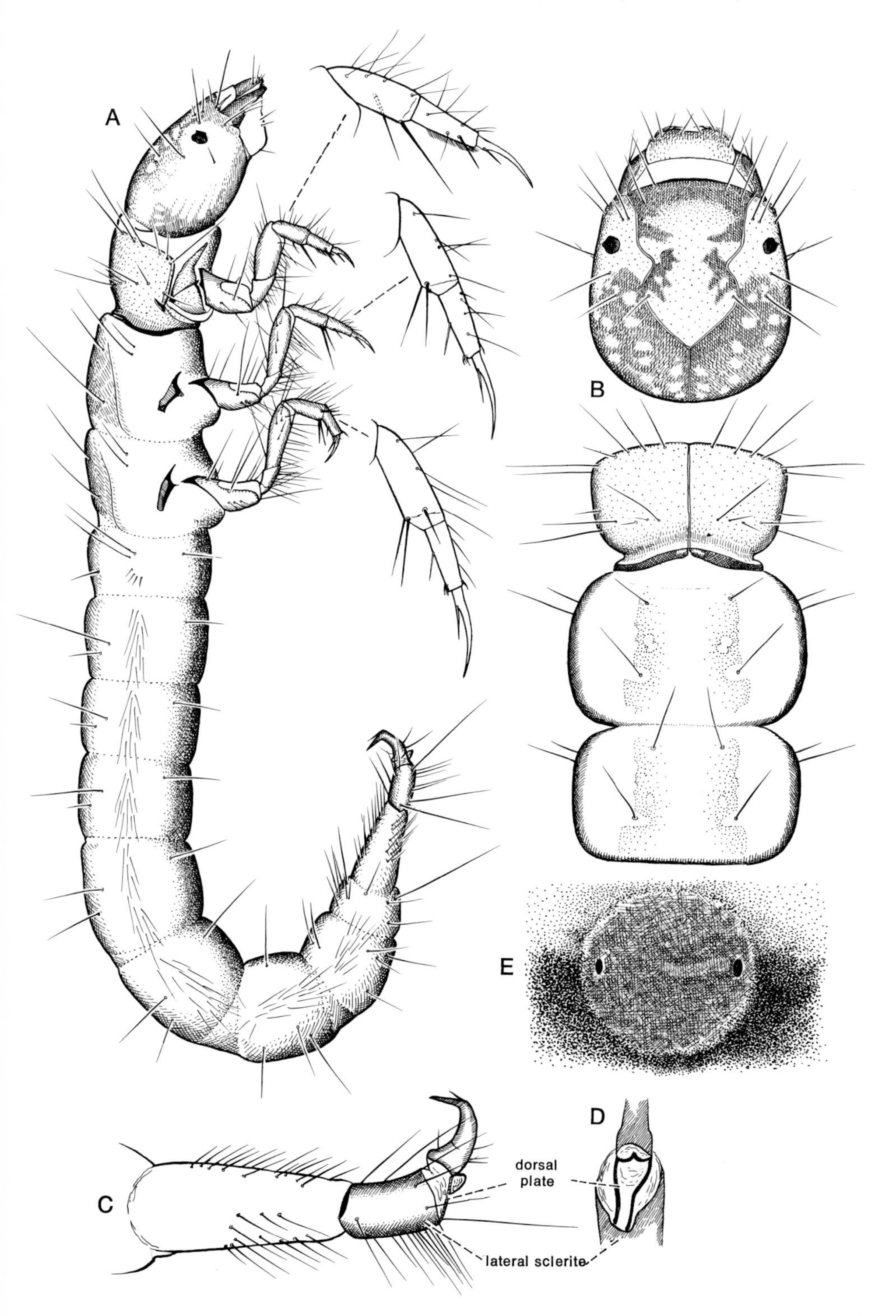
A
B
E
C
D
dorsal plate
lateral sclerite

15.2 Genus Neureclipsis

DISTRIBUTION AND SPECIES *Neureclipsis* is a Holarctic genus. Five species are known in North America, and all occur in the eastern half of the continent; but one of them, *N. bimaculata* (L.), also extends to British Columbia and through northern Europe and Asia.

Larvae have been described for *N. crepuscularis* (Walker) by Ross (1944) and for *N. bimaculata* (L.) by Lepneva (1964). We have associated larval material for both these species and for *N. valida* (Walker).

MORPHOLOGY *Neureclipsis* larvae are distinguished by a basal segment of the anal proleg that is approximately the same length as the distal segment, and without setae (A,C) although a few may arise at the distal end where the basal and distal segments join. The concave, ventral edge of the anal claw bears a row of fine spines (A). Larvae of *N. bimaculata* have a pair of stout bristles on the venter of segment IX (C). Length of larva up to 21 mm.

RETREAT The larval retreat in this genus is a distinctive trumpet-shaped net of silk (D) that tapers to a slender tube where the larva is concealed. Guy lines of silk support the large opening that is exposed to the current, and food particles are filtered from the water. Net-spinning behaviour of *N. bimaculata* in Germany was studied in detail by Brickenstein (1955). Length of nets at least up to 12 cm, excluding recurved posterior end, 3 to 4 cm diameter at the anterior end.

BIOLOGY Filter feeding of *Neureclipsis* larvae obliges them to live in running waters, but generally they are in slower currents since the nets are relatively large and would probably be difficult to stabilize in stronger currents. On rooted aquatic plants and submerged tree branches, the nets are often abundant and staged one below the other from just beneath the surface to the bottom substrate, exploiting the full potential of the water column. Larvae feed on animals and algae (Lepneva 1964; Noyes 1914); each of three larvae we examined contained arthropod remains.

Neureclipsis bimaculata (South Dakota, Yankton Co., 1968, ROM)
A, larva, lateral x9, propleuron, fore tarsus, and anal claw enlarged; B, head, pro- and meso-notum, dorsal; C, segment IX and anal prolegs, ventral; D, *N. bimaculata* larval capture net and retreat (after Brickenstein 1955), arrow indicating direction of current

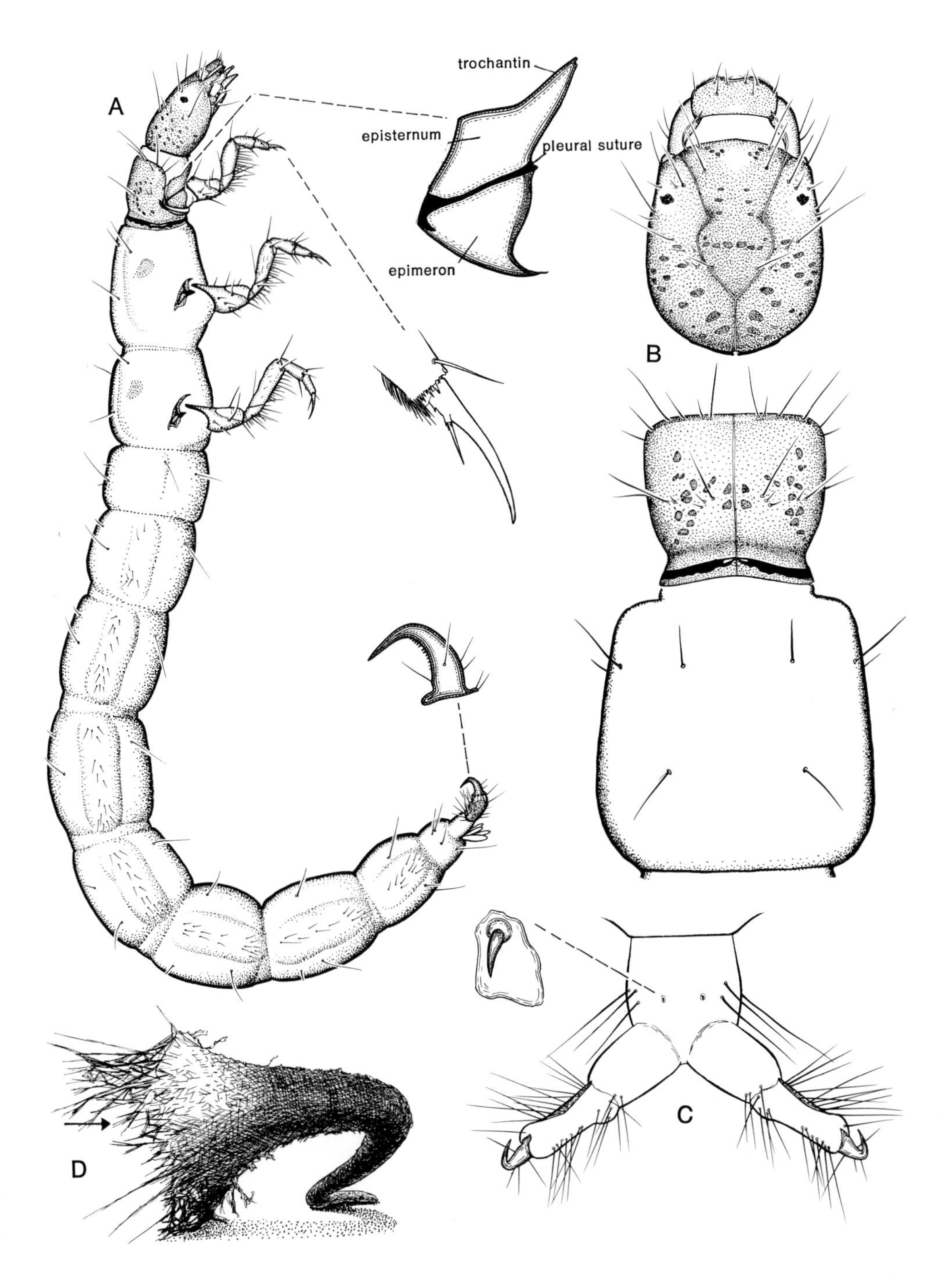
A
trochantin
episternum
pleural suture
epimeron
B
C
D

15.3 Genus Nyctiophylax

DISTRIBUTION AND SPECIES Judging from current listings, *Nyctiophylax* is recorded from all faunal regions except the European Palaearctic (Fischer 1960-73). Eight species have been identified in North America, some confined to the southeastern United States, others widespread through the eastern half of the continent extending westward to British Columbia and Oregon. Taxonomy of the North American species has been extensively revised by Morse (1972).

Generic characters were summarized by Flint (1964a), and larvae have been described for three North American species: *N. celta* Denning by Ross (1944, as Psychomyiid Genus B) and by Flint (1964a, as *N. vestitus* (Hagen)); *N. moestus* Banks by Noyes (1914, as *Cyrnus pallidus*?), by Ross (1944, 1959, as Psychomyiid Genus A), and by Flint (1964a, as *Nyctiophylax* sp. A); *N. nephophilus* Flint by Flint (1964a). We have associated larvae for *N. affinis* (Banks).

MORPHOLOGY Larvae of this genus are distinguished by short teeth on the ventral, concave surface of the anal claw (A), often best seen on the mesial face; and by a stout lateral bristle on each side of the pronotum (C). Small spines on venter IX are arranged in a T-shaped field, and these also occur on the venter of the basal segment of the anal prolegs (D). The anal claw bears an accessory hook. Length of larva up to 10 mm.

RETREAT Larval retreats of *Nyctiophylax* species (E) consist of a silken roof over a depression in a piece of wood or a rock, enclosing a cylindrical chamber open at each end; the silken floor is extended beyond the roof at each end as a threshold of threads. A covering of silt and diatoms makes the shelter difficult to detect. According to Noyes (1914) each opening can be closed by a flap-like extension of the silken roof, and a loose network of threads floats up from the threshold; the larva darts out from the retreat to capture any small creature causing the threshold threads to move. Larvae pupate within the retreat. Length of larval retreat up to at least 15 mm.

BIOLOGY *Nyctiophylax* larvae live in lakes and streams, but in running waters they are generally to be found in pools and sections with reduced current. Larvae are almost entirely carnivorous (Coffman et al. 1971).

Nyctiophylax sp. (Arkansas, Washington Co., 5 May 1970, ROM 700306)
A, larva, lateral x15, anal claw enlarged; B, head and thorax, dorsal; C, right margin of pronotum and pleuron, dorsal; D, segment IX and bases of anal prolegs, ventral
E, *N. affinis* (Ontario, Muskoka Dist., May 1975) larval retreat on wood substrate

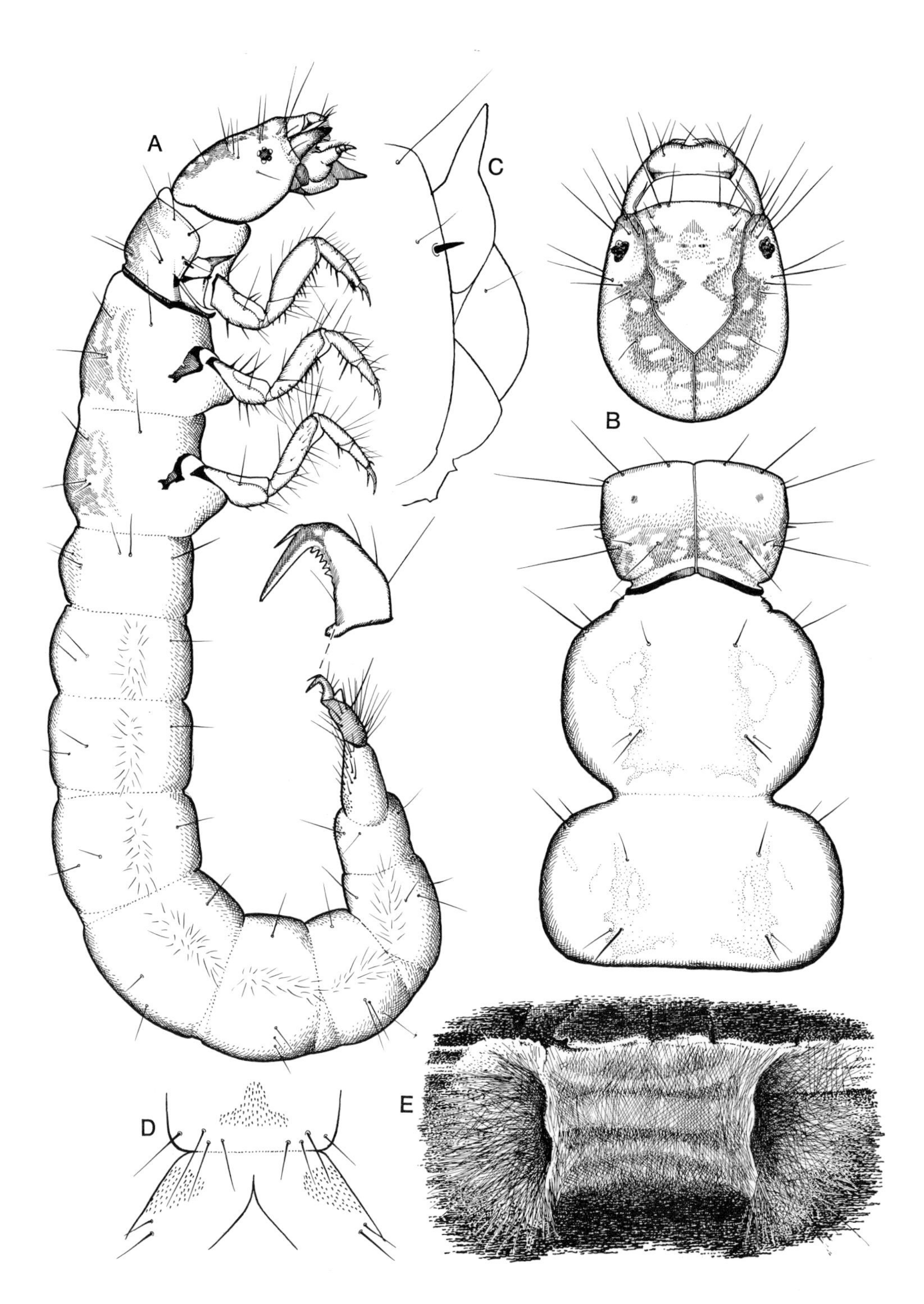
A
B
C
D
E

15.4 Genus **Phylocentropus**

DISTRIBUTION AND SPECIES The genus *Phylocentropus* comprises five species in North America and a single one from Malaysia; the North American species occur within the eastern half of the continent (Root 1965; Ross 1965).

Larvae have been described for two species: *P. lucidus* (Hagen) by Sibley (1926); and *P. placidus* (Banks) by Vorhies (1909, as *P. maximus*). We have associated material for these two and for *P. carolinus* Carpenter.

MORPHOLOGY *Phylocentropus* larvae are chiefly distinguished by characters of the legs (A): all tarsi are flattened and paddle-like; tarsal claws of the middle and hind legs are short and stout; tibiae of all legs are short. The tip of the labium is very long, probably an asset in applying silk to the dwelling tubes. Length of larva up to 20 mm.

RETREAT Larvae of this genus construct branching tubes of silk covered with sand and small pieces of detritus (D) in sedimentary deposits along margins of sandy streams and lakes. The upstream end of the tube extends above the substrate, and descends more or less vertically into the sand. Diverging from this main tube at an angle of approximately 90° is a lateral branch that opens at the surface, usually downstream from the first opening, and bears a bulbous portion near its junction with the main tube; within the bulb is a sack-shaped capture net of silk strands randomly arranged (Wallace et al. 1976). Tubes are often found with more than one bulbous side-branch, but usually only the upper one is functional and the others are sealed off at the base of their junction with the main tube. Total length of a single main tube up to 16 cm.

BIOLOGY The larva is positioned within the main vertical tube between the open end and the lateral branch (Wallace et al. 1976). Observations on similar tubes of the African *Protodipseudopsis* (Gibbs 1968) indicate that a current of water enters through the upstream opening and passes through the bulb to the downstream surface opening; abdominal undulation by the larva probably provides much of the current propulsion within the tube. *Phylocentropus* larvae ingest mainly fine particles with some vascular plant pieces and diatoms (Wallace et al. 1976); it is presumed that the larva enters the side branch periodically to remove these particles from the net, and setal brushes on the mandibles (C) and maxillae are probably adapted for this function.

Phylocentropus carolinus (South Carolina, Oconee Co., 18-19 May 1970, ROM 700350)
A, larva, lateral x14, fore and middle tarsi and claws enlarged; B, head, pro- and mesonotum, dorsal; C, mandibles, ventral; D, dwelling tube in sediment, ends projecting, x2, arrows indicating direction of current

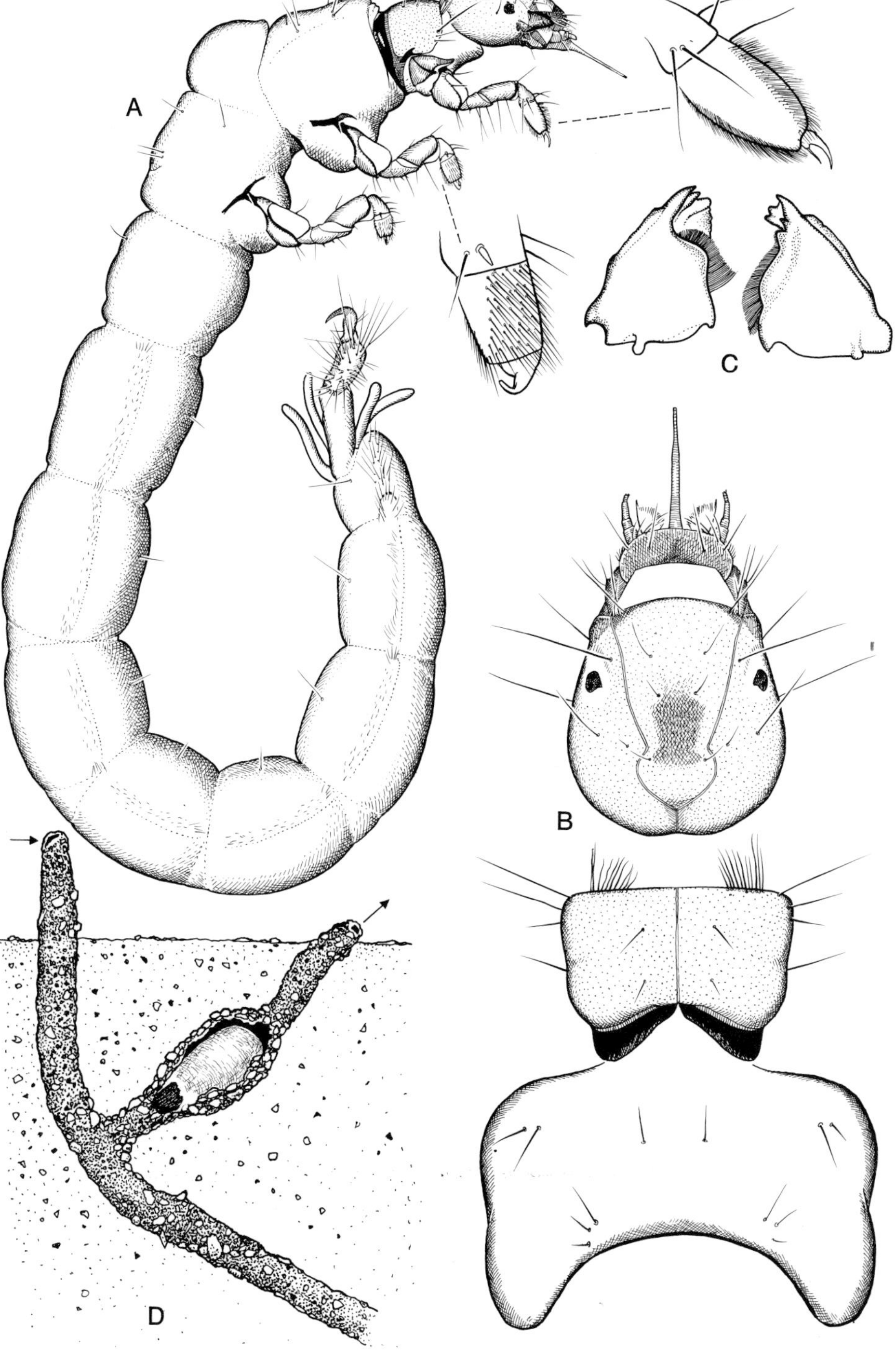
A
B
C
D

15.5 Genus **Polycentropus**

DISTRIBUTION AND SPECIES Limits applied to this genus in North America are somewhat broader than elsewhere, for together with *Polycentropus* sens. str., there are also on this continent species in groups recognized by European and some other workers as the genera *Holocentropus* and *Plectrocnemia*. *Polycentropus* sens. lat. occurs in all faunal regions except the Australian. Approximately 40 species are now known north of Mexico, and the group occurs throughout most of the continent.

Larval stages have been described for six species by Ross (1944); we have evidence for larval identity in an additional six species.

MORPHOLOGY Mandibular characters (Ross 1944, 1959) were found to be inadequate for diagnosis of *Polycentropus* sens. lat. (Flint 1964a); and some of our collections from western North America indicate that muscle scars of the head can be lighter than the base colour in addition to darker as stipulated in some diagnoses (e.g. Flint 1964a). Characters of tibial spurs in the specimen illustrated also differ in other species. Proposed tentatively here as a diagnostic character for *Polycentropus* is the X-shaped conformation of the two dark bands of the dorsal plate of the anal proleg (D,E). Length of larva up to 25 mm.

RETREAT For North American larvae assigned to *Polycentropus* sens. lat., we have found two types of retreats. One is a loosely constructed silken tube with one or both ends flared out into an irregular maze of silken threads; in captivity, larvae rest in the tube, quickly darting out to seize any small creature that moves across the maze. In a similar net, illustrated and described by Noyes (1914) (F), the tube was 21 mm long. The other type is a bag-like structure expanded by the current, similar to that illustrated for *P. flavomaculatus* (Pictet) by Wesenburg-Lund (1911).

BIOLOGY North American larvae of *Polycentropus* live in most types of freshwater habitats, and are better able to cope with warm lentic conditions than others in the family. This is the only genus of retreat-makers represented in temporary vernal pools (Wiggins 1973a); I have also taken them from lake sediments at depths of 5 m in association with *Chironomus* larvae. They are primarily predacious (Coffman et al. 1971; Winterbourn 1971a).

Polycentropus sp. (Ontario, Muskoka Dist., 16 Sept. 1967, ROM)
A, larva, lateral x12, fore tibia and tarsus enlarged; B, head, pro- and meso-notum, dorsal; C, maxillae and labium, ventral, labial palp enlarged; D, anal proleg, lateral; E, dorsal plate of anal proleg, dorsal
F, *Polycentropus* sp. larval retreat, approx. x1.6 (after Noyes 1914)

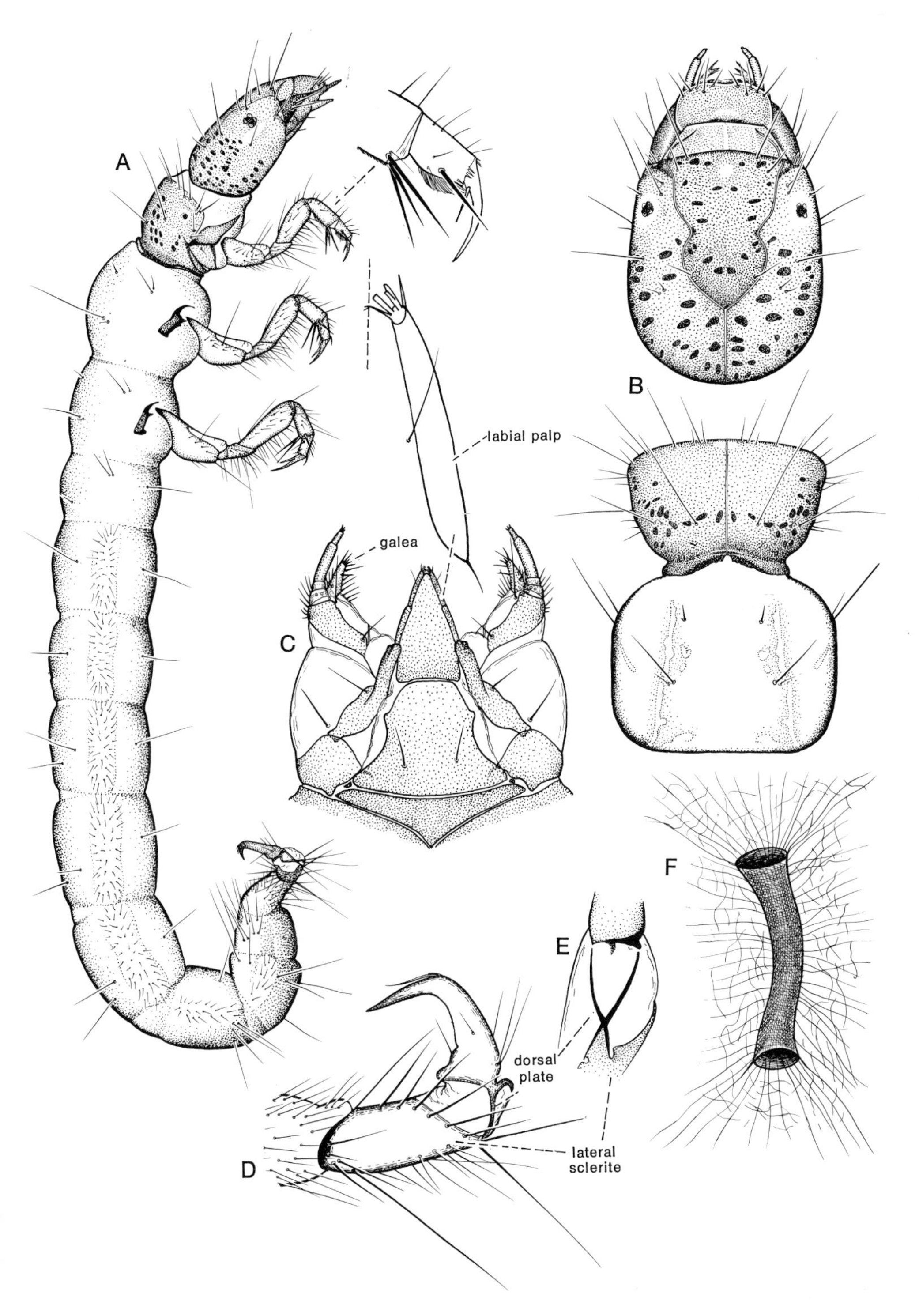
A
B
C
D
E
F
labial palp
galea
dorsal plate
lateral sclerite

15.6 Genus **Polyplectropus**

DISTRIBUTION AND SPECIES In addition to species from North and South America, forms from Africa and the Orient have also been assigned to *Polyplectropus* although many of the Old World species may not be congeneric (Flint 1968a). Several species occur in Mexico and southward, but only *P. charlesi* (Ross) and *P. proditus* (Edwards) from Texas are yet known north of Mexico.

The larva described by Flint (1964a) as Genus C is now believed to belong to *Polyplectropus* (Flint 1968a), and probably to *P. charlesi* since it was collected at the type locality of that species in Texas.

MORPHOLOGY Larvae are clearly distinguished from all other North American polycentropodids by the long teeth on the ventral concave margin of the anal claw (A); the claw lacks an accessory hook. A central dark patch on the dorsum of the head, broken by pale muscle scars, is also characteristic (B); the basal segment of the anal proleg is somewhat longer than the distal and bears a few setae. Length of larva up to 8 mm.

RETREAT The retreat of *Polyplectropus* larvae is reported to be similar to that of *Nyctiophylax* (Flint 1964a).

BIOLOGY Larvae live on rocks in small, cool streams (Flint 1968a).

REMARKS The larva illustrated, collected in San Felipe Spring near Del Rio, Texas, was made available from the United States National Museum by O.S. Flint.

Polyplectropus sp. (prob. *charlesi*) (Texas, Val Verde Co., 21 Sept. 1960, USNM)
A, larva, lateral x26, anal claw enlarged; B, head, pro- and meso-notum, dorsal; C, head, ventral

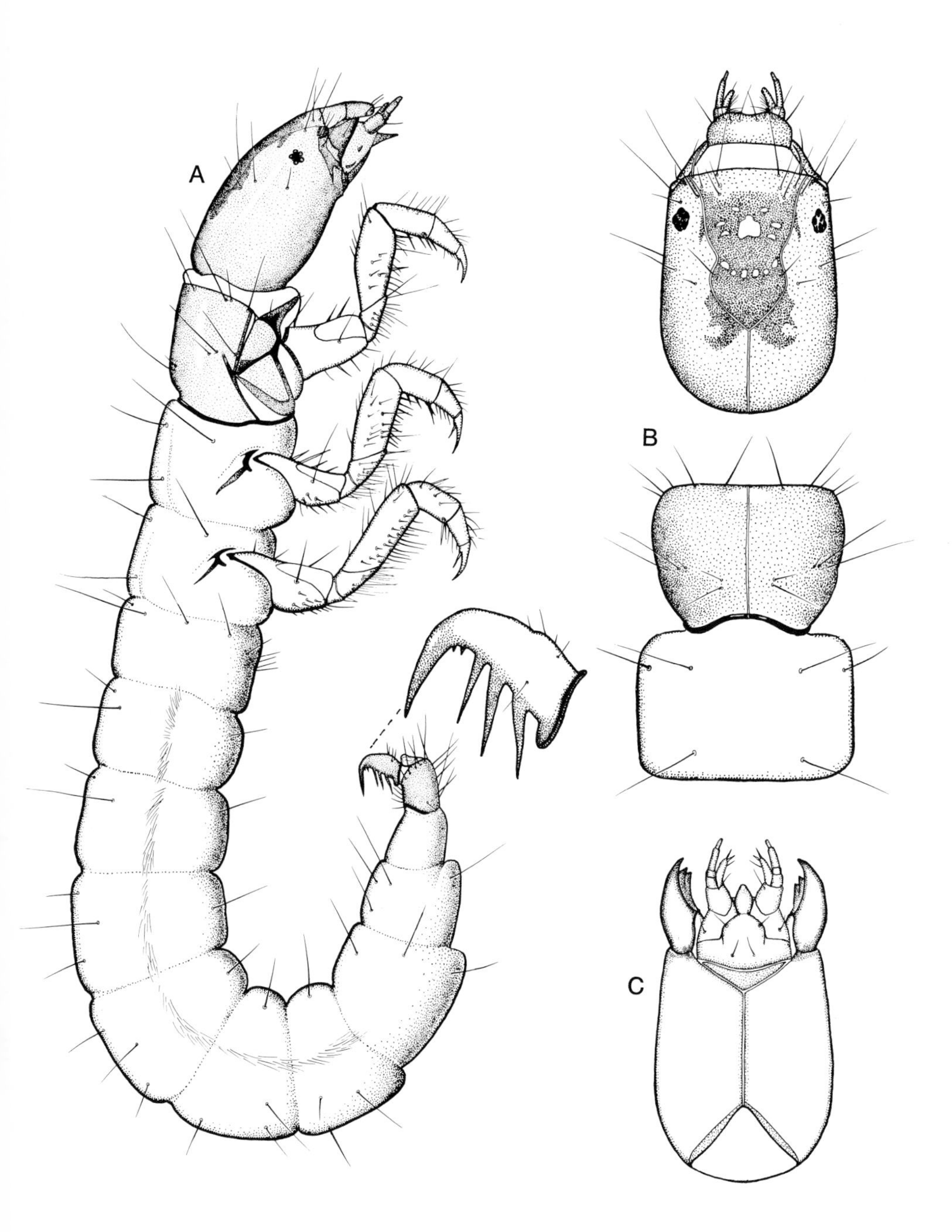
A
B
C

16
Family Psychomyiidae

The Psychomyiidae are widely distributed over much of the world, but are nowhere as dominant as the other net-spinners, the Hydropsychidae or Polycentropodidae. Five genera are recognized in the Nearctic region, with a total of about 15 species north of Mexico; representatives occur in most parts of the continent except for the far north.

The larvae differ from those of the Polycentropodidae in several features. A precise diagnostic character for the Psychomyiidae is the broad trochantin of the prothorax which is separated from the propleuron by a well-marked suture (Fig. 16.3c). The labium is extended far beyond the anterior margin of the head, probably to facilitate application of silk to the interior of the dwelling tube; labial palpi are absent. The galeae are broad and flattened, the submental sclerites usually paired and separate (Fig. 16.2c). The pronotum alone is sclerotized, the prothoracic legs stouter than those of the other segments; tarsal claws are short and stout, bearing a stout basal process from which arises a long seta. The abdomen lacks a lateral fringe, and anal papillae are usually present. The basal segment of the anal proleg is much shorter than the distal segment that bears the lateral sclerite; the anal claw is well developed.

Psychomyiids construct tubes of silk covered with sand and debris on rock and wood substrates. For the most part, larvae live in cool, running water, but some *Tinodes* live in isolated stream pools in western North America and along lake margins in Europe. This evidence of independence from stream current in *Tinodes* is noteworthy because the larvae lack a lateral abdominal fringe of setae which is held to be an asset for larvae generating their own respiratory current (e.g. Lepneva 1964). Information available indicates that psychomyiid larvae feed mainly on detritus and associated microflora (Lepneva 1964) and algae.

There are three subfamilies in the Psychomyiidae, all of them represented in North America. The typical subfamily is the Psychomyiinae, and includes *Lype, Psychomyia,* and *Tinodes.* The Xiphocentroninae, comprising only the genus *Xiphocentron*, are given family status by some authors. The Paduniellinae are represented in North America only by *Paduniella nearctica* Flint from Arkansas (Flint 1967d); no larva has yet been identified in this genus.

16 Family **Psychomyiidae**

Key to Genera*

1 Mesopleuron extended anteriorly as lobate process, tibiae and tarsi fused together on all legs (Fig. 16.4D). Known from southern Texas
(subfamily Xiphocentroninae) **16.4 Xiphocentron**

Mesopleuron not extended anteriorly, tibiae and tarsi separate on all legs (Fig. 16.2A) (subfamily Psychomyiinae) **2**

2 (1) Anal claw with well-developed teeth arising from ventral, concave margin (Fig. 16.2A); ventral surface of labium with paired submental sclerites much longer than wide (Fig. 16.2C). Widespread **16.2 Psychomyia**

Anal claw without teeth on ventral, concave margin (Fig. 16.3A); ventral surface of labium with submental sclerites wider than long (Fig. 16.1D) 3

3 (2) Lateral border of mandibles smoothly curved and without protuberance, pair of lateral setae arising approximately one-third of distance from base of each mandible (Fig. 16.1C). Eastern **16.1 Lype**

Lateral border of mandible with protuberance, pair of lateral setae arising near middle of each mandible (Fig. 16.3D,E). Western **16.3 Tinodes**

*Larvae are not yet known for the genus *Paduniella*; see above.

16.1 Genus Lype

DISTRIBUTION AND SPECIES Although largely a genus of the Palaearctic, *Lype* is known also from the Ethiopian, Oriental, and Nearctic regions. Only a single species, *L. diversa* (Banks), is known from North America; it is confined to the eastern half of the continent, extending as far north as Wisconsin, Ontario, Quebec, and Maine.

Diagnostic characters at the generic level were given by Ross (1959) and Flint (1964a); the larva of *L. diversa* was described in detail by Flint (1959).

MORPHOLOGY There is little that is distinctive about the larva of *Lype diversa*, and it is very much like that of the western genus *Tinodes*; shape of the mandibles offers the best means of distinguishing between the two. In *Lype* the lateral margin of the mandibles is evenly curved, there are two setal brushes on the inner edge of the left mandible, and the mandibles are approximately as long as they are wide; the pair of lateral setae on each mandible is situated near the base (C). Length of larva up to 8.5 mm.

RETREAT Larval retreats of *L. diversa* (E) are exceedingly well camouflaged. A slightly arched roof of silk and small pieces of detritus cover a groove in a piece of submerged wood, forming a chamber in which the larva is concealed. The retreat is open at each end. Length of retreat illustrated approximately 8 mm.

BIOLOGY Larvae of *L. diversa* live in small, cool streams, where their retreats are usually constructed on submerged logs and branches. The retreat illustrated was found on small pieces of wood incorporated into the case of a living larva of *Limnephilus*.

Lype diversa (South Carolina, Oconee Co., 17–18 May 1970, ROM 700347)
A, larva, lateral x19, trochantin enlarged; B, head, pro- and meso-notum, dorsal; C, mandibles, dorsal; D, head, ventral; E, retreat, dorsal approx. x12

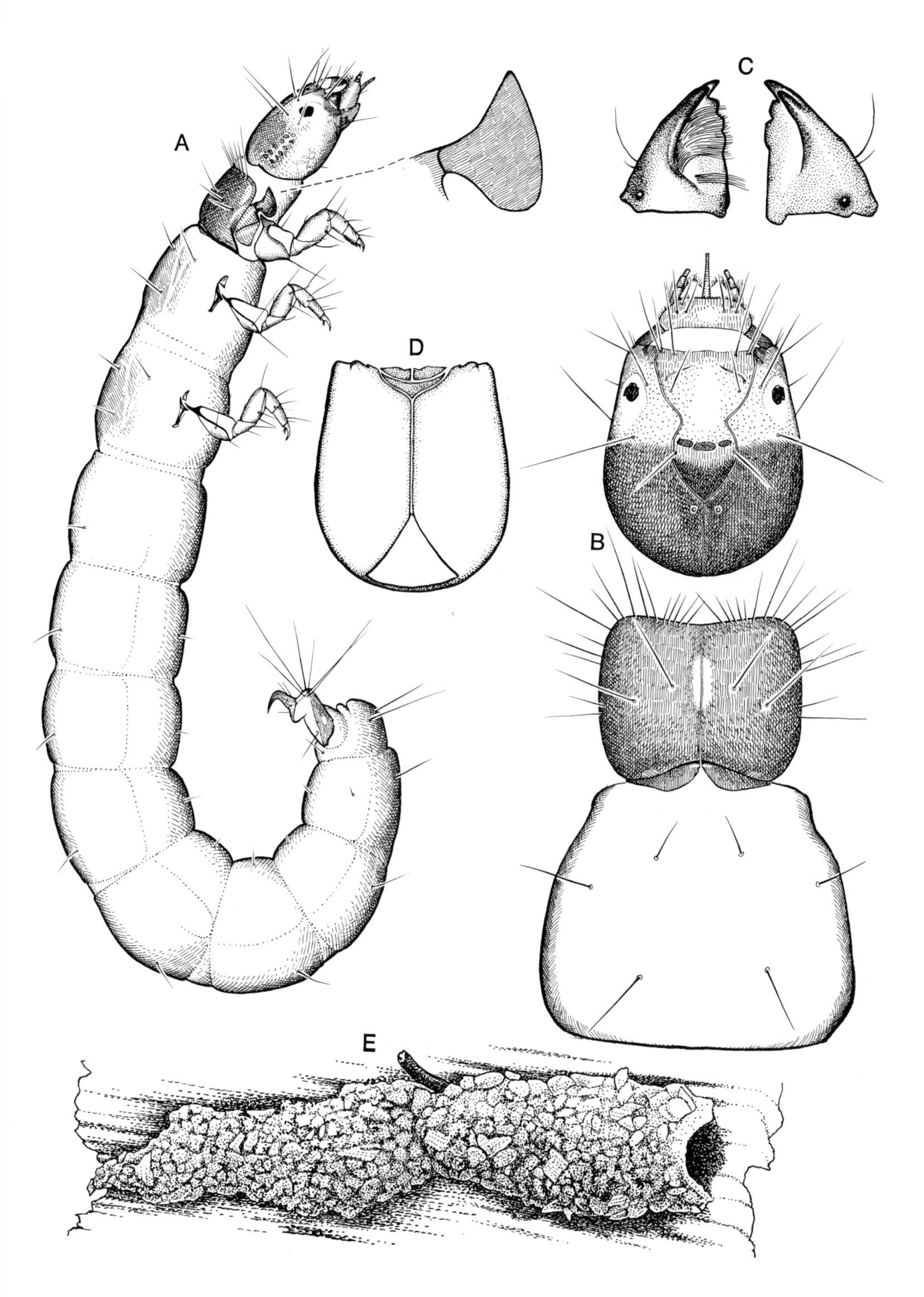
A
B
C
D
E

16.2 Genus **Psychomyia**

DISTRIBUTION AND SPECIES *Psychomyia* is of Holarctic and Oriental distribution. Three species are known in North America: *P. flavida* Hagen throughout most of the continent, and recorded also from Siberia (Schmid 1965); *P. lumina* (Ross) in Oregon; and *P. nomada* (Ross) in the Appalachian area.

The larva of *P. flavida* was described by Ross (1944); and larvae of *P. flavida* and *P. nomada* were described by Flint (1964a). The larva of a western species illustrated here from material we collected in Oregon is probably *P. lumina.*

MORPHOLOGY These are the only psychomyiid larvae known in North America with teeth on the ventral, concave margin of the anal claw (A). The submental sclerites are very large and longer than wide, and the ventral apotome is reduced to a small equilateral triangle (C). In the series illustrated, the anterior margin of the frontoclypeal apotome (B) lacks the median notch characteristic of the other two species; larvae of all three Nearctic species can probably be separated on the basis of different development of this notch (cf. Flint 1964a, fig. 5). Length of larva up to 9 mm.

RETREAT The larvae illustrated construct meandering tubes of silk several centimetres long covered with sand grains on rocks (D). In eastern species these tubes are little more than 1 cm in length.

BIOLOGY In North America, *Psychomyia* larvae are usually found in rivers and streams, although European species are also reported from the littoral zone of lakes (Lepneva 1964). Gut contents of *P. flavida* in a Pennsylvania woodland stream were found to be largely algae with smaller proportions of detritus and animals (Coffman et al. 1971). Guts (3) from the larval series illustrated all contained some vascular plant tissue, as well as fine organic particles.

REMARKS *P. flavida* is probably facultatively parthenogenetic (Corbet 1966); females are often abundant in light traps, and males, though infrequently collected, are sometimes found in equal proportions.

Psychomyia (prob. *lumina*) (Oregon, Benton Co., 7 April 1964, ROM)
A, larva, lateral x18, tarsi and anal claw enlarged; B, head, pro- and meso-notum, dorsal; C, ventral apotome of head and mouthparts, ventral; D, rock with sand-covered tubes, approx. x0.25, section of tube enlarged

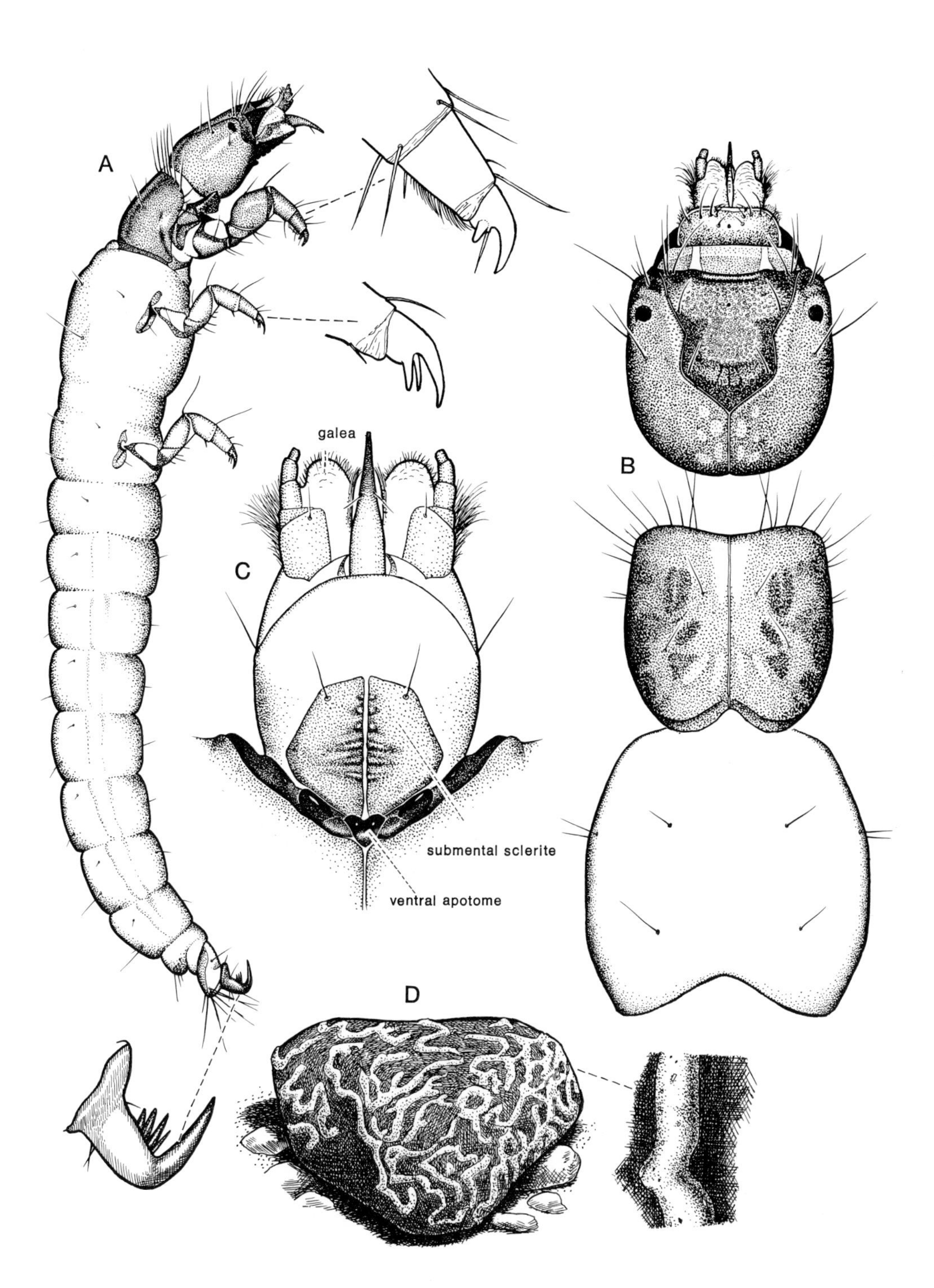
A
B
C
D
galea
submental sclerite
ventral apotome

16.3 Genus Tinodes

DISTRIBUTION AND SPECIES Species assigned to *Tinodes* are characteristic of all faunal regions except the Australian and Neotropical. In North America north of Mexico nine species are recognized, all in western montane areas.

On this continent, generic diagnostic characters for *Tinodes* larvae were given by Ross (1959) and Flint (1964a). In the ROM collection we have larvae from two sites where *Tinodes* adults were taken.

MORPHOLOGY In our material of *Tinodes*, both mandibles have a dorsal bump on the lateral edge near the base. The left mandible bears only a single setal brush on the inner edge, and both mandibles are somewhat longer than wide; the pair of lateral setae on each mandible is situated at about its mid-point. Some European species have small teeth on the ventral, concave edge of the anal claw, but all North American larvae known are without teeth in this position. Length of larva up to 15 mm.

RETREAT *Tinodes* larvae construct flattened, silken tubes covered with small rock fragments, usually on rock surfaces (F); the tube has a thin silken floor. Tubes are variable in length, but often reach several centimetres.

BIOLOGY Larvae of the two species we have were collected in rather warm streams of desert country. Those assumed to be *T. powelli* Denning came from isolated pools of a desert stream of widely fluctuating flow in California (Deep Creek, P.L. Boyd Desert Research Center, Riverside Co., provided by S.I. Frommer); adults were collected in June. Water temperatures in a series of these pools in Deep Creek ranged from 18 to 27.5°C (Frommer and Sublette 1971). Larvae assumed to be *T. provo* Ross came from a spring-fed stream in Idaho (Deep Creek, Oneida Co., provided by R.L. Newell) with a constant temperature of approximately 18°C; adults were collected in January, but some other aquatic insects, normally bivoltine, had multivoltine life cycles at this station (R.L. Newell, pers. comm.). Larvae feed largely on detritus and algae.

REMARKS Taxonomy of adults was reviewed by Denning (1956).

Tinodes (prob. *powelli*) (California, Riverside Co., 13 June 1969, ROM)
A, larva, lateral x12, claws of tarsus and anal proleg enlarged; B, head, pro- and mesonotum, dorsal; C, pleuron and trochantin with fore coxa, lateral; D, mandibles, dorsal; E, right mandible, lateral; F, section of retreat, dorsal approx. x4

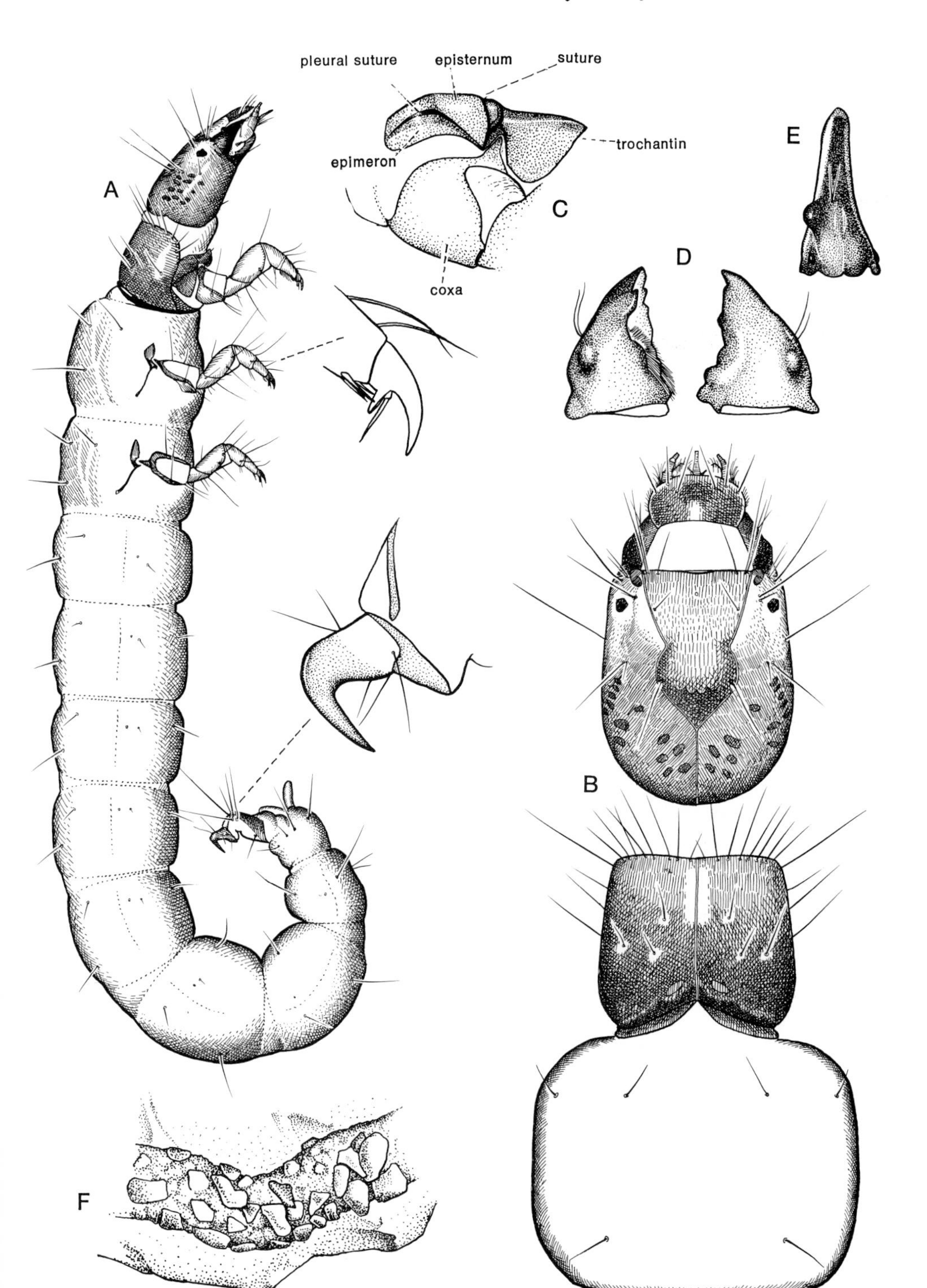
pleural suture
episternum
suture
trochantin
epimeron
coxa
A
B
C
D
E
F

16.4 Genus **Xiphocentron**

DISTRIBUTION AND SPECIES Species of this genus occur in the Antilles, Mexico, and Central and South America (Flint 1964b); one of them, *X. mexico* Ross, extends into southern Texas. A Chinese species may also belong to this genus (Ross 1949).

The larva and pupa of *X. mexico* were described by Edwards (1961).

MORPHOLOGY *Xiphocentron* is unique among all North American genera in having a lobate process extending anteriorly from the mesopleuron, its base invaginated beneath the mesopleuron (A,B,D); the tibiae and tarsi of all legs are fused. The prothoracic trochantin (D,E) is relatively smaller than in the other psychomyiid genera, but is separated from the pleuron by a dark brown line apparently homologous with the trochanteral suture of other genera such as *Tinodes* (Fig. 16.3C). Anal papillae are present. Length of larva up to 8 mm.

RETREAT According to Edwards (1961) and Flint (1964b), *Xiphocentron* larvae build tubes of fine sand grains on rocks below the water surface, frequently extending several centimetres above the surface on wet substrates. Tubes up to 5 cm long and 2.5 mm wide are recorded. At the ends of tubes constructed by a Colombian species on the undersides of logs overhanging streams, larvae construct pendant ovoid vesicles of silk in which they pupate out of water (Sturm 1960).

BIOLOGY The Texas larvae of *X. mexico* were found in the outflow of a small spring (Edwards 1961), and the Antilles species live in small streams (Flint 1964b).

REMARKS Larvae from the United States National Museum were provided by O.S. Flint.

Xiphocentron mexico (Texas, Hays Co., 1 July 1960, USNM)
A, larva, lateral x24, anal claw enlarged; B, head, pro- and meso-notum, dorsal; C, head, ventral; D, prothorax and mesopleuron, right side, lateral; E, fore trochantin, right side, ventral

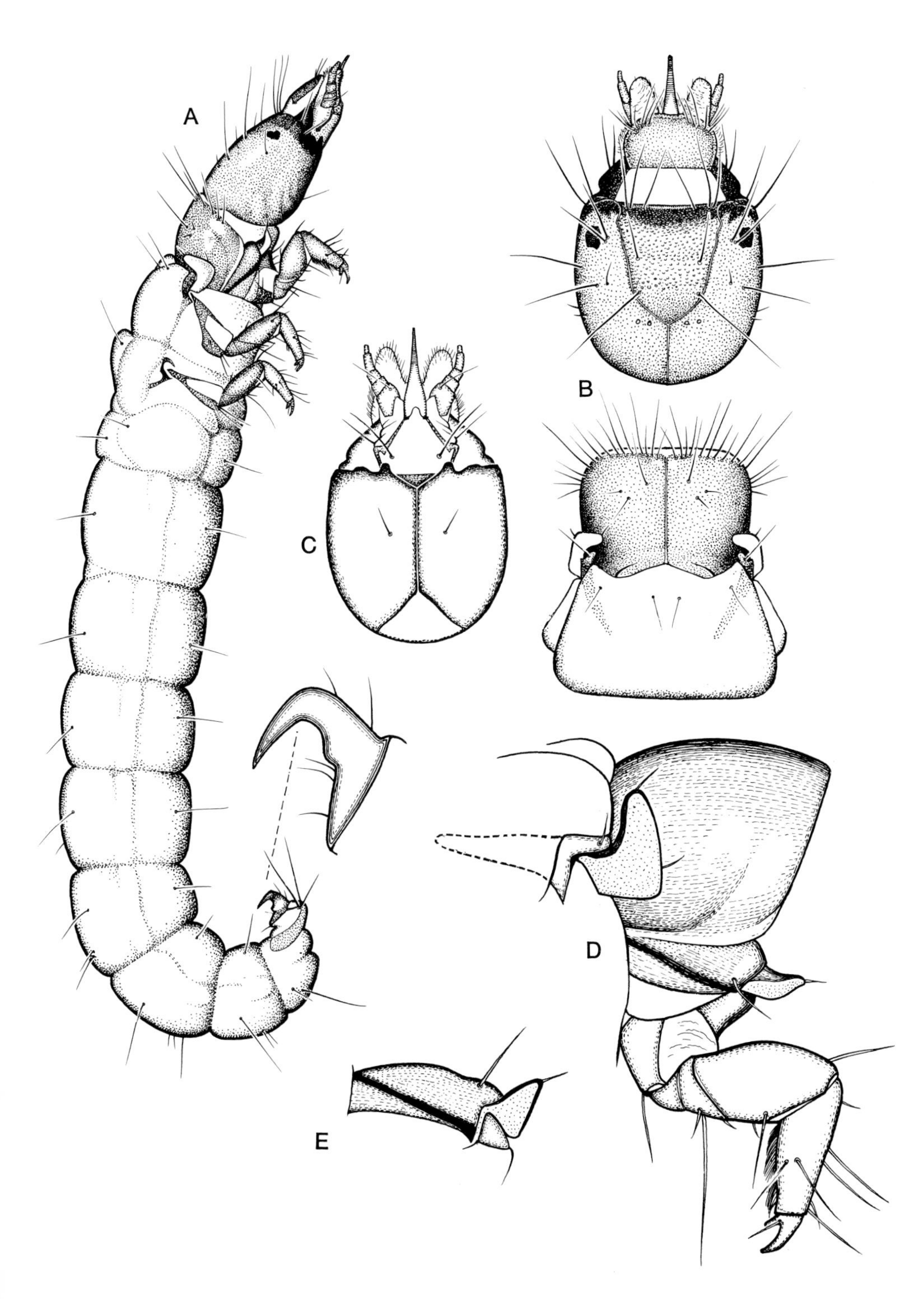
A
B
C
D
E

17

Family Rhyacophilidae

The Rhyacophilidae are a large and important family in cool, running waters of all faunal regions except the Ethiopian. Three genera are recognized in North America with well over 100 species, almost all of which are in the single genus *Rhyacophila.*

Rhyacophilid larvae contrast strongly with larvae of other families of the Trichoptera because they are free-living, passing their larval existence without constructing either a portable case or a fixed retreat. Larvae crawl actively over rocks, and are reputed to leave a silken thread wherever they go. Prior to pupation, the larva fashions a crude enclosure of small stones or sand, either dome-shaped on a larger rock surface or ring-like between two rocks; within this enclosure it spins a dark brown, silken cocoon and undergoes metamorphosis.

Rhyacophilids are typical campodeiform caddis larvae with head and mouth parts prognathous and a body of strong muscular appearance with deep constrictions between segments. On the thorax only the pronotum is sclerotized, and all three pairs of legs are similar in size. Gills are present in some species but unlike the long filamentous gills of other families, those of the Rhyacophilidae are often stiff and in dense tufts. A sclerite is always present on the dorsum of segment IX. Anal prolegs are strongly developed, projecting freely from the body for their entire length, and presumably help to maintain stability of the larva on rocks in currents.

Although generally regarded as one of the most primitive families of the Trichoptera (Ross 1967), the Rhyacophilidae are among the most successful; evidently they monopolize the free-living lotic predator niche among the Trichoptera. Taxonomy of adults and phylogeny of all genera were reviewed by Ross (1956).

There are two subfamilies and both are represented on this continent. The Rhyacophilinae, comprising only *Himalopsyche* and *Rhyacophila* in North America, are Holarctic and Oriental. The Hydrobiosinae, with several genera but only *Atopsyche* in North America, occur in cool streams of the Australian and Neotropical regions; the group is raised to family status by some authors (Schmid 1970).

17 Family **Rhyacophilidae**

Key to Genera

1 Tibia, tarsus, and claw of fore leg articulating against stout ventral lobe of femur to form chelate leg (Fig. 17.1C). Southwestern
(subfamily Hydrobiosinae) **17.1 Atopsyche**

Fore leg normal, not chelate as above (Fig. 17.2A)
(subfamily Rhyacophilinae) **2**

2 (1) Dense tuft of stout gills on side of meso- and meta-thorax and on segments I–VIII (Fig. 17.2A); final-instar larva up to 32 mm long. Western
17.2 Himalopsyche

Tufts of gills usually absent (Fig. 17.3A), but if present, tufts neither as dense nor on as many segments as above; final-instar larvae not more than 30 mm long. Widespread **17.3 Rhyacophila**

17.1 Genus **Atopsyche**

DISTRIBUTION AND SPECIES *Atopsyche* is a large Neotropical genus with many species in South America through Central America, Mexico, and the Antilles. Three of the species known in Mexico have been recorded from Texas and Arizona: *A. erigia* Ross, *A. sperryi* Denning, and *A. tripunctata* Banks. We have *Atopsyche* larvae from Nevada.

North American larvae in *Atopsyche* are well known (Ross 1959), but only *A. erigia* has been described at the species level (Edwards and Arnold 1961). We have larvae for *A. sperryi* from Arizona.

MORPHOLOGY *Atopsyche* larvae have very characteristic chelate front legs (A,C), in which the shortened tibia, tarsus, and claw close against a concave extension of the femur. They are further distinguished from those of other Nearctic rhyacophilids by a broad prosternal plate (D). Length of larva up to 22 mm.

RETREAT Pupal enclosures constructed just before pupation are crudely made of small stones fastened to rocks; larvae are often found within the enclosure sealed in the typical rhyacophilid silk cocoon.

BIOLOGY *Atopsyche* larvae live in cool streams, where they are predacious; gut contents of larvae (3) we examined all contained arthropod remains. One can only agree with Hinton (1950), in describing the unusual fore legs and their musculature, that the chelae are almost certainly used in securing prey; but that has yet to be observed.

REMARKS Taxonomy of adults and phylogeny for *Atopsyche* species were reviewed by Ross and King (1952).

Atopsyche sp. (Arizona, Coconino Co., 9 April 1968, ROM)
A, larva, lateral x12; B, head, pro- and meso-notum, dorsal; C, fore leg, lateral; D, prothorax, ventral

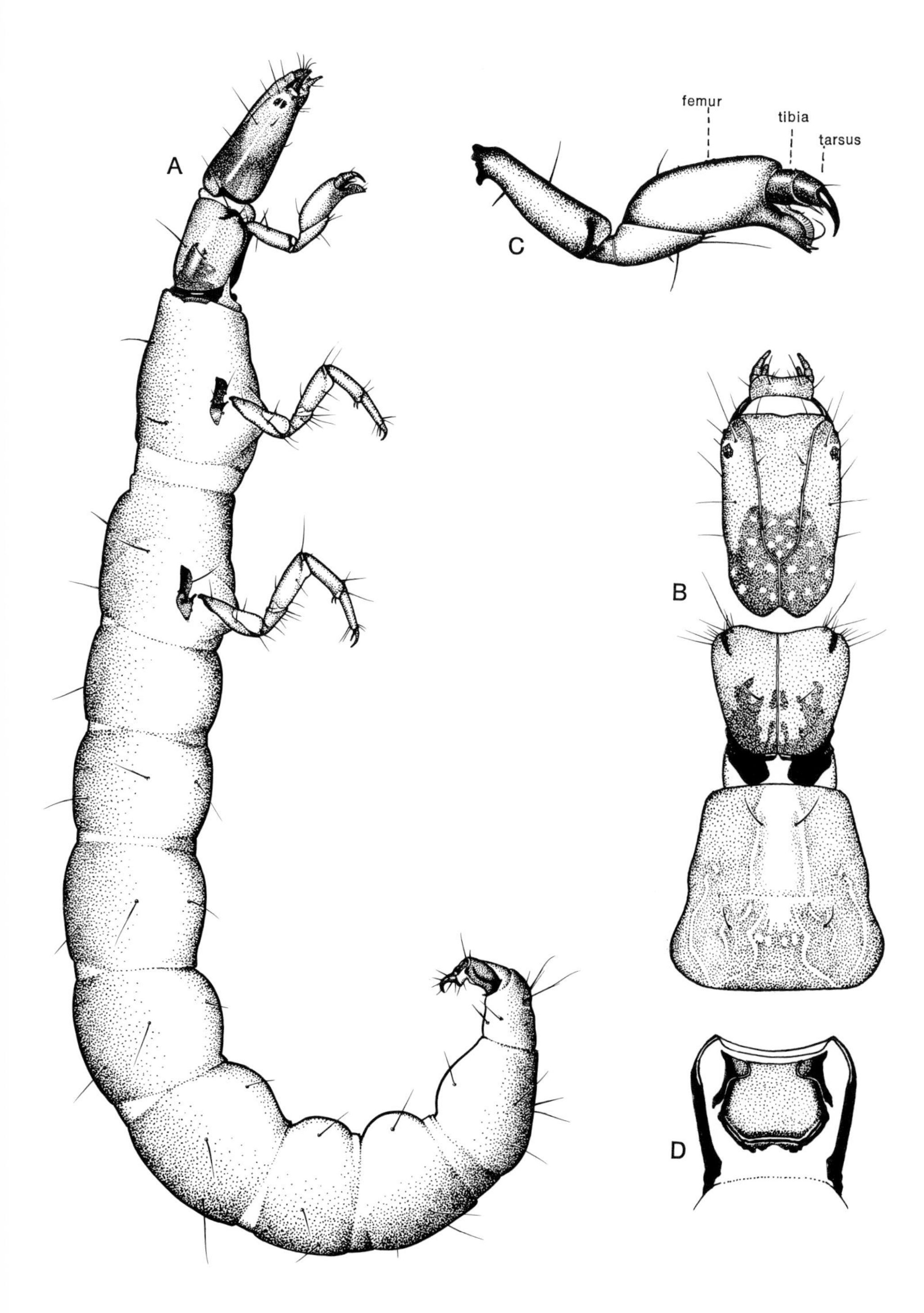
A
femur
tibia
tarsus
C
B
D

17.2 Genus **Himalopsyche**

DISTRIBUTION AND SPECIES This is largely a genus of Oriental and Asian Palaearctic regions with one species, *H. phryganea* (Ross), recorded from the western Nearctic; this single North American species is known from California, Oregon, and Washington in both Cascade and Coast ranges.

The larva of *H. phryganea* was described by Flint (1961b); we have collected several series of them.

MORPHOLOGY *Himalopsyche* larvae are large and stout-bodied (A). Although Old World larvae of the genus frequently have lateral processes bearing short gill filaments on each abdominal segment (Lepneva 1964, fig. 310; Schmid and Botoşăneanu 1966, pl.VI, fig. 1), there are no lateral processes in *H. phryganea.* A dense tuft of stout gill filaments arises from the sides of abdominal segments I–VIII and from the meso- and meta-thorax; these tufts are probably denser and present on more segments than in any North American *Rhyacophila*, but since so much of the structural diversity of *Rhyacophila* larvae remains undocumented, no firm diagnosis can yet be made. An ovoid sclerite is present on the venter of most abdominal segments (A). Length of larva up to 32 mm.

RETREAT Pupal enclosures of *H. phryganea* are elongate domes constructed of rather uniform small stones fastened together in such a way that spaces are left for water circulation around each stone. Length at least 25 mm.

BIOLOGY Larvae of *Himalopsyche* are recorded from mountain streams of many kinds including torrents, and collections of *H. phryganea* have been made at elevations between 2,000 and 5,000 ft (600 and 1,500 m) (Schmid and Botoşăneanu 1966); our collections of larvae of this species were made for the most part in rather small mountain streams. Guts of larvae (3) which we examined all contained arthropod parts.

REMARKS Taxonomy of adults and phylogeny of *Himalopsyche* species have been reviewed by Ross (1956) and by Schmid and Botoşăneanu (1966).

Himalopsyche phryganea (Oregon, Benton Co., 3 June 1968, ROM)
A, larva, lateral x6, detail of anal claw and ventral sclerite; B, dorsal sclerite of segment IX

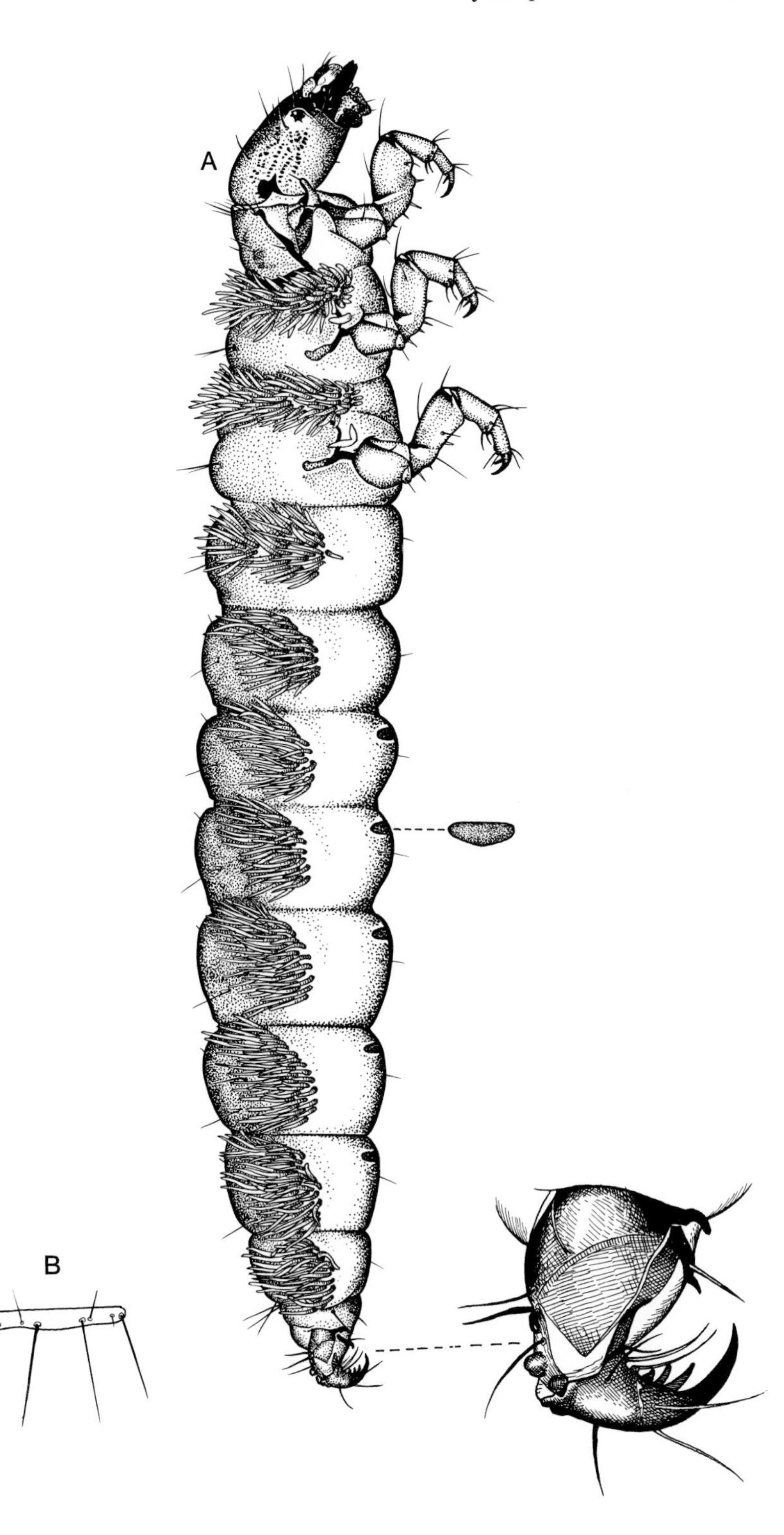
A
B

17.3 Genus **Rhyacophila**

DISTRIBUTION AND SPECIES *Rhyacophila* is the largest genus in the Trichoptera, with close to 500 species widely distributed through the Holarctic and Oriental regions. More than 100 species are known in North America, and they occur in most parts of the continent where surface relief is sufficient to sustain lotic habitats.

Larvae for North American species have been described by Vorhies (1909), Lloyd (1921), Ross (1944), Flint (1962c), and Smith (1968b).

MORPHOLOGY Taxonomy of *Rhyacophila* larvae is not yet sufficiently explored for diagnostic limits to be fixed for this genus. In some species the lateral sclerite of the anal proleg is developed into a curved spike, as in the common eastern *R. fuscula* (Walker) (A,C); but in most species the lateral sclerite is unmodified (D). Abdominal segments usually lack filamentous gills, as in the larva illustrated, but gills of several types are present in some species; gills as stout and densely clustered as in *Himalopsyche* are at least very unusual, and perhaps not represented, in North American *Rhyacophila.* Larvae differ considerably in conformation of the head, in colour markings, and in details of the mandibles and anal prolegs. Ventral abdominal sclerites similar to those in *Himalopsyche* are present in at least some species of *Rhyacophila.* Length of larva up to 23 mm.

RETREAT Larvae build only a crude pupal shelter of small stones.

BIOLOGY *Rhyacophila* larvae live in a wide range of running-water habitats, and some species are evidently adapted to temporary streams (Ross 1944). Species frequently occur together in the same stream; temporal and spatial separation in such instances was documented by Mackay (1969) and Thut (1969). In general, *Rhyacophila* larvae are free-living predators, and some species exhibit a preference for certain groups of prey organisms (Malicky 1973). Larvae of a few species are herbivorous, feeding on vascular plant tissue both living and dead, and on algae (Smith 1968b; Thut 1969).

REMARKS Taxonomy of adults and phylogeny have been reviewed by Ross (1956) and Schmid (1970).

Rhyacophila fuscula (Ontario, Algonquin Prov. Park, 20 May 1959, ROM)
A, larva, dorsal x7, detail of anal proleg; B, mouthparts, ventral; C, anal proleg, lateral
D, *R. banksi* Ross (Quebec, Rouville Co., ROM) anal prolegs, dorsal

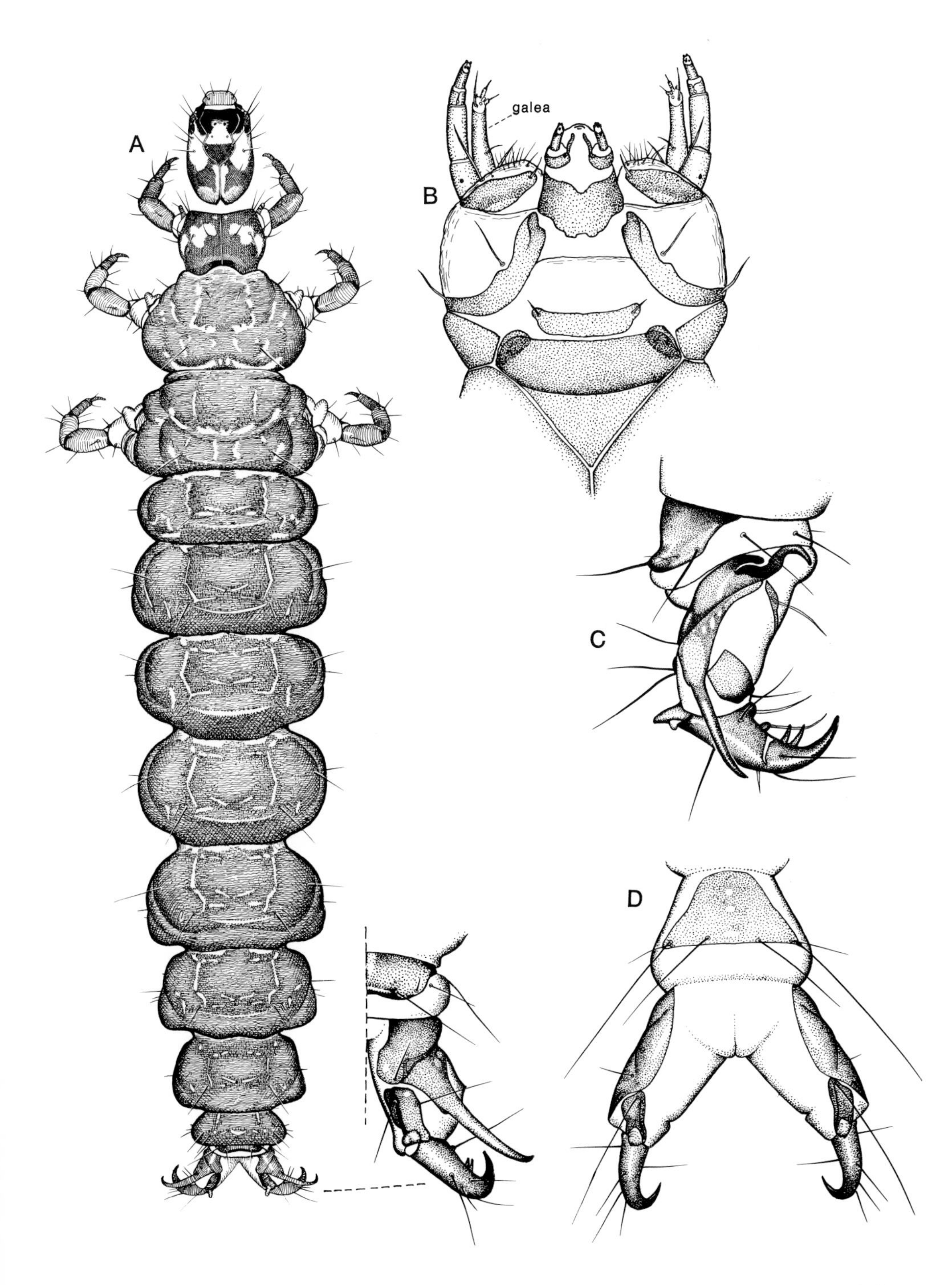
A
B
galea
C
D

18

Family Sericostomatidae

The Sericostomatidae are represented in all faunal regions, usually by a small number of genera and species. Twelve North American species are now known and were assigned to three genera in a recent revision (Ross and Wallace 1974). Representatives are widespread and somewhat local through the eastern half of the continent and in the west.

Larvae are generally found in running waters, often in small springs, but also along lake shores where there is wave action. Observations indicate that larvae of at least the two eastern genera, *Agarodes* and *Fattigia*, are burrowers in sandy deposits; those of the western *Gumaga* do not burrow to the same extent. Sericostomatid larvae feed mainly on plant materials, and appear to be principally detritivorous. Six larval instars were attributed to the European species *Sericostoma personatum* (Spence) (Elliott 1969), compared to five in most families.

Nearctic sericostomatid larvae have a number of distinctive features. The head lacks much of the ventral pigmentation, and bears a dorsolateral ridge, variously developed on each side. The labrum is rather narrow, and the mandibles have several tooth-like points; the ventral apotome is roughly triangular, separating the genae incompletely. The pronotum lacks the prominent posterior transverse ridge of the Limnephilidae; prosternal sclerites are absent and there is no prosternal horn; the fore trochantin is relatively large and hook-shaped, the fore femur broad and flat. The mesonotum tends to be sclerotized posterior to the primary sclerites, and in *Gumaga* the *sa*3 sclerite is separated from the plate representing the fused *sa*1 and *sa*2. Sclerites of the metanotal setal areas are indistinct and difficult to interpret. On abdominal segment I the lateral humps are oblique and flat, provided with a dense patch of short setae, and ringed basally by a more or less complete, weak sclerite (Fig. 18.2A); setae on segment I are sparse, with only the lateral hump seta and usually one on each side of the dorsal hump and a pair on the venter. Gills, usually single, are present on segment I, and on other segments most gills of the lateral series are short and single, those of the dorsal and ventral series either branched or single; segment VIII bears a row of sclerotized tubercles on each side, segment IX lacks a well-defined dorsal sclerite but does bear setae on the posterodorsal border; the lateral fringe of the abdomen is absent. On the dorsum of each anal proleg there is a characteristic cluster of approximately 30 or more setae mesiad from the somewhat reduced lateral sclerite (Fig. 18.1D); the anal claw bears a stout accessory hook which in some genera bears a small accessory hook itself (Fig. 18.1A).

Larval cases are usually made of rock pieces, and are variously tapered and curved. The posterior opening is considerably reduced with silk.

Two subfamilies are currently recognized; the Sericostomatinae to which the Holarctic genera belong, and the Conoesucinae which comprise a number of southern hemisphere genera. It should be noted, though, that in some earlier works on the North American fauna (e.g. Betten 1934) and in many European works even of fairly recent origin (e.g. Hickin 1967), the Lepidostomatinae, Brachycentrinae, and Goerinae were also treated as subfamilies of the Sericostomatidae. Association of these groups with the Sericostomatidae sens. str. is an inheritance from the time when classification of the Trichoptera was based mainly on data derived from adult stages.

Taxonomy of Nearctic sericostomatid adults was reviewed by Ross (1948b) and Ross and Wallace (1974).

Key to Genera

1 Anterolateral corner of pronotum extended into sharp point (Fig. 18.1A); metanotal *sa*2 comprising many setae in transverse band on pair of sclerites (Figs. 18.1B, 18.2B) — 2

Anterolateral corner of pronotum not extended into sharp point as above (Fig. 18.3A); each metanotal *sa*2 consisting of only single seta (Fig. 18.3B). Western — **18.3 Gumaga**

2 (1) Dorsum of abdominal segment IX with approximately 40 setae (Fig. 18.2E); head flat dorsally, prominent carina at each side (Fig. 18.2A,B). Southeastern — **18.2 Fattigia**

Dorsum of abdominal segment IX with approximately 15 setae (Fig. 18.1D); head rounded dorsally, lateral carina less prominent than above (Fig. 18.1A,B). Widespread through eastern half of continent — **18.1 Agarodes**

18.1 Genus Agarodes

DISTRIBUTION AND SPECIES The genus *Agarodes* is confined to the eastern half of North America from Louisiana and Florida to Minnesota and Maine; nine species are recognized.

Some characters for the larva of *A. griseus* Banks were given by Ross and Wallace (1974). The larva of *A. distinctus* Ulmer illustrated here is one from series we reared in Ontario; and we have larvae of other species from the south.

MORPHOLOGY The anterolateral corner of the pronotum is extended anteriorly as a sharp point (A), and each of the three lightly sclerotized setal areas of the metanotum bears many setae (B), the *sa*1 setae arranged in more of a transverse line than those in *Fattigia.* Most abdominal gills of the dorsal and ventral series are branched, but those of the lateral series are mainly single (A). The dorsum of segment IX bears four major setae and approximately ten shorter ones along its posterior edge (D). Length of larva up to 15 mm.

CASE Larval cases in the genus are made of moderately coarse but uniform rock fragments, and generally are not elongate as in *Gumaga.* Length of larval case up to 24 mm.

BIOLOGY In Ontario, we have collected larvae of *A. distinctus* in sand and gravel deposits along the edges of lakes and rivers; but we have taken larvae of other species of the genus in the southeast in the sand of cool springs. Since *Agarodes* larvae are rarely seen other than by sifting sand and gravel, I presume they are chiefly burrowing detritivores in these materials; guts of larvae (3) we examined contained pieces of vascular plants and fine organic particles for the most part.

REMARKS Taxonomy of adults of *Agarodes* was reviewed by Ross and Scott (1974).

Agarodes distinctus (Ontario, Algonquin Prov. Park, 31 May 1960, ROM)
A, larva, lateral x9, anal claw enlarged; B, head and thorax, dorsal; C, case x8; D, segment IX and anal prolegs, dorsal

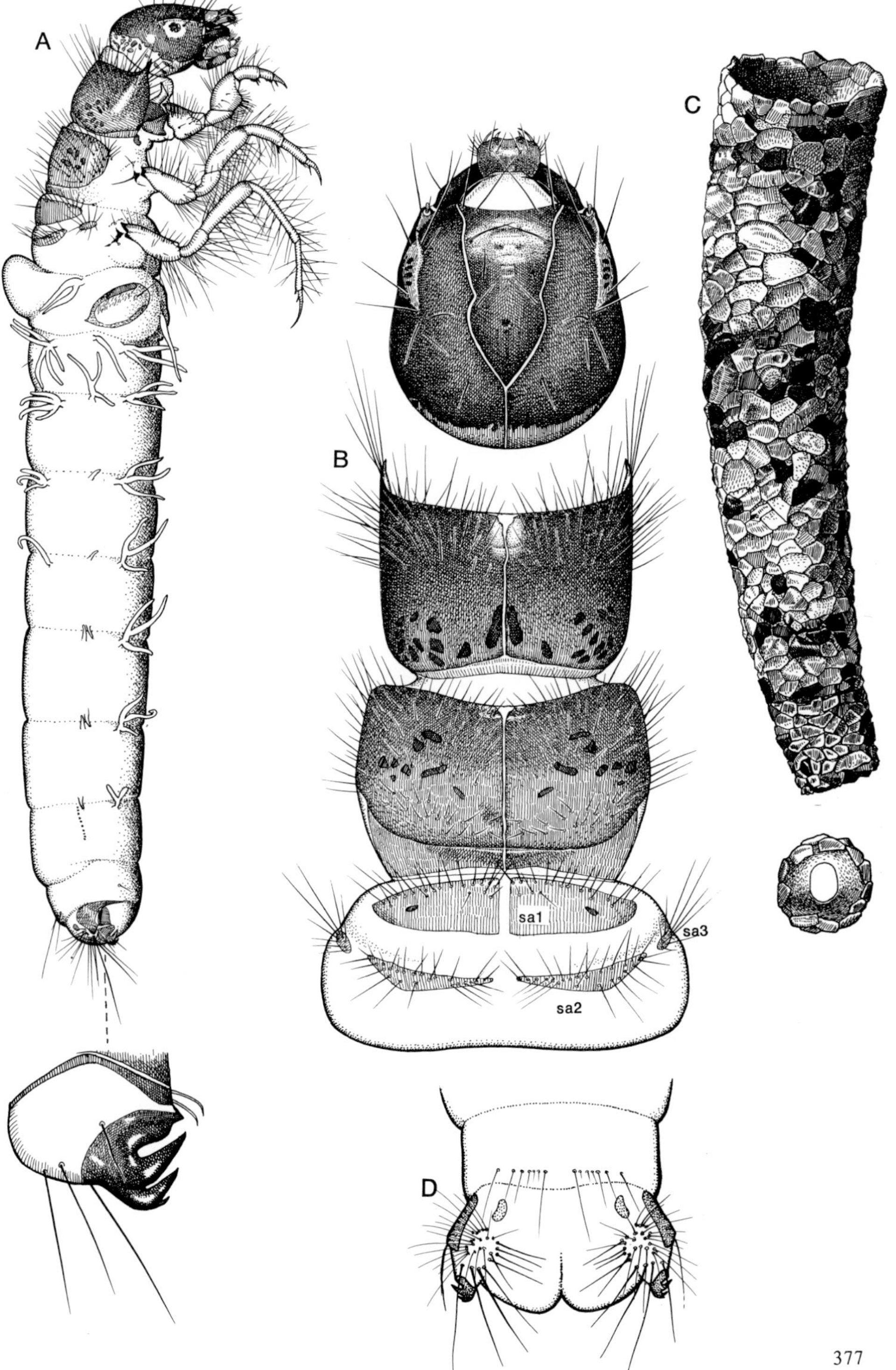
A
B
C
D
sa1
sa2
sa3

18.2 Genus **Fattigia**

DISTRIBUTION AND SPECIES This genus comprises only the species *F. pele* (Ross) of the southeastern United States. Diagnostic characters of the larva were given by Ross and Wallace (1974); we have associated larvae in several localities.

MORPHOLOGY The dorsum of the head of *F. pele* is distinctly flattened and the carina prominent laterally (A). As in *Agarodes*, the anterolateral corner of the pronotum is extended as a sharp point (A), but is directed more ventrally than in that genus; the metanotal *sa*1 setae are more scattered than in *Agarodes*. Most abdominal gills of the dorsal and ventral series are branched, but gills of the lateral series are mostly single. This genus is distinct from the other two in having many setae (approx. 40) on the dorsum of segment IX (E), and several setae (approx. 15) on the mesial surface of the base of the anal claw. Length of larva up to 15 mm.

CASE The larval case of *F. pele* (C) is similar to that of *Agarodes* species in material and general proportions. Length of larval case up to 18 mm.

BIOLOGY Our collections of *F. pele* larvae were made in sand deposits of cool springs and streams, mostly by sifting the sand. Gut contents of larvae (3) we examined were largely vascular plant tissue and fine organic particles.

Fattigia pele (North Carolina, Macon Co., 20 May 1970, ROM 700363)
A, larva, lateral x11, lateral hump enlarged; B, head and thorax, dorsal; C, case x8; D, fore trochantin, lateral; E, posterior edge of segment IX, dorsal

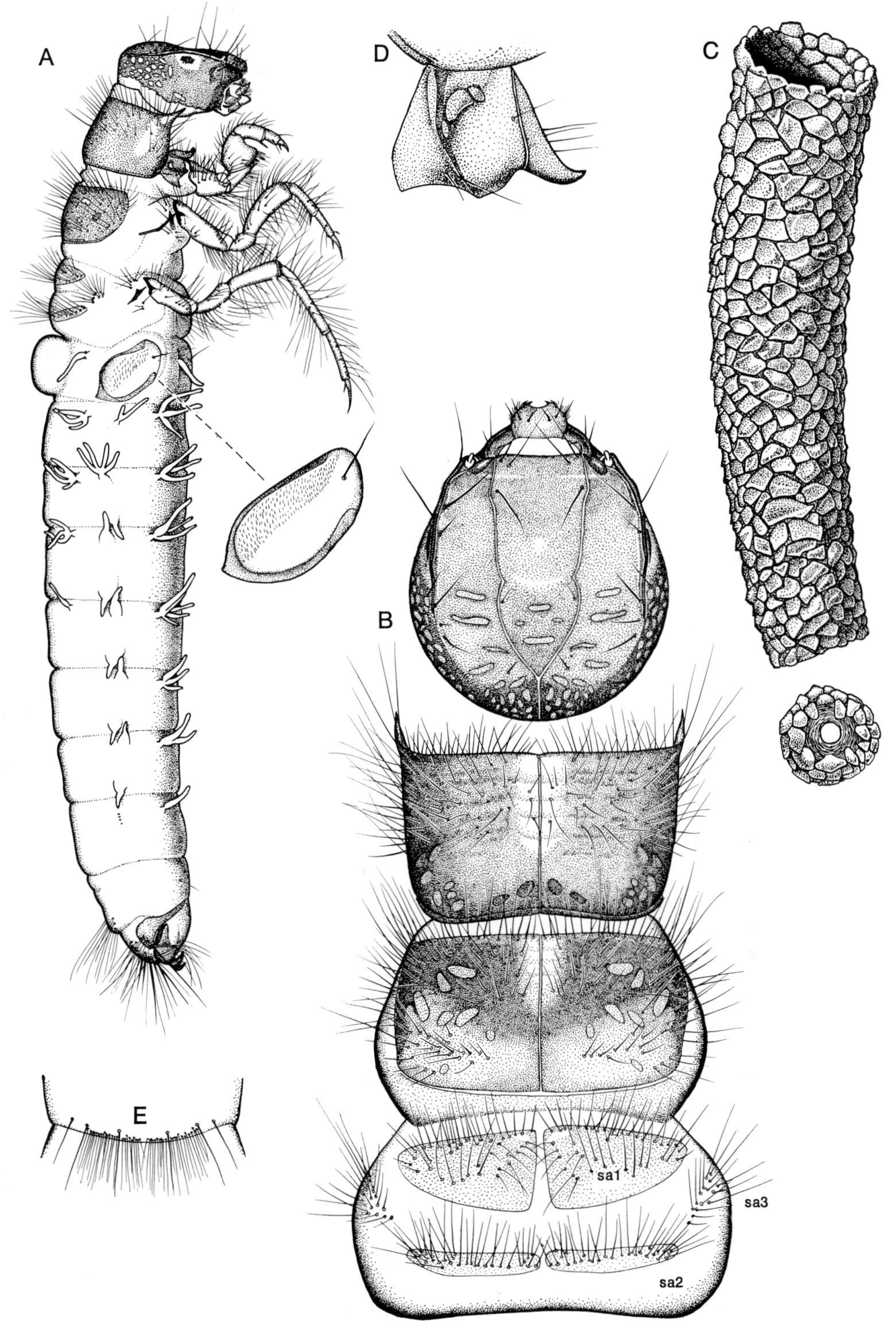
A
B
C
D
E
sa1
sa2
sa3

18.3 Genus Gumaga

DISTRIBUTION AND SPECIES As recently constituted, *Gumaga* is a genus of Japan and North America (Ross and Wallace 1974). On this continent the group occurs from New Mexico and Utah (Ross and Wallace 1974) through Arizona and California to southern Oregon. There are two species, *G. nigricula* (McL.) and *G. griseola* (McL.).

Diagnostic characters of *Gumaga* larvae were given by Ross and Wallace (1974) and by Ross (1959, as *Sericostoma*). We have associated larvae for both species, and have collected larvae in many localities.

MORPHOLOGY The North American larvae of *Gumaga* are distinguished from the larvae of other sericostomatids on this continent by several features; the most readily apparent are the lack of a sharp anterolateral point on the pronotum, the separation of mesonotal *sa*3 as a separate sclerite (A,B), and the reduction of metanotal *sa*2 to a single seta (B). Abdominal gills are always single in *Gumaga* (A). The dorsum of segment IX bears approximately six major setae and about the same number of smaller ones (E), similar to *Agarodes*. Length of larva up to 19 mm.

CASE Larval cases in *Gumaga* (C) are frequently long and slender, and made of sand grains, rather smaller pieces than in cases of the other two genera. Length of larval case up to 29 mm.

BIOLOGY Our collections of *Gumaga* larvae were made in running water, in both cold springs and warmer streams. Larvae in cases were usually inactive, lying exposed on the gravel substrate or beneath rocks, and evidently less inclined to burrow than those of the other two genera. Larvae of the European *Sericostoma personatum* (Spence), found under rocks by day, moved actively about on the stream bed at night (Elliott 1969); the inactivity of *Gumaga* larvae where we have collected them during the day suggests the possibility of similar nocturnal activity. In guts of *Gumaga* larvae (3) that we examined there was a much higher proportion of fine particles than in the other sericostomatid genera.

REMARKS Taxonomy of the males of Nearctic *Gumaga* was reviewed by Kimmins and Denning (1951).

Gumaga nigricula (Arizona, Coconino Co., 24 May 1961, ROM)
A, larva, lateral x12; B, head and thorax, dorsal; C, case x5; D, head, ventral; E, posterior edge of segment IX, dorsal

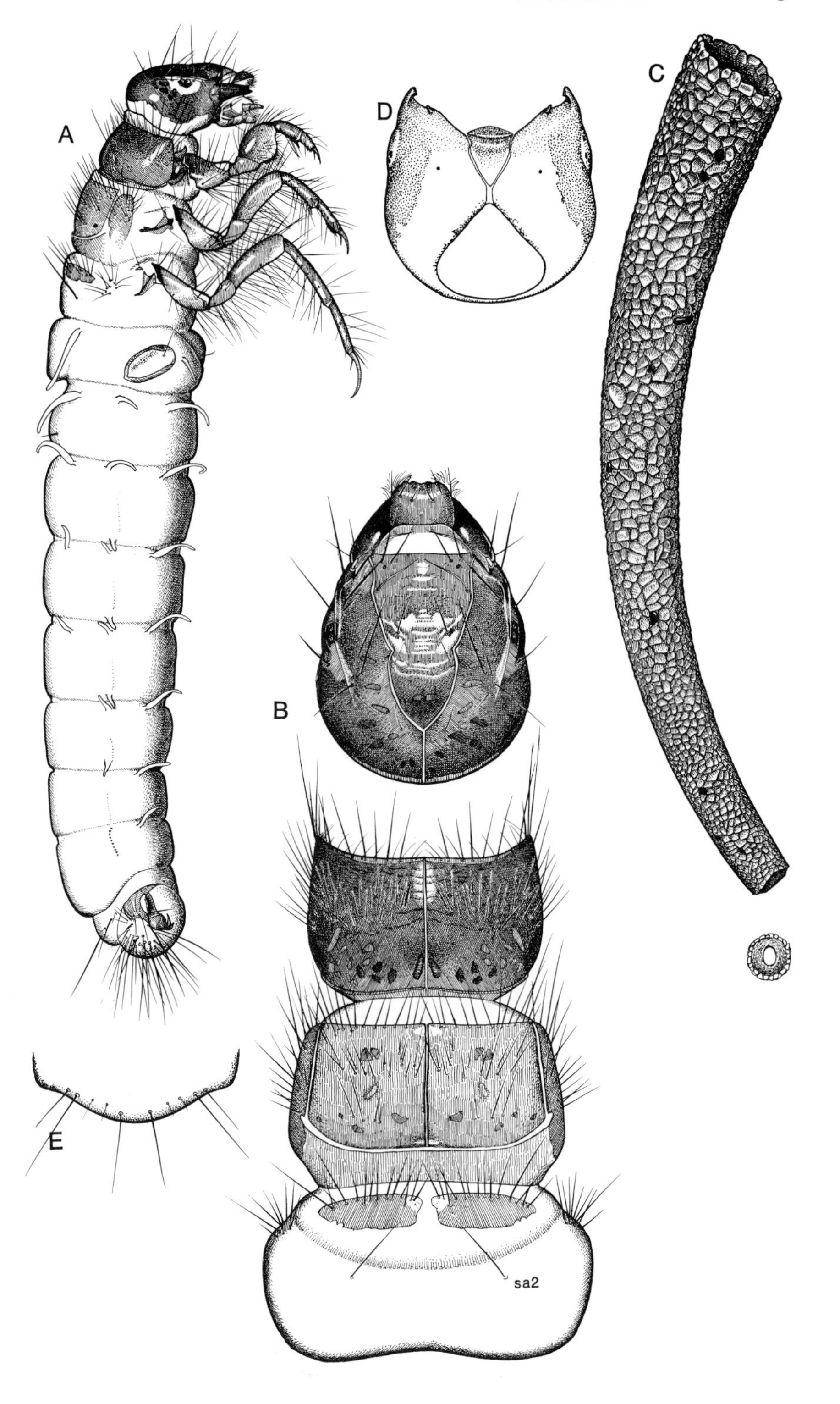
A
B
C
D
E
sa2

Literature Cited

Anderson, N.H. 1967a. Life cycle of a terrestrial caddisfly, *Philocasca demita* (Trichoptera: Limnephilidae), in North America. Ann. Ent. Soc. Am. 60(2): 320-23

- 1967b. Biology and downstream drift of some Oregon Trichoptera. Can. Ent. 99(5): 507-21

- 1974a. Observations on the biology and laboratory rearing of *Pseudostenophylax edwardsi* (Trichoptera: Limnephilidae). Can. J. Zool. 52(1): 7-13

-1974b. The eggs and oviposition behaviour of *Agapetus fuscipes* Curtis (Trich., Glossosomatidae). Ent. Mon. Mag. 109: 129-31

Anderson, N.H., and Bourne, J.R. 1974. Bionomics of three species of glossosomatid caddis flies (Trichoptera: Glossosomatidae) in Oregon. Can. J. Zool. 52(3): 405-11

Anderson, N.H., and Grafius, E. 1975. Utilization and processing of allochthonous material by stream Trichoptera. Verh. Int. Ver. Limnol. 19: 3083-88

Anderson, N.H., and Wold, J.L. 1972. Emergence trap collections of Trichoptera from an Oregon stream. Can. Ent. 104(2): 189-201

Balduf, W.V. 1939. The Bionomics of Entomophagous Insects, pt. II. St Louis: John S. Swift

Bärlocher, F., and Kendrick, B. 1973. Fungi and food preferences of *Gammarus pseudolimnaeus*. Arch. Hydrobiol. 72(4): 501-16

- 1975. Assimilation efficiency of *Gammarus pseudolimnaeus* (Amphipoda) feeding on fungal mycelium or autumn-shed leaves. Oikos, 26: 55-59

Berg, C.O. 1949. Limnological relations of insects to plants of the genus *Potamogeton*. Trans. Am. Microsc. Soc. 68(4): 279-91

Bernhardt, S.A. 1966. Observations on case-building by *Nemotaulis hostilis* (Hagen) larvae (Trichoptera: Limnephilidae). Bull. Brooklyn Ent. Soc. 59-60: 63-76

Betten, C. 1902. The larva of the caddis fly, *Molanna cinerea* Hagen. J. N.Y. Ent. Soc. 10: 147-54

- 1934. The caddis flies or Trichoptera of New York State. Bull. N.Y. St. Mus. 292

- 1950. The genus *Pycnopsyche* (Trichoptera). Ann. Ent. Soc. Am. 43(4): 508-22

Literature Cited

Betten, C., and Mosely, M.E. 1940. The Francis Walker types of Trichoptera in the British Museum. London: British Museum (Natural History)

Bjarnov, N., and Thorup, J. 1970. A simple method for rearing running-water insects, with some preliminary results. Arch. Hydrobiol. 67(2): 201-9

Botoşăneanu, L. 1956. Le développement postembryonnaire, la biologie et la position systématique d'un des Trichoptères les plus intéressants de la faune européenne: *Helicopsyche bacescui* Orghidan et Botoşăneanu. Acta Soc. Zool. Bohemoslovenicae, 20(4): 285-312

Botoşăneanu, L., and Sýkora, J. 1973. Sur quelques Trichoptères (Insecta: Trichoptera) de Cuba. *In* Résultats des expéditions biospéologiques cubano-roumaines à Cuba, pp. 380-407. Bucharest: Editura Academiei Republicii Socialiste România

Bray, R.P. 1966. A technique to aid taxonomic studies on Trichoptera larvae. Entomologist's Mon. Mag. 102(1223-25): 134

Brickenstein, C. 1955. Über den Netzbau der Larve von *Neureclipsis bimaculata* L. (Trichopt., Polycentropidae). Abh. Bayer. Akad. Wiss., n.s. 69: 1-44

Brusven, M.A., and Scoggan, A.C. 1969. Sarcophagous habits of Trichoptera larvae on dead fish. Ent. News, 80: 103-5

C.S.I.R.O. 1970. The Insects of Australia. Melbourne: Melbourne University Press

Chapman, D.W., and Demory, R.L. 1963. Seasonal changes in the food ingested by aquatic insect larvae and nymphs in two Oregon streams. Ecology, 44(1): 140-46

Cloud, T.J., and Stewart, K.W. 1974. Seasonal fluctuations and periodicity in the drift of caddisfly larvae (Trichoptera) in the Brazos River, Texas. Ann. Ent. Soc. Am. 67(5): 805-11

Coffman, W.P., Cummins, K.W., and Wuycheck, J.C. 1971. Energy flow in a woodland stream ecosystem: I. Tissue support trophic structure of the autumnal community. Arch. Hydrobiol. 68(2): 232-76

Corbet, P.S. 1966. Parthenogenesis in caddisflies (Trichoptera). Can. J. Zool. 44(5): 981-82

Corbet, P.S., Schmid, F., and Augustin, C.L. 1966. The Trichoptera of St. Helen's Island, Montreal. I. The species present and their relative abundance at light. Can. Ent. 98(12): 1284-98

Cummins, K.W. 1964. Factors limiting the microdistribution of larvae of the caddisflies *Pycnopsyche lepida* (Hagen) and *Pycnopsyche guttifer* (Walker) in a Michigan stream (Trichoptera: Limnephilidae). Ecol. Monogr. 34(3): 271-95

- 1973. Trophic relations of aquatic insects. A. Rev. Ent. 18: 183-206

Décamps, H., and Lafont, M. 1974. Cycles vitaux et production des *Micrasema* pyrénéennes dans les mousses d'eau courante. Ann. Limnol. 10(1): 1-32

Denning, D.G. 1937. The biology of some Minnesota Trichoptera. Trans. Am. Ent. Soc. 63: 17-43

- 1956. Trichoptera. *In* Aquatic Insects of California, edited by R.L. Usinger, pp. 237-70. Berkeley and Los Angeles: University of California Press

- 1958a. New western Trichoptera. Pan-Pacif. Ent. 34(2): 93-98

- 1958b. The genus *Farula* (Trichoptera: Limnephilidae). Ann. Ent. Soc. Am. 51(6): 531-35

- 1964. The genus *Homophylax* (Trichoptera: Limnephilidae). Ann. Ent. Soc. Am. 57(2): 253-60
- 1970. The genus *Psychoglypha* (Trichoptera: Limnephilidae). Can. Ent. 102(1): 15-30
- 1973. New species of Trichoptera. Pan-Pacif. Ent. 49(2): 132-43
- 1975. New species of Trichoptera from western North America. Pan-Pacif. Ent. 51(4): 318-26

Denning, D.G., and Blickle, R.L. 1972. A review of the genus *Ochrotrichia* (Trichoptera: Hydroptilidae). Ann. Ent. Soc. Am. 65(1): 141-51

Denning, D.G., and Sýkora, J. 1966. New North American Trichoptera. Can. Ent. 98(11): 1219-26

Dodds, G.S., and Hisaw, F.L. 1924. Ecological studies of aquatic insects. II. Size of respiratory organs in relation to environmental conditions. Ecology, 5(3): 262-71
- 1925. Ecological studies on aquatic insects. III. Adaptations of caddisfly larvae to swift streams. Ecology, 6(2): 123-37

Edwards, S.W. 1956. Two new species of Trichoptera from Tennessee. J. Tenn. Acad. Sci. 31(1): 3-7
- 1961. The immature stages of *Xiphocentron mexico* (Trichoptera). Tex. J. Sci. 13(1): 51-56
- 1966. An annotated list of the Trichoptera of middle and west Tennessee. J. Tenn. Acad. Sci. 41(4): 116-28

Edwards, S.W., and Arnold, C.R. 1961. The caddis flies of the San Marcos River. Tex. J. Sci. 13(4): 398-415

Elkins, W.A. 1936. The immature stages of some Minnesota Trichoptera. Ann. Ent. Soc. Am. 29(4): 656-81

Elliott, J.M. 1969. Life history and biology of *Sericostoma personatum* Spence (Trichoptera). Oikos, 20: 110-18
- 1970. The diel activity patterns of caddis larvae (Trichoptera). J. Zool. Lond. 160(3): 279-90
- 1971. The life history and biology of *Apatania muliebris* McLachlan (Trichoptera). Entomologist's Gaz. 22: 245-51

Feldmeth, C.R. 1970. The respiratory energetics of two species of stream caddis fly larvae in relation to water flow. Comp. Biochem. Physiol. 32: 193-202

Fischer, F.C.J. 1960-73. Trichopterorum catalogus. Amsterdam: Nederlandse Entomologische Vereeniging (vol. 1, 1960; vol. 2, 1961; vol. 3, 1962; vol. 4, 1963; vol. 5, 1964; vol. 6, 1965; vol. 7, 1966; vol. 8, 1967; vol. 9, 1968; vol. 10, 1969; vol. 11, 1970; vol. 12, 1971; vol. 13, 1972a; vol. 14, 1972b; vol. 15, 1973)

Flint, O.S. 1956. The life history and biology of the genus *Frenesia* (Trichoptera: Limnephilidae). Bull. Brooklyn Ent. Soc. 51(4, 5): 93-108
- 1957. Description of the immature stages of *Drusinus uniformis* Betten (Trichoptera: Limnephilidae). Bull. Brooklyn Ent. Soc. 52(1): 1-4
- 1958. The larva and terrestrial pupa of *Ironoquia parvula* (Trichoptera, Limnephilidae). J. N.Y. Ent. Soc. 66: 59-62
- 1959. The immature stages of *Lype diversa* (Banks) (Trichoptera, Psychomyiidae). Bull. Brooklyn Ent. Soc. 54(2): 44-47

- 1960. Taxonomy and biology of Nearctic limnephilid larvae (Trichoptera), with special reference to species in eastern United States. Entomologica Am., n.s. 40: 1-117
- 1961a. The immature stages of the Arctopsychinae occurring in eastern North America (Trichoptera: Hydropsychidae). Ann. Ent. Soc. Am. 54(1): 5-11
- 1961b. The presumed larva of *Himalopsyche phryganea* (Ross) (Trichoptera: Rhyacophilidae). Pan-Pacif. Ent. 37(4): 199-202
- 1962a. The immature stages of *Paleagapetus celsus* Ross (Trichoptera: Hydroptilidae). Bull. Brooklyn Ent. Soc. 57(2): 40-44
- 1962b. The immature stages of *Matrioptila jeanae* (Ross) (Trichoptera: Glossosomatidae). J. N.Y. Ent. Soc. 70: 64-67
- 1962c. Larvae of the caddis fly genus *Rhyacophila* in eastern North America (Trichoptera: Rhyacophilidae). Proc. U.S. Nat. Mus. 113(3464): 465-93
- 1963. Studies of Neotropical caddis flies, I: Rhyacophilidae and Glossosomatidae (Trichoptera). Proc. U.S. Nat. Mus. 114(3473): 453-78
- 1964a. Notes on some Nearctic Psychomyiidae with special reference to their larvae (Trichoptera). Proc. U.S. Nat. Mus. 115(3491): 467-81
- 1964b. The caddisflies (Trichoptera) of Puerto Rico. Univ. Puerto Rico, Agric. Exp. Sta., Tech. Pap. 40
- 1965. New species of Trichoptera from the United States. Proc. Ent. Soc. Wash. 67(3): 168-76
- 1966. Notes on certain Nearctic Trichoptera in the Museum of Comparative Zoology. Proc. U.S. Nat. Mus. 118(3530): 373-89
- 1967a. Studies of Neotropical caddis flies, IV. New species from Mexico and Central America. Proc. U.S. Nat. Mus. 123(3608): 1-24
- 1967b. Studies of Neotropical caddis flies, V. Types of the species described by Banks and Hagen. Proc. U.S. Nat. Mus. 123(3619): 1-37
- 1967c. Studies of Neotropical caddis flies, VI: on a collection from northwestern Mexico. Proc. Ent. Soc. Wash. 69(2): 162-76
- 1967d. The first record of the Paduniellini in the New World (Trichoptera: Psychomyiidae). Proc. Ent. Soc. Wash. 69(4): 310-11
- 1968a. Bredin-Archbold-Smithsonian biological survey of Dominica. 9. The Trichoptera (caddisflies) of the Lesser Antilles. Proc. U.S. Nat. Mus. 125(3665): 1-86
- 1968b. The caddisflies of Jamaica (Trichoptera). Bull. Inst. Jamaica, Sci. Ser. 19
- 1970. Studies of Neotropical caddisflies, X: *Leucotrichia* and related genera from North and Central America (Trichoptera: Hydroptilidae). Smithson. Contr. Zool. 60
- 1972. Studies of Neotropical caddisflies, XIII: the genus *Ochrotrichia* from Mexico and Central America (Trichoptera: Hydroptilidae). Smithson. Contr. Zool. 118
- 1973. Studies of Neotropical caddisflies, XVI: the genus *Austrotinodes* (Trichoptera: Psychomyiidae). Proc. Biol. Soc. Wash. 86(11): 127-42
- 1974a. Studies of Neotropical caddisflies, XVII: the genus *Smicridea* from North and Central America (Trichoptera: Hydropsychidae). Smithson. Contr. Zool. 167
- 1974b. The genus *Culoptila* Mosely in the United States, with two new combinations (Trichoptera: Glossosomatidae). Proc. Ent. Soc. Wash. 76(3): 284
- 1974c. The Trichoptera of Surinam; studies of Neotropical caddisflies, XV. Stud. Fauna Suriname, 55

Flint, O.S., and Wiggins, G.B. 1961. Records and descriptions of North American species in the genus *Lepidostoma*, with a revision of the *Vernalis* group (Trichoptera: Lepidostomatidae). Can. Ent. 93(4): 279-97

Fox, H.M., and Sidney, J. 1953. The influence of dissolved oxygen on the respiratory movements of caddis larvae. J. Exp. Biol. 30(2): 235-37

Fremling, C.R. 1960. Biology and possible control of nuisance caddisflies of the upper Mississippi River. Iowa State Univ. Sci. Tech., Res. Bull. 483: 856-79

Frommer, S.I., and Sublette, J.E. 1971. The Chironomidae (Diptera) of the Philip L. Boyd Deep Canyon Desert Research Center, Riverside Co., California. Can. Ent. 103(3): 414-23

Gallepp, G.W. 1974a. Diel periodicity in the behaviour of the caddisfly, *Brachycentrus americanus* (Banks). Freshwat. Biol. 4(2): 193-204

- 1974b. Behavioral ecology of *Brachycentrus occidentalis* Banks during the pupation period. Ecology, 55(6): 1283-94

Gibbs, D.G. 1968. The larva, dwelling-tube and feeding of a species of *Protodipseudopsis* (Trichoptera: Dipseudopsidae). Proc. Roy. Ent. Soc. Lond. (A), 43(4-6): 73-79

Glime, J.M. 1968. Ecological observations on some bryophytes in Appalachian mountain streams. Castanea, 33: 300-25

Gordon, A.E. 1974. A synopsis and phylogenetic outline of the Nearctic members of *Cheumatopsyche.* Proc. Acad. Nat. Sci. Philad. 126(9): 117-60

Hanna, H.M. 1960. Methods of case-building and repair by larvae of caddis flies. Proc. Roy. Ent. Soc. Lond. (A), 35(7-9): 97-106

Hansell, M.H. 1968a. The selection of house building materials by the caddis fly larva, *Agapetus fuscipes* (Curtis). Rev. Comp. Anim. 2: 91-102

- 1968b. The house building behaviour of the caddis-fly larva *Silo pallipes* Fabricius, pts. 1-3. Anim. Behav. 16(4): 558-84

- 1972. Case building behaviour of the caddis fly larva, *Lepidostoma hirtum.* J. Zool. Lond. 167(2): 179-92

- 1974. Regulation of building unit size in the house building of the caddis larva *Lepidostoma hirtum.* Anim. Behav. 22(1): 133-43

Hickin, N.E. 1967. Caddis Larvae. London: Hutchinson

Hiley, P.D. 1969. A method of rearing Trichoptera larvae for taxonomic purposes. Entomologist's Mon. Mag. 105(1265-67): 278-79

Hill-Griffin, A.L. 1912. New Oregon Trichoptera. Ent. News, 23(1): 17-21

Hinton, H.E. 1949. On the function, origin, and classification of pupae. Proc. Trans. S. Lond. Ent. Nat. Hist. Soc. 1947-48: 111-54

- 1950. A trichopterous larva with a chelate front leg. Proc. Roy. Ent. Soc. Lond. (A), 25(4-6): 62-65

- 1971. Some neglected phases in metamorphosis. Proc. Roy. Ent. Soc. Lond. (C), 35: 55-64

Hodkinson, I.D. 1975. A community analysis of the benthic insect fauna of an abandoned beaver pond. J. Anim. Ecol. 44(2): 533-51

Hynes, H.B.N. 1970. The Ecology of Running Waters. Toronto: University of Toronto Press

Jaag, O., and Ambühl, H. 1964. The effect of the current on the composition of biocoenoses in flowing water streams. *In* Advances in Water Pollution Research; Proceedings of the International Conference, London, 1962, edited by B.A. Southgate, pp. 31-44. Oxford: Pergamon Press

Johnstone, G.W. 1964. Stridulation by larval Hydropsychidae (Trichoptera). Proc. Roy. Ent. Soc. Lond. (A), 39(10-12): 146-50

Kimmins, D.E., and Denning, D.G. 1951. The McLachlan types of North American Trichoptera in the British Museum. Ann. Ent. Soc. Am. 44(1): 111-40

Krivda, W.V. 1961. Notes on the distribution and habitat of *Chilostigma areolatum* (Walker) in Manitoba (Trichoptera: Limnephilidae). J. N.Y. Ent. Soc. 69: 68-70

Lea, I. 1834. Observations on the Naiades and descriptions of new species of that and other families. Trans. Am. Phil. Soc., n.s. 4: 63-121

Lehmkuhl, D.M. 1970. A North American trichopteran larva which feeds on freshwater sponges (Trichoptera: Leptoceridae; Porifera: Spongillidae). Am. Midl. Nat. 84(1): 278-80

Lepneva, S.G. 1964. Fauna SSSR, Rucheiniki, vol. 2, no.1. Lichinki i kukolki podotryada kol'chatoshchupikovykh. Zoologicheskii Institut Akademii Nauk SSSR, n.s. 88. [In Russian. Translated into English as: Fauna of the U.S.S.R.; Trichoptera, vol. 2, no.1. Larvae and Pupae of *Annulipalpia.* Published by the Israel Program for Scientific Translations, 1970]

- 1966. Fauna SSSR, Rucheiniki, vol. 2, no.2. Lichinki i kukolki podotryada tsel'nosh-chupikovykh. Zoologicheskii Institut Akademii Nauk SSSR, n.s. 95. [In Russian. Translated into English as Fauna of the U.S.S.R.; Trichoptera, vol. 2, no.2. Larvae and Pupae of *Integripalpia.* Published by the Israel Program for Scientific Translations, 1971]

Levanidova, I.M. 1967. Materialy po faune rucheinikov (Trichoptera) Sibiri i Dalnego Vostoka. Entomologicheskoe Obozrenie, 46: 793-98 [In Russian. Translated into English as: Data on the caddisfly (Trichoptera) fauna of Siberia and the Far East. Ent. Rev. 46(4): 467-70 (1967)]

- 1975. Rucheiniki Kamchatskogo Poluostrova (ekologo-faunisticheskii obzor). Izvestia Tikhookeanskogo Nauchno-Issledovatelskogo Instituta Rybnogo Khozyaistva i Okea-nografii 97: 83-114. [In Russian. Translated as the Caddisflies (Trichoptera) of Kamchatka (an ecological-faunistic outline). Bulletin of the Pacific Scientific Institute of Fisheries and Oceanography]

Lloyd, J.T. 1915a. Notes on *Astenophylax argus* Harris (Trichoptera). J. N.Y. Ent. Soc. 23(1): 57-60

- 1915b. Notes on *Ithytrichia confusa* Morton. Can. Ent. 47(4): 117-21.

- 1921. The biology of North American caddis fly larvae. Bull. Lloyd Libr. 21

Mackay, R.J. 1968. Seasonal variation in the structure of stream insect communities. M.Sc. thesis, McGill University

- 1969. Aquatic insect communities of a small stream on Mont St. Hilaire, Quebec. J. Fish. Res. Bd. Can. 26(5): 1157-83

- 1972. Temporal patterns in life history and flight behaviour of *Pycnopsyche gentilis, P. luculenta*, and *P. scabripennis* (Trichoptera: Limnephilidae). Can. Ent. 104(11): 1819-35

Mackay, R.J., and Kalff, J. 1973. Ecology of two related species of caddis fly larvae in the organic substrates of a woodland stream. Ecology, 54(3): 499-511

Mackereth, J.C. 1956. Taxonomy of the larvae of the British species of the sub-family Glossosomatinae (Trichoptera). Proc. Roy. Ent. Soc. Lond. (A), 31(10-12): 167-72

Malicky, H. 1973. Trichoptera (Köcherfliegen). Handb. Zool. 4(2)

Marlier, G. 1964. Trichoptères de l'Amazonie recueillis par le Professeur H. Siolo. Mém. Inst. Roy. Sci. Nat. Belg., deux. sér., fasc. 76

Martynov, A.V. 1924. Praklicheskaya entomologie, vyp.5. Rucheiniki. [Practical entomology, vol. 5. Trichoptera.] Leningrad: Gosudarstvennoe Izdatelstvo [in Russian]

- 1930. On the trichopterous fauna of China and eastern Tibet. Proc. Zool. Soc. Lond. 1930(5): 65-112

McGaha, Y.J. 1952. The limnological relations of insects to certain aquatic flowering plants. Trans. Am. Microsc. Soc. 71(4): 355-81

Mecom, J.O. 1972a. Feeding habits of Trichoptera in a mountain stream. Oikos, 23: 401-7

- 1972b. Productivity and distribution of Trichoptera larvae in a Colorado mountain stream. Hydrobiologia, 40: 151-76

Mecom, J.O., and Cummins, K.W. 1964. A preliminary study of the trophic relationships of the larvae of *Brachycentrus americanus* (Banks) (Trichoptera: Brachycentridae). Trans. Am. Microsc. Soc. 83(2): 233-43

Merrill, D. 1965. The stimulus for case-building activity in caddis-worms (Trichoptera). J. Exp. Zool. 158(1): 123-30

- 1969. The distribution of case recognition in ten families of caddis larvae (Trichoptera). Anim. Behav. 17(3): 486-93

Merrill, D., and Wiggins, G.B. 1971. The larva and pupa of the caddisfly genus *Setodes* in North America (Trichoptera: Leptoceridae). Life Sci. Occ. Pap. Roy. Ont. Mus. 19

Mickel, C.E., and Milliron, H.E. 1939. Rearing the caddice fly, *Limnephilus indivisus* Walker and its hymenopterous parasite *Hemiteles biannulatus* Grav. Ann. Ent. Soc. Am. 32(3): 575-80

Milne, L.J. 1934. Studies in North American Trichoptera, 1. Cambridge, Mass.

Milne, M.J. 1938. Case-building in Trichoptera as an inherited response to oxygen deficiency. Can. Ent. 70(9): 177-80

Minckley, W.L. 1963. The ecology of a spring stream, Doe Run, Meade County, Kentucky. Wildl. Monogr. 11

Moretti, G. 1942. Studii sui tricotteri: XV. Comportamento del *Triaenodes bicolor* Curt. (Trichoptera - Leptoceridae). Boll. Lab. Zool. Agr. Bachic., Milano, 11: 89-131

Morse, J.C. 1972. The genus *Nyctiophylax* in North America. J. Kans. Ent. Soc. 45(2): 172-81

- 1975. A phylogeny and revision of the caddisfly genus *Ceraclea* (Trichoptera, Leptoceridae). Contr. Am. Ent. Inst. 11(2)

Murphy, H.E. 1919. Observations on the egg-laying of the caddice-fly *Brachycentrus nigrisoma* Banks, and on the habits of the young larvae. J. N.Y. Ent. Soc. 27: 154-59

Muttkowski, R.A., and Smith, G.M. 1929. The food of trout stream insects in Yellowstone National Park. Roosevelt Wild Life Ann. 2(2): 241-63

Neave, F. 1933. Ecology of two species of Trichoptera in Lake Winnipeg. Int. Rev. Hydrobiol. 29(1/2): 17-28

Neboiss, A. 1958. Larva and pupa of an Australian Limnephilid (Trichoptera). Proc. Roy. Soc. Vict. 70(2): 163-68

Needham, J.G. 1902. A probable new type of hypermetamorphosis. Psyche, 9(316): 375-78

Nielsen, A. 1942. Über die Entwicklung und Biologie der Trichopteren mit besonderer Berücksichtigung der Quelltrichopteren Himmerlands. Arch. Hydrobiol., Suppl. Bd. 17: 255-631

- 1943a. *Apatidea auricula* Forsslund from a Norwegian mountain lake; description of the imago and notes on the biology. Saertr. af Ent. Medd. 23: 18-30

- 1943b. Trichopterologische Notizen. Vidensk. Medd. fra Dansk naturh. Foren. 107: 105-20

- 1948. Postembryonic development and biology of the Hydroptilidae. Kgl. Danske Vidensk. Selsk. Biol. Skr. 5(1)

- 1950. Notes on the genus *Apatidea* MacLachlan, with descriptions of two new and possibly endemic species from the springs of Himmerland. Saertr. af Ent. Medd. 25: 384-404

- 1957. A comparative study of the genital segments and their appendages in male Trichoptera. Biol. Skr. Danske Vidensk. Selsk. 8(5)

Nimmo, A.P. 1965. A new series of *Psychoglypha* Ross from western Canada, with notes on several other species of Limnephilidae (Trichoptera). Can. J. Zool. 43(5): 781-87

- 1966. A list of Trichoptera taken at Montreal and Chambly, Quebec, with descriptions of three new species. Can. Ent. 98(7): 688-93

- 1971. The adult Rhyacophilidae and Limnephilidae (Trichoptera) of Alberta and eastern British Columbia and their post-glacial origin. Quaest. Ent. 7: 3-234

Novák, K. 1952. Tracheáln í soustava larev našich Trichopter. Acta Soc. Zool. Bohemoslovenicae, 16(3-4): 249-70

- 1960. Entwicklung und Diapause der Köcherfliegenlarven *Anabolia furcata* Br. (Trichopt.). Čas. Čs. Spol. Ent. 57(3): 207-12

Novák, K., and Sehnal, F. 1963. The development cycle of some species of the genus *Limnephilus* (Trichoptera). Čas. Čs. Spol. Ent. 60(1-2): 68-80

- 1965. Imaginaldiapause bei den in periodischen Gewässern lebenden Trichopteren. Proc. XII Int. Congr. Ent., London, 1964, p. 434

Noyes, A.A. 1914. The biology of the net-spinning Trichoptera of Cascadilla Creek. Ann. Ent. Soc. Am. 7(4): 251-72

Nüske, H., and Wichard, W. 1971. Die Analpapillen der Köcherfliegenlarven. I. Feinstruktur und histochemischer Nachweis von Natrium und Chlorid bei *Philopotamus montanus* Donov. Cytobiologie, 4(3): 480-86 [in German, English abstract]

- 1972. Die Analpapillen der Köcherfliegenlarven. II. Feinstruktur des ionentransportierenden und respiratorischen Epithels bei Glossosomatiden. Cytobiologie, 6(2): 243-49 [in German, English abstract]

Pearson, W.D., and Kramer, R.H. 1972. Drift and production of two aquatic insects in a mountain stream. Ecol. Monogr. 42(3): 365-85

Percival, E., and Whitehead, H. 1929. A quantitative study of the fauna of some types of stream-bed. J. Ecol. 17: 282-314

Philipson, G.N. 1953a. A method of rearing trichopterous larvae collected from swift-flowing waters. Proc. Roy. Ent. Soc. Lond. (A), 28(1-3): 15-16

- 1953b. The larva and pupa of *Wormaldia subnigra* McLachlan (Trichoptera: Philopotamidae). Proc. Roy. Ent. Soc. Lond. (A), 28(4-6): 57-62

- 1954. The effect of water flow and oxygen concentration on six species of caddis fly (Trichoptera) larvae. Proc. Zool. Soc. Lond. 124(3): 547-64

Resh, V.H. 1972. A technique for rearing caddisflies (Trichoptera). Can. Ent. 104(12): 1959-61

- 1976. The biology and immature stages of the caddisfly genus *Ceraclea* in eastern North America (Trichoptera: Leptoceridae). Ann. Ent. Soc. Am. 69(6): 1039-61

Resh. V.H., Morse, J.C., and Wallace, I.D. 1976. The evolution of the sponge feeding habit in the caddisfly genus *Ceraclea* (Trichoptera: Leptoceridae). Ann. Ent. Soc. Am. 69(5): 937-41

Resh, V.H., and Unzicker, J.D. 1975. Water quality monitoring and aquatic organisms: the importance of species identification. J. Wat. Pollut. Control Fed. 47(1): 9-19

Riek, E.F. 1970. Trichoptera. *In* The Insects of Australia, pp. 741-64. Melbourne: C.S.I.R.O. and Melbourne University Press

Root, D.W. 1965. A new northeastern caddisfly species of the genus *Phylocentropus* (Trichoptera: Psychomyiidae). Bull. Brooklyn Ent. Soc. 59-60: 85-87

Ross, H.H. 1938. Descriptions of Nearctic caddis flies (Trichoptera) with special reference to the Illinois species. Bull. Ill. Nat. Hist. Surv. 21(4): 101-83

- 1944. The caddis flies, or Trichoptera, of Illinois. Bull. Ill. Nat. Hist. Surv. 23(1)

- 1946. A review of the Nearctic Lepidostomatidae (Trichoptera). Ann. Ent. Soc. Am. 39(2): 265-91

- 1947. Descriptions and records of North American Trichoptera, with synoptic notes. Trans. Am. Ent. Soc. 73: 125-68

- 1948a. Notes and descriptions of Nearctic Hydroptilidae (Trichoptera). J. Wash. Acad. Sci. 38(6): 201-6

- 1948b. New species of sericostomatoid Trichoptera. Proc. Ent. Soc. Wash. 50(6): 151-57

- 1949. Xiphocentronidae, a new family of Trichoptera. Ent. News, 60(1): 1-7

- 1950. Synoptic notes on some Nearctic limnephilid caddisflies (Trichoptera, Limnephilidae). Am. Midl. Nat. 43(2): 410-29

- 1951a. The caddisfly genus *Anagapetus* (Trichoptera: Rhyacophilidae). Pan-Pacif. Ent. 27(3): 140-44

- 1951b. Phylogeny and biogeography of the caddisflies of the genus *Agapetus* and *Electragapetus* (Trichoptera: Rhyacophilidae). J. Wash. Acad. Sci. 41(11): 347-56

- 1951c. New American species of *Cernotina* (Trichoptera). Rev. Ent. 22(1-3): 343-48

- 1952. The caddisfly genus *Molannodes* in North America. Ent. News, 63(4): 85-87

- 1956. Evolution and Classification of the Mountain Caddisflies. Urbana: University of Illinois Press

- 1959. Trichoptera. *In* Fresh-water Biology, 2nd ed., edited by W.T. Edmondson, pp. 1024-49. New York: Wiley

- 1964. Evolution of caddisworm cases and nets. Am. Zoologist, 4: 209-20

- 1965. The evolutionary history of *Phylocentropus* (Trichoptera: Psychomyiidae). J. Kans. Ent. Soc. 38(4): 398-400
- 1967. The evolution and past dispersal of the Trichoptera. A. Rev. Ent. 12: 169-206
- 1970. Hydropsychid genus A, *Diplectrona* (Trichoptera: Hydropsychidae). J. Georgia Ent. Soc. 5(4): 229-31
Ross, H.H., and Gibbs, D.G. 1973. The subfamily relationships of the Dipseudopsinae (Trichoptera, Polycentropodidae). J. Georgia Ent. Soc. 8(4): 312-16
Ross, H.H., and King, E.W. 1952. Biogeographic and taxonomic studies in *Atopsyche* (Trichoptera, Rhyacophilidae). Ann. Ent. Soc. Am. 45(2): 177-204
Ross, H.H., and Merkley, D.R. 1952. An annotated key to the Nearctic males of *Limnephilus* (Trichoptera, Limnephilidae). Am. Midl. Nat. 47(2): 435-55
Ross, H.H., and Scott, D.C. 1974. A review of the caddisfly genus *Agarodes*, with descriptions of new species (Trichoptera: Sericostomatidae). J. Georgia Ent. Soc. 9(3): 147-55
Ross, H.H., and Unzicker, J.D. 1965. The *Micrasema rusticum* group of caddisflies (Brachycentridae, Trichoptera). Proc. Biol. Soc. Wash. 78: 251-57
Ross, H.H., and Wallace, J.B. 1974. The North American genera of the family Sericostomatidae (Trichoptera). J. Georgia Ent. Soc. 9(1): 42-48
Roy, D., and Harper, P.P. 1975. Nouvelles mentions de trichoptères du Québec et description de *Limnephilus nimmoi* sp. nov. (Limnephilidae). Can. J. Zool. 53(8): 1080-88
Sattler, W. 1958. Beiträge zur Kenntnis von Lebensweise und Körperbau der Larve und Puppe von *Hydropsyche* Pict. (Trichoptera) mit besonderer Berücksichtigung des Netzbaues. Z. Morph. Ökol. Tiere, 47(2): 115-92
- 1963a. Die Larven- und Puppenbauten von *Diplectrona felix* McLach. (Trichoptera). Zool. Anz. 170(1/2): 53-55
- 1963b. Über den Körperbau, die Ökologie und Ethologie der Larve und Puppe von *Macronema* Pict. (Hydropsychidae), ein als Larve sich von 'Mikro-Drift' ernährendes Trichopter aus dem Amazonasgebiet. Arch. Hydrobiol. 59(1): 26-60
- 1968. Weitere Mitteilungen über die Ökethologie einer neotropischen *Macronema*-Larve (Hydropsychidae, Trichoptera). Amazoniana, 1: 211-29
Schmid, F. 1950a. Le genre *Anabolia* Steph. (Trichoptera, Limnophilidae). Rev. Suisse Hydrol. 12(2): 300-39
- 1950b. Monographie du genre *Grammotaulius* Kolenati (Trichoptera Limnophilidae). Rev. Suisse Zool. 57(7): 317-52
- 1950c. Le genre *Halesochila* Banks (Trichopt. Limnophilid.). Mitt. Schweiz. Ent. Ges. 23(1): 55-60
- 1950d. Le genre *Hydatophylax* Wall (Trichopt. Limnophilidae). Mitt. Schweiz. Ent. Ges. 23(3): 265-96
- 1951. Le genre *Ironoquia* Bks. (Trichopt. Limnophilid.). Mitt. Schweiz. Ent. Ges. 24(3): 317-28
- 1952a. Les genres *Glyphotaelius* Steph. et *Nemotaulius* Bks. (Trichopt. Limnophil.). Bull. Soc. Vaud. Sci. Nat. 65(280): 213-44
- 1952b. Le groupe de *Lenarchus* Mart. (Trichopt. Limnophil.). Mitt. Schweiz. Ent. Ges. 25(3): 157-210

- 1952c. Le groupe de *Chilostigma.* Arch. Hydrobiol. 47(1): 75-163
- 1953. Contribution à l'étude de la sous-famille des Apataniinae (Trichoptera, Limnophilidae). I. Tijdschr. Ent. 96(1-2): 109-67
- 1954a. Contribution à l'étude de la sous-famille des Apataniinae (Trichoptera, Limnophilidae). II. Tijdschr. Ent. 97(1-2): 1-74
- 1954b. Le genre *Asynarchus* McL. (Trichopt., Limnoph.). Mitt. Schweiz. Ent. Ges. 27(2): 57-96
- 1955. Contribution à l'étude des Limnophilidae (Trichoptera). Mitt. Schweiz. Ent. Ges. 28
- 1964a. Quelques Trichoptères asiatiques. Can. Ent. 96(6): 825-40
- 1964b. Some Nearctic species of *Grammotaulius* Kol. (Trichoptera: Limnephilidae). Can. Ent. 96(6): 914-17
- 1965. Ergebnisse der zoologischen Forschungen von Dr. Z. Kaszab in der Mongolei. 63. Trichoptera. Reichenbachia, 7(22): 201-3
- 1968a. La famille des Arctopsychides (Trichoptera). Mem. Ent. Soc. Quebec, 1
- 1968b. Quelques Trichoptères néarctiques nouveaux ou peu connus. Naturaliste Can. 95(3): 673-98
- 1970. Le genre *Rhyacophila* et la famille des Rhyacophilidae (Trichoptera). Mem. Ent. Soc. Can. 66

Schmid, F., and Botoşăneanu, L. 1966. Le genre *Himalopsyche* Banks (Trichoptera, Rhyacophilidae). Ann. Ent. Soc. Quebec, 11(2): 123-76

Schmitz, M., and Wichard, W. 1975. Ionenabsorption an Chloridepithelien von Köcherfliegenlarven (Trichoptera). Ent. Germ. 2(1): 30-34
- In press. Der Ort der osmoregulatorischen Salzaufnahme bei Phryganeidae-Larven (Trichoptera). Mitt. Schweiz. Ent. Ges.

Scott, K.M.F. 1961. Some new caddis flies (Trichoptera) from the Western Cape Province - III. Ann. S. Afr. Mus. 46(2): 15-33

Selgeby, J.H. 1974. Immature insects (Plecoptera, Trichoptera, and Ephemeroptera) collected from deep water in western Lake Superior. J. Fish. Res. Bd. Can. 31(1): 109-11

Sherberger, F.F., and Wallace, J.B. 1971. Larvae of the southeastern species of *Molanna.* J. Kans. Ent. Soc. 44(2): 217-24

Sibley, C.K. 1926. Trichoptera. *In* A preliminary biological survey of the Lloyd-Cornell Reservation. Bull. Lloyd Libr. 27: 102-8, 181-221

Silfvenius, A.J. 1904. Über die Metamorphose einiger Phryganeiden und Limnophiliden III. Acta Soc. pro Fauna et Flora Fennica, 27(2): 3-74
- 1905. Beiträge zur Metamorphose der Trichopteren. Acta Soc. pro Fauna et Flora Fennica, 27(6): 3-168

Siltala, A.J. 1907a. Trichopterologische Untersuchungen. No. 2. Über die postembryonale Entwicklung der Trichopteren-Larven. Zool. Jahrb., suppl. 9(2): 309-626
- 1907b. Über die Nahrung der Trichopteren. Acta Soc. pro Fauna et Flora Fennica, 29(5): 3-34

Smirnov, N.N. 1962. On nutrition of caddis worms *Phryganea grandis* L. Hydrobiologia, 19: 252-61

Smith, S.D. 1968a. The Arctopsychinae of Idaho (Trichoptera: Hydropsychidae). Pan-Pacif. Ent. 44(2): 102-12

- 1968b. The *Rhyacophila* of the Salmon River drainage of Idaho with special reference to larvae. Ann. Ent. Soc. Am. 61(3): 655-74
- 1969. Two new species of Idaho Trichoptera with distributional and taxonomic notes on other species. J. Kans. Ent. Soc. 42(1): 46-53
Snodgrass, R.E. 1954. Insect metamorphosis. Smithson. Misc. Coll. 122(9)
Solem, J.O. 1970. Contributions to the knowledge of the larvae of the family Molannidae (Trichoptera). Norsk. Ent. Tidsskr. 17(2): 97-102
Stauffer, R.C., ed. 1975. Charles Darwin's Natural Selection, being the Second Part of His Big Species Book Written from 1856 to 1858. Cambridge, Eng.: Cambridge University Press
Sturm, H. 1960. Die terrestrischen Puppengehäuse von *Xiphocentron sturmi* Ross (Xiphocentronidae, Trichoptera). Zool. Jahrb. Syst. 87(4/5): 387-94
Svensson, B.W., and Tjeder, B. 1975. Check-list of the Trichoptera of north-western Europe. Ent. Scand. 6: 261-74
Thut, R.N. 1969. Feeding habits of larvae of seven *Rhyacophila* (Trichoptera: Rhyacophilidae) species with notes on other life-history features. Ann. Ent. Soc. Am. 62(4): 894-98
Tindall, A.R. 1960. The larval case of *Triaenodes bicolor* Curtis (Trichoptera: Leptoceridae). Proc. Roy. Ent. Soc. Lond. (A), 35(7-9): 93-96
- 1963. Some observations on the biology of caddis-fly larvae. Entomologist's Gaz., 14: 28-30
Tomaszewski, C. 1973. Studies on the adaptive evolution of the larvae of Trichoptera. Acta Zool. Cracoviensia, 18(13): 311-98
Ulmer, G. 1957. Köcherfliegen (Trichopteren) von den Sunda-Inseln. III. Arch. Hydrobiol., suppl. 23(2/4): 109-470
Vaillant, F. 1965. Les larves de Trichoptères hydroptilides mangeuses de substrat. Proc. XII Int. Congr. Ent., London, 1964, p. 165
Van Dam, L. 1938. On the Utilization of Oxygen and Regulation of Breathing in Some Aquatic Animals. Groningen: Volharding
Vorhies, C.T. 1905. Habits and anatomy of the larva of the caddis-fly, *Platyphylax designatus*, Walker. Trans. Wis. Acad. Sci. Arts Lett. 15: 108-23
- 1909. Studies on the Trichoptera of Wisconsin. Trans. Wis. Acad. Sci. Arts Lett. 16: 647-738
Wallace, J.B. 1971. A new species of *Brachycentrus* from Georgia with two unusual larval characters (Trichoptera: Brachycentridae). Ent. News 82: 313-21
- 1975a. The larval retreat and food of *Arctopsyche*; with phylogenetic notes on feeding adaptations in Hydropsychidae larvae (Trichoptera). Ann. Ent. Soc. Am. 68(1): 167-73
- 1975b. Food partitioning in net-spinning Trichoptera larvae: *Hydropsyche venularis, Cheumatopsyche etrona*, and *Macronema zebratum* (Hydropsychidae). Ann. Ent. Soc. Am. 68(3): 463-72
Wallace, J.B., and Malas, D. 1976. The fine structure of capture nets of larval Philopotamidae (Trichoptera), with special emphasis on *Dolophilodes distinctus.* Can. J. Zool. 54(10): 1788-1802
Wallace, J.B., and Ross, H.H. 1971. Pseudogoerinae: a new subfamily of Odontoceridae (Trichoptera). Ann. Ent. Soc. Am. 64(4): 890-94

Wallace, J.B., and Sherberger, F.F. 1970. The immature stages of *Anisocentropus pyraloides* (Trichoptera: Calamoceratidae). J. Georgia Ent. Soc. 5(4): 217-24
- 1974. The larval retreat and feeding net of *Macronema carolina* Banks (Trichoptera: Hydropsychidae). Hydrobiologia, 45(2-3): 177-84
Wallace, J.B., Sherberger, S.R., and Sherberger, F.F. 1976. Use of the diatom *Terpsinoë musica* Ehrenb. (Biddulphiales: Biddulphiaceae) as casemaking material by *Nectopsyche* larvae (Trichoptera: Leptoceridae). Am. Midl. Nat. 95(1): 236-39
Wallace, J.B., Woodall, W.R., and Staats, A.A. 1976. The larval dwelling-tube, capture net and food of *Phylocentropus placidus* (Trichoptera: Polycentropodidae). Ann. Ent. Soc. Am. 69(1): 149-54
Walton, I. 1653. The Compleat Angler. London: printed by T. Maxey for R. Marriot
Waters, T.F. 1962. Diurnal periodicity in the drift of stream invertebrates. Ecology, 43(2): 316-20
- 1968. Diurnal periodicity in the drift of a day-active stream invertebrate. Ecology, 49(1): 152-53
Wesenberg-Lund, C. 1911. Biologische Studien über netzspinnende, campodeoide *Trichopteren*larven. Int. Rev. Hydrobiol., biol. suppl. ser. 3, 1: 1-64
Wichard, W. 1973. Zur Morphogenese des respiratorischen Epithels der Tracheenkiemen bei Larven der Limnephilini Kol. (Insecta, Trichoptera). Z. Zellforsch. 144: 585-92
- 1974. Zur morphologischen Anpassung von Tracheenkiemen bei Larven der Limnephilini Kol. (Insecta, Trichoptera). II. Adaptationsversuche unter verschiedenen O_2-Bedingungen während der larvalen Entwicklung. Oecologia, 15: 169-75
- 1976. Morphologische Komponenten bei der Osmoregulation von Trichopterenlarven. *In* Proc. of the First Int. Symp. on Trichoptera, Lunz am See (Austria), 1974, edited by H. Malicky, pp. 171-77. The Hague: Junk
Wichard, W., and Komnick, H. 1973. Fine structure and function of the abdominal chloride epithelia in caddisfly larvae. Z. Zellforsch. 136: 579-90
Wiggins, G.B. 1954. The caddisfly genus *Beraea* in North America (Trichoptera). Contr. Roy. Ont. Mus. Zool. Palaeont. 39
- 1956. A revision of the North American caddisfly genus *Banksiola* (Trichoptera: Phryganeidae). Contr. Roy. Ont. Mus. Zool. Palaeont. 43
- 1958. A study of the systematics of the Phryganeidae of the world (Insecta: Trichoptera). Ph.D. thesis, University of Toronto
- 1959. A method of rearing caddisflies (Trichoptera). Can. Ent. 91(7): 402-5
- 1960a. The unusual pupal mandibles in the caddisfly family Phryganeidae (Trichoptera). Can. Ent. 92(6): 449-57
- 1960b. A preliminary systematic study of the North American larvae of the caddisfly family Phryganeidae (Trichoptera). Can. J. Zool. 38(6): 1153-70
- 1961. The rediscovery of an unusual North American phryganeid, with some additional records of caddisflies from Newfoundland (Trichoptera). Can. Ent. 93(8): 695-702
- 1962. A new subfamily of phryganeid caddisflies from western North America (Trichoptera: Phryganeidae). Can. J. Zool. 40(5): 879-91
- 1963. Larvae and pupae of two North American limnephilid caddisfly genera (Trichoptera: Limnephilidae). Bull. Brooklyn Ent. Soc. 58(4): 103-12
- 1965. Additions and revisions to the genera of North American caddisflies of the family

Brachycentridae with special reference to the larval stages (Trichoptera). Can. Ent. 97(10): 1089-1106
- 1966. The critical problem of systematics in stream ecology. *In* Symposium on Organism-Substrate Relationships in Streams, pp. 52-58. University of Pittsburgh, Pymatuning Laboratory of Ecology
- 1968. Contributions to the systematics of the caddisfly family Molannidae in Asia (Trichoptera). Life Sci. Contr., Roy. Ont. Mus. 72
- 1972. The caddisfly family Phryganeidae: classification and phylogeny for the world fauna (Trichoptera). Symposium A. Systematics, Ecology and Phylogeny of Odonata, Ephemeroptera, Plecoptera and Trichoptera. Proc. XIII Int. Congr. Ent., Moscow, 1968, p. 342
- 1973a. A contribution to the biology of caddisflies (Trichoptera) in temporary pools. Life Sci. Contr., Roy. Ont. Mus. 88
- 1973b. New systematic data for the North American caddisfly genera *Lepania, Goeracea* and *Goerita* (Trichoptera: Limnephilidae). Life Sci. Contr., Roy. Ont. Mus. 91
- 1973c. Contributions to the systematics of the caddisfly family Limnephilidae (Trichoptera). I. Life Sci. Contr., Roy. Ont. Mus. 94
- 1975. Contributions to the systematics of the caddisfly family Limnephilidae (Trichoptera). II. Can. Ent. 107(3): 325-36
- 1976. Contributions to the systematics of the caddis-fly family Limnephilidae (Trichoptera). III: The genus *Goereilla*. *In* Proc. of the First Int. Symp. on Trichoptera, Lunz am See (Austria), 1974, edited by H. Malicky, pp. 7-19. The Hague: Junk

Wiggins, G.B., and Anderson, N.H. 1968. Contributions to the systematics of the caddisfly genera *Pseudostenophylax* and *Philocasca* with special reference to the immature stages (Trichoptera: Limnephilidae). Can. J. Zool. 46(1): 61-75

Wiggins, G.B., and Kuwayama, S. 1971. A new species of the caddisfly genus *Oligotricha* from northern Japan and Sakhalin, with a key to the adults of the genus (Trichoptera: Phryganeidae). Kontyu, 39(4): 340-46

Williams, D.D., and Hynes, H.B.N. 1974. The occurrence of benthos deep in the substrature of a stream. Freshwat. Biol. 4(3): 233-56

Williams, D.D., and Williams, N.E. 1975. A contribution to the biology of *Ironoquia punctatissima* (Trichoptera: Limnephilidae). Can. Ent. 107(8): 829-32

Williams, N.E., and Hynes, H.B.N. 1973. Microdistribution and feeding of the net-spinning caddisflies (Trichoptera) of a Canadian stream. Oikos, 24: 73-84

Winterbourn, M.J. 1971a. The life histories and trophic relationships of the Trichoptera of Marion Lake, British Columbia. Can. J. Zool. 49(5): 623-35
- 1971b. An ecological study of *Banksiola crotchi* Banks (Trichoptera, Phryganeidae) in Marion Lake, British Columbia. Can. J. Zool. 49(5): 637-45

Yamamoto, T., and Ross, H.H. 1966. A phylogenetic outline of the caddisfly genus *Mystacides* (Trichoptera: Leptoceridae). Can. Ent. 98(6): 627-32

Yamamoto, T., and Wiggins, G.B. 1964. A comparative study of the North American species in the caddisfly genus *Mystacides* (Trichoptera: Leptoceridae). Can. J. Zool. 42(6): 1105-26

Taxonomic Index

All taxa of subgeneric to subordinal status mentioned in this book are listed here. Entries after each name refer to pages on which there is some information concerning that group; page entries are in numerical order, the principal reference, if any, for the group in boldface type. In groups for which illustrations are provided, the plate reference is found at the end of the page references. All genera listed as valid in the checklist of Nearctic Trichoptera (Ross 1944: 291–303) are included in this index.

Taxonomic Index